Socially Responsible Engineering

Socially Responsible Engineering
Justice in Risk Management

Daniel A. Vallero
and
P. Aarne Vesilind

JOHN WILEY & SONS, INC.

Library of Congress Cataloging-in-Publication Data:
Vallero, Daniel A.
  Socially responsible engineering: justice in risk management/Daniel A.
Vallero and P. Aarne Vesilind.
        p. cm.
    Includes index.
    ISBN-13: 978-0-471-78707-5 (pbk.)
    ISBN-10: 0-471-78707-8 (pbk.)
  1. Environmental engineering—Moral and ethical aspects. 2. Environmental
ethics. 3. Engineering ethics. I. Vesilind, P. Aarne. II. Title.
    TD153.V34 2006
    174′.9628—dc22
                                                            2005033604

Printed in the United States of America

10 9 8 7 6 5 4 3 2 1

Contents

v

Foreword

Issues concerning ethical practices in research and technological application have been publicized in the last decade. Although these problems originally were concentrated in biological and medical science areas, problems with work done at an industrial laboratory and at a US national laboratory showed that ethical lapses occur in other disciplines. Another type of ethical issue was demonstrated in the shuttle Challenger destruction, an example of how politically-driven pressures can cause engineers to repress crucial analyses.

A great American engineer, Norman Augustine[1] wrote "Engineers who make bad decisions often don't realize they are confronting ethical issues." He went on to write that "it is important that ethics courses also deal with the pragmatic issues that confront engineers in the rough-and-tumble, everyday world in which they live and work."[2]

This text by Vallero and Vesilind brings together two authors who have written extensively on issues that do relate to the work-day world of the engineer. Vallero, a senior engineer and environmental researcher in the Executive and Legislative Branches and an adjunct professor at Duke University, has incorporated in this text many examples of practical engineering problems. These are the strength of this text. Vesilind, a long time engineering professor at Duke University and now at Bucknell, has published many articles on the ethics and engineering. Their perspective is seen in that the dominant theme of this text is the first rule of practice from the National Society of Professional Engineers. The First Fundamental Canon is "Engineers, in the fulfillment of their professional duties, shall hold paramount the safety, health, and welfare of the public."[3] The many examples in the text illustrate application of this criterion.

For several decades Edward Obert was on the mechanical engineering faculty of the University of Wisconsin-Madison. He sprinkled his lecture notes with ethical guidelines. and wanted to be remembered by a quote from Socrates: "When my sons grow up, I would ask you, my friends, to punish them if they care about anything more than virtue."[4]

Those words are guidance for the engineering profession.

JOHN F. AHEARNE

Director, Ethics Program, Sigma Xi, the Scientific Research Society
Member, National Academy of Engineering

REFERENCES AND NOTES

1. Former chairman of Lockheed Martin, former Under Secretary of the Army, and former lecturer with rank of professor at Princeton University. He is a former chairman of the National Academy of Engineering and the chair of the National Academy of Engineering Committee on Engineering Ethics and Society.

2. "Ethics and the Second Law of Thermodynamics," Norman R. Augustine, *The Bridge,* Volume 32, No. 3, Fall 2002.
3. http://www.nspe.org/ethics/Code-2006-Jan.pdf
4. From the Dedication to *The Responsible Researcher: Paths and Pitfalls,* J.F. Ahearne, Sigma Xi, The Scientific Research Society, 1999.

Preface

Few technical books begin with questions of how we can be virtuous in our duties as professionals, although many engineers, planners, and scientists are concerned with what it means to be a "good" engineer, planner, or scientist. Most books used by technical professionals are limited to discussions of practical questions (or more correctly, questions of professional practice), and it is perfectly appropriate and absolutely essential to be concerned with concepts such as physical integrity of structures, sources and exposures to pollutants, and the applications of physics, chemistry, and biology to solve problems. But only a limited number of texts delve into the philosophical underpinnings of science and engineering professions. Of those that do, most mainly address issues in professional ethics.

One of the challenges in dealing with social issues in a way that informs and interests most engineers and other design professionals is to cover concepts and to use language familiar to a technical audience without wandering into the vernacular of the social sciences. In fact, many of us are "compartmental" in how we think about the world. All engineers are very interested in solving problems by applying the laws of science and the principles of mathematics. The researchers among us are equally interested in the theoretical foundations of these laws and principles. So the first division among those of us who are technical types is between basic and applied science. Next, as educated people, engineers are very interested in the other "nonscientific"* aspects of the world around us, but that is an entirely different compartment from what we do for a living. A growing number of design professions pull all of these compartments together. They do not see the need to draw bright lines between the physical, chemical, and biological sciences and the social sciences and humanities, as long as they affect one another. In fact, we believe that there is a growing need for engineers to become more comfortable with social issues. This book is one step in that direction.

We are not naïve in thinking that all engineering students and practicing engineers will simply embrace what some may perceive to be "soft" subject matter.† In fact, we believe that the approaches applied to date in raising our awareness of societal issues have been woefully inadequate, especially in their seeming disregard for how engineers teach and learn. In particular, they seem to treat issues like justice as sidebars where the engineer is asked to suspend engineering realities to think about social issues. This approach is unfortunate and ignores the essence of engineering since issues like ethics and social justice are often even more mathematically challenging than "typical" engineering

---

*We use this term advisedly. In this context it distinguishes engineering and physical sciences from the social sciences and humanities. We recognize this line of distinction is frequently blurred and that rigorous adherence to the scientific method is not limited to the physical sciences.

† Vallero recalls a colleague's somewhat sarcastic contention that any study whose textbook requires more than one page to explain "why" it is a science is indeed *not* a science. We are a bit less strident in our viewpoint.

problems like those solved using the Bernoulli and Navier–Stokes equations, Fick's and Newton's laws, and Bohr's concept of the atom. In fact, social concepts may often be described as "ill-posed" and nonlinear, just as are many of the most challenging physical concepts being studied in engineering classes today. Compared to many design problems, social issues frequently have more variables, exhibit initial and boundary conditions that are extremely difficult to define, require sophisticated mathematical approaches, call for creative optimization schemes, and hardly ever have a singular solution.

Thus, the structure and tenor of this book address a profound challenge to the engineering profession—social justice—in language and concepts familiar to engineers. We do not shy away from the mathematics, although in light of the large range of engineering disciplines, all equations and notation are explained. In this way, this book is very different from most environmental justice texts. We also do not avoid discussing the intricate details of economic, sociological, and philosophical subjects. Since not all of our technical audience is familiar with these topics, we explain them in detail. This book also differs from most environmental engineering texts in that in addition to explaining the basic and applied sciences, we use case studies, examples, sidebars, and biographies to drive home important concepts.

The authors of this book are both environmental engineers. Among the other important experiences that we have in common, we have both directed the Program in Science, Technology and Human Values at Duke. As the name implies, the program calls for a new way of looking at the world. Leading such a program demands that science and engineering be approached as a social construct. This may bother those who hold that "real science" is unaffected by the norms and influences of the culture in which it resides. How science is conducted and who is able to conduct science are affected by culture. Some advocates erroneously argue, however, that the actual meaning of results differs. Therefore, we agree with the need for objectivity and adherence to the scientific methods prescribed by Bacon, Boyle, and other giants in the history of science, but we hold that ethics is an integral part of this construct, as is appreciation of the myriad policy and analytical approaches for studying the role of science and engineering in shaping and in being shaped by society. An essential part of an engineer's social contract is that we be trusted as professionals. Such trust must be based on both sound science and the appropriate societal application of that science.*

The dearth of virtue-related textbooks is particularly ironic for the fields of environmental science and engineering. After all, the environmentalism movement came to the fore in the 1960s, a time when Western society was rethinking its systems of values. The values long held by the "establishment" were being tested. Everything from capitalism to the law to politics was up for review. While Martin Luther King was leading marches in Selma and Washington, Rachel Carson was challenging the petrochemical revolution

---

*Incidentally, since we mentioned Robert Boyle, we should also mention Thomas Hobbes. Boyle is recognized as one the key proponents of *a posteriori* (experimentally based) science, whereas Hobbes argued for *a priori* science, the idea that knowledge is independent of experience. Since the Renaissance, modern science has embraced Boyle's view. However, as mentioned here, Hobbes' concept of the social contract is widely held by many social theorists as an important normative force in society. According to contractarian theory, all professionals, including engineers and physicians, have a special place of trust within society. In fact, public trust is a common feature of all "professions."

in her famous book *Silent Spring*. Notable examples of reason and stability have emerged since these times, as codified in the Voting Rights, the Civil Rights, the National Environmental Policy Acts, as well as the successes in establishing the arsenal of federal and state environmental laws, but we have yet to formulate a consensus on which virtues are absolutely requisite for our environmental professions.

Leafing through the book, practicing professionals or students might decide that the book may not be relevant to them and their interests. Like many of us, they are less comfortable with "soft" considerations such as philosophy than with the physical sciences, engineering, and other "hard" intellectual matter. But please bear with us! The face of engineering is changing. We must remain highly competent in our application of mathematics and the natural sciences, but engineering is more than that. The engineer today and in the coming decades must adapt to a changing world. The National Academy of Engineering puts it this way:

> [O]ne thing is clear: engineering will not operate in a vacuum separate from society in 2020 any more than it does now. Both on a macro scale, where the world's natural resources will be stressed by population increases, to the micro scale, where engineers need to work in teams to be effective, consideration of social issues is central to engineering. Political and economic relations between nations and their peoples will impact engineering practice in the future, probably to a greater extent than now.*

There has been no better example of the societal responsibilities of the engineer than that demonstrated by the events that led to and followed Hurricane Katrina on the U.S. Gulf coast in 2005. Engineers were clearly part of the problem. Some of this culpability was technical malfeasance, but much of it was a failure to recognize risks beyond math and science. Justice is a big part of what engineers and scientists are all about. In fact, the audience for social justice is not only those of us immersed in either the environmental aspects (e.g., ecologists, environmental engineers, environmental scientists) or the justice considerations (e.g., lawyers, policymakers), but all engineers and most scientists. For example, a structural engineer who is designing a bridge is wise to consider not only the structural elements (e.g., stress and strain, elasticity, weight loading), but the context of the bridge in terms of what it means to the fabric of the community and whether the design is benefiting from sufficient input and participation regarding the placement, aesthetics, and flow patterns to and from the bridge. Certainly, most of the case studies we consider here and throughout the environmental justice literature are specifically "environmental," such as hazardous wastes sites and landfills, but every design must consider potential justice issues.

It is interesting to contrast the role of town engineer *versus* that of engineers in a large urban bureaucracy. The town engineer may have duties that would be represented by whole departments in a large city, such as public works, transportation, housing, and environmental protection. Thus the town engineer is the only professional who can ensure that just decisions are made. That is, the engineer is the only one at the table (literally *and* figuratively) with the technical expertise and power of position to represent those

---

*National Academy of Engineering, *The Engineer of 2020: Visions of Engineering in the New Century,* National Academies Press, Washington, DC, 2004.

with little or no "voice." The same goes for an engineer in a small firm who advises clients *versus* an engineer in a highly specialized job in a large firm.

In this book we begin by addressing questions that concern all of us in asking what is meant by such terms as *just science, just engineering, virtue, environmental justice,* and *environmental racism.* In the second chapter we inquire into our justification for invoking the principles of environmental justice. In the third chapter we look back at the engineering profession and consider what it means to be a professional, especially in light of such issues as ethics and justice. In Chapter 4 we delve into the question of risk from harm due to environmental effects and suggest a new paradigm for thinking about how risk should be estimated. In Chapter 5 we follow up this discussion by noting that risk assessment is merely the first step, and that action is needed to protect the public health and welfare. The sixth chapter focuses on the concept of sustainability and how this ties in with environmental justice. Finally, Chapter 7 is a discussion of how engineers interact with society, engaging the reader in various issues of applied engineering in the context of environmental issues. In this way, the engineer is given a tool kit to help to prepare for the coming expectations of the new century.

In the writing of this book, we have assumed that the reader believes that he or she, through the skill of professional engineering, has a responsibility to try to make the world a better place. To this end, we hope that our work will assist practicing engineers and engineering students alike in both choosing a career path and in gaining a deeper understanding of how engineering affects society, particularly those segments of society that are underrepresented and that have little political or economic clout. In short, this book is about how an engineer can be fair to one's clients while being fair to one's own career.

DANIEL A. VALLERO
P. AARNE VESILIND

Socially Responsible Engineering

1

And Justice for All

Justice is a universal human value. It is a concept that is built into every code of practice and behavior, including the codes of ethics of all engineering and other professional disciplines—and it is at the heart of environmental protection. It is the linchpin of social responsibility. An interesting aspect of justice in a society is that it is found in different venues and stated in many ways. Much understanding of justice is passed from one generation to the next. Although history has shown that human beings can be highly moral agents, it has also shown that we can be very unfair. As theologian Reinhold Niebuhr[1] puts it:

> Man's capacity for justice makes democracy possible, but man's inclination to injustice makes democracy necessary.

Thus, society must not only have norms, it must enforce such norms. In the United States, normative justice is articulated by the U.S. Constitution and encapsulated in the "equal protection" clause of the fourteenth amendment:

> All persons born or naturalized in the United States, and subject to the jurisdiction thereof, are citizens of the United States and of the state wherein they reside. No state shall make or enforce any law which shall abridge the privileges or immunities of citizens of the United States; nor shall any state deprive any person of life, liberty, or property, without due process of law; nor deny to any person within its jurisdiction the equal protection of the laws.

The recently contested *Pledge of Allegiance* pairs the cherished value of freedom with the principle of fairness articulated throughout the Constitution when its ending affirms that the flag represents ". . . liberty and justice for all." Whereas the phrase "under God" has received much scrutiny of late, the concept of fairness has not.

Justice is also a key fixture of the U.S. Declaration of Independence. The Declaration's second paragraph states:

> We hold these truths to be self-evident, that all men are created equal, that they are endowed by their Creator with certain unalienable Rights, that among these are Life, Liberty and the pursuit of Happiness. . . . That whenever any Form of Government becomes destructive of these ends, it is the Right of the People to alter or to abolish it, and to institute new Government, laying its foundation on such principles and organizing its powers in such form, as to them shall seem most likely to effect their Safety and Happiness.

These unalienable rights of life, liberty, and the pursuit of happiness depend upon a livable environment. The Declaration warns against a destructive government. Arguably, the government holds a central role in overcoming the forces that will militate against equity in environmental protection. Democracy and freedom are at the core of achieving fairness, and Americans rightfully take great pride in these foundations of our Republic. The framers of our Constitution wanted to make sure that life, liberty, and the pursuit of happiness were available to all: first with the protection of property rights and later, with the Bill of Rights, by granting human and civil rights to all the people.

Engineers may be surprised to know it, but we are agents of justice. As a profession, we have arguably done more than any other in the past century to improve the quality and length of life in developed nations and are making similar strides in developing countries. The treatment of wastes, provision of potable water, controlling air pollution, handling of solid and hazardous wastes, safer modes of transportation, reliable energy sources, improved communication networks, safer buildings, and improved disaster response are examples of how engineers have enhanced people's pursuit of happiness.

Certainly, a modern connotation of "safety and happiness" is that of risk reduction. As our codes of ethics mandate, the socially responsible engineer and design practitioner must be "faithful agents." But faithful to whom? What has become evident only in the past few decades is that without a clean environment, life is threatened by toxic substances, liberty is threatened by the loss of resources, and happiness is less likely in an unhealthful and unappealing place to live.

Justice must be universalized and applied to everyone. This may seem obvious, but so few things are distributed evenly, we may be tempted to assume that systems are fair simply because "most" are satisfied with the current situation. However, the only way to preserve public health and to protect the environment is to ensure that *all* persons are adequately protected. In the words of Reverend Martin Luther King, "Injustice anywhere is a threat to justice everywhere."[2] Extending this logic means that if any group is disparately exposed to an unhealthy environment, the entire nation is subjected to inequity and injustice; we are all "at risk." An optimistic view (and most engineers are by nature optimistic) is that our projects and products can advance the opportunities for a safe and livable environment by including everyone, leaving no one behind. This mandate has a name, *environmental justice,* and in this book we argue that equal protection can be extended intellectually (if not legally) to matters of public health and environmental quality.

ENVIRONMENTAL JUSTICE

The concept of environmental justice has evolved over time. In the early 1980s, the first name for the movement was *environmental racism,* followed by *environmental equity.* These transitional definitions reflect more than changes in jargon. When attention began to be paid to the particular incidents of racism, the focus was logically placed on eradicating the menace at hand (i.e., blatant acts of willful racism). This was a necessary but not completely sufficient component in addressing the environmental problems of minority communities and economically disadvantaged neighborhoods, so the concept of equity was employed more assertively. *Equity* implies the need not only to eliminate the overt problems associated with racism, but to initiate positive change to achieve more evenly distributed environmental protection.

We now use the term *environmental justice,* which is usually applied to social issues, especially as they relate to neighborhoods and communities. *Environmental justice* (EJ) *communities* possess two basic characteristics:

- Environmental justice communities have suffered historical exposures to disproportionately high doses of potentially harmful substances[3] (the *environmental* part of the definition). Such exposures have often occurred for several decades. These communities are home to numerous pollution sources, including heavy industry and pollution control facilities, which may be obvious by their stacks and outfall structures, or which may be more subtle, such as long-buried wastes with little evidence on the surface of their existence. These sites increase the likelihood of exposure to dangerous substances. Exposure is preferred to *risk,* since risk is a function of the hazard and the exposure to that hazard. Even a substance with a very high toxicity (one type of hazard) that is confined to a laboratory of a manufacturing operation may not pose much of a risk, due to the potentially low levels of exposure.
- Environmental justice communities have certain, specified socioeconomic and demographic characteristics. EJ communities must have a majority representation of low socioeconomic status (SES), racial, ethnic, and historically disadvantaged people (the *justice* part of the definition).

These definitions point to the importance of an integrated response to ensure justice. The first component of this response is a sound scientific and engineering underpinning to decisions. The technical quality of designs and operations is vital to addressing the needs of any group. However, the engineering codes' call that we be faithful agents lends an added element of social responsibility to environmental practitioners.[4] For example, we cannot assume a "blank slate" for any design. Historic disenfranchisement and even outright bias may well have put certain neighborhoods at a disadvantage.

Thus, the responsibility of professionals cannot stop at sound science but should consider the social milieu, especially possible disproportionate impacts. The determination of disproportionate impacts, especially pollution-related diseases and other health endpoints, is a fundamental step in ensuring environmental justice. But even this step relies on the application of sound physical science. Like everything else that technical professionals do, we must first assess the situation to determine what needs to be done to improve it. At a first step in assessing environmental insult, epidemiologists look at clusters and other indications of elevated exposures and effects in populations. For example, certain cancers, as well as neurological, hormonal, and other chronic diseases have been found to be significantly higher in minority communities and in socioeconomically depressed areas. Acute diseases, as indicated by hospital admissions, may also be higher in certain segments of society, such as pesticide poisoning in migrant workers.[5] These are examples of *disparate effects.* In addition, each person responds to an environmental insult uniquely, and that person is affected differently at various life stages. For example, young children are at higher risk to neurotoxins. This is an example of *disparate susceptibility.* However, subpopulations also can respond differently than the entire population, meaning that developmental and genetic differences seem to affect people's susceptibility to contaminant exposure. Scientists are very interested in genetic variation, so that genomic techniques[6] (e.g., identifying certain polymorphisms associated with disease susceptibility) are a growing area of inquiry.

In a sense, historical characteristics constitute the "environmental" aspects of EJ communities, and socioeconomic characteristics entail the "justice" considerations. The two sets of criteria are mutually inclusive, so for a community to be defined as an EJ community, both of these sets of criteria must be present.

One of the National Academies, the Institute of Medicine,[7] has found that numerous EJ communities experience a "certain type of double jeopardy." Not only must these communities tolerate elevated levels of exposure to contaminants, but they are usually ill equipped to deal with these exposures because so little is known about the exposure scenarios in EJ communities. The first problem (i.e., higher concentrations of contaminants) is an example of *disparate exposure*. The latter problem is exacerbated by the disenfranchisement from the political process that is endemic to EJ community members. This is a problem of *disparate opportunity* or even *disparate protection*. (This harkens back to the Constitution's requirement of equal protection.) The report also found large variability among communities as to the type and amount of exposure to toxic substances. Each contaminant has its own type of toxicity. For example, one of the most common exposures in EJ communities is to the metal lead (Pb) and its compounds. The major health problem associated with Pb is diseases of the brain as well as central and peripheral nervous system harm, including learning and behavioral problems. Another common contaminant in EJ communities is benzene, as well as other organic solvents. These contaminants can also be neurotoxic, but also have very different toxicity profiles from neurotoxic metals such as Pb. For example, benzene is a potent carcinogen, having been linked to leukemia and lymphatic tumors as well as to severe types of anemia. The two contaminants also have very different exposure profiles. For example, Pb exposure is often in the home and yard, whereas benzene exposures often result from breathing air near a source (e.g., at work or near an industry, such as an oil refinery or pesticide manufacturer). The Institute of Medicine's findings point to the need for improved approaches for characterizing human exposures to toxicants in EJ communities.

One of the first places to recognize the disparate exposures was in Warren County, North Carolina, but numerous other communities have experienced uneven, and arguably unjust, disparities in environmental protection. However, there is little consensus as to what defines an environmental injustice and whether, in fact, an injustice has occurred in many of these communities.

CASE STUDY: THE WARREN COUNTY, NORTH CAROLINA, PCB LANDFILL[8]

A rural county in North Carolina is recognized as the birthplace of the environmental justice movement. The story began in the late 1970s when the Raleigh-based Ward Transfer Company needed to get rid of more than 30,000 gallons of oil contaminated with polychlorinated biphenyls (PCBs). PCBs were first marketed in 1929 and were manufactured in various countries and with different trade names (e.g., Aroclor, Phenoclor). They were used as a heat transfer fluid in electrical transformers and, as such, were the "perfect" solution to the generation of heat during electricity transmission.

PCBs were nonreactive, did not biodegrade, were easy to use, and most important, were cheap. Millions of gallons of PCBs were used in electrical transformers all over the world.

Concern over the toxicity and persistence in the environment of PCBs caused the U.S. Congress in 1976 to enact a specific section, 6(e), of the Toxic Substances Control Act (TSCA) to address PCB contamination. This included prohibitions on the manufacture, processing, and distribution in commerce of PCBs. This is the "cradle to grave" (i.e., from manufacture to disposal) management of PCBs in the United States. Similar prohibitions and management measures were adopted worldwide.

Like all environmental problems, or any engineering problem for that matter, the first step is to understand the scientific facts of the case. To begin, what are PCBs, and why do they elicit such concern? PCBs all have the structure of $C_{12}H_{10-n}Cl_n$, where n is within the range 1 to 10:

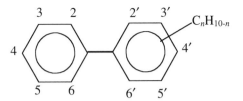

polychlorinated biphenyl structure

Although all PCBs have this arrangement, they differ from each other by the number and location of chlorine atoms at each of the numbered positions. These different arrangements are known as *congeners* (i.e., a single, unique, well-defined chemical compound in the PCB category). The name of a congener specifies the total number of chlorine substituents[9] and the position of each chlorine. For example: 4,4'-dichlorobiphenyl is a congener comprising the biphenyl structure with two chlorine substituents, one on each of the two carbons at the 4 (also known as *para*) positions of the two rings. PCBs can exist as 209 possible chlorinated biphenyl congeners, although only about 130 of these are generally found commercially.

The U.S. Environmental Protection Agency (EPA) decided to control the manufacture, transportation, and use of PCBs, and included in this control was making the resale of PCB-contaminated transformer oil illegal. The Ward Transfer Company's problem was a foreshadowing of what companies handling PCBs would face in the coming decades. They had an enormous amount of PCB-contaminated oil to dispose of, so they asked another company, owned by Robert J. Burns, to remove the soil. According to Robert D. Bullard (see his biographical sketch), this latter company "chose the cheap way out," and Burns's trucks ended up getting rid of the contaminated oil by spraying it along North Carolina roadsides. By the time this crime was discovered, 200 miles of highways had been contaminated, and the soil alongside the highways became hazardous.[10] PCBs adhere tightly to soil particles, so there was little danger of the chemical leaching into water, but it could not be just left there either, so the decision was made to dig up the

Figure 1.1 The confluence of two social upheavals, the civil rights movement and environmental protection, was apparent in the demonstrations in Warren County, North Carolina. Waste transport was interrupted as one of the protests against the proposed siting of a PCB landfill in the county. (Photo credit: Jenny Labalme, used with permission.)

soil along the highways and take it to a controlled landfill facility where it could be properly managed.

After an extensive search, a site in Warren County was chosen for the PCB landfill. This did not sit well with the predominantly African American residents of the county, and the spontaneous and large demonstration (see Figure 1.1) opposing the siting of a landfill made international news.

The site is located in Shocco Township, which has a population of approximately 1300. Sixty-nine percent of the township residents are nonwhite, and 20% of the residents have incomes below the federal poverty level. Residents of Warren County and civil rights leaders passionately protested the location of the landfill in Warren County. These protests are considered the watershed event that brought environmental justice to the national level. In 1982, during construction of the landfill, then-Governor Jim Hunt made a commitment to the people of Warren County. He stated that if appropriate and feasible technology became available, the state would explore detoxification of the landfill.

An environmental response is often precipitated first by a complaint. But to complain, one must have a "voice." If a certain group of people has had little or no voice in the past, they are likely to feel and be disenfranchised. Although there have been recent examples to the contrary, African American communities have had little success in voicing concerns about environmentally unacceptable conditions in their neighborhoods. Hispanic Americans may have even less voice in environmental matters since their perception of government, the final arbiter in many environmental disagreements, is one of skepticism and outright fear of reprisal in the form of being deported or being "profiled." Many of the most adversely affected communities are not likely to complain.

Biographical Sketch: Robert D. Bullard

Robert D. Bullard is Ware Professor of Sociology and Director of the Environmental Justice Resource Center at Clark Atlanta University (CAU). Prior to joining the faculty at CAU in 1994, he served as a professor of sociology at the University of California–Riverside and visiting professor in the Center for African American Studies at UCLA.

Bullard served on President Clinton's Transition Team in the Natural Resources and Environment Cluster (i.e., Departments of Energy, Interior, and Agriculture, and the Environmental Protection Agency) and on the U.S. EPA National Environmental Justice Advisory Council (NEJAC), where he chaired the Health and Research Subcommittee. He is a widely published and read author, with one of his latest books, *Dumping on Dixie: Class and Environmental Quality* (Westview Press, Boulder, CO, 2000), becoming a standard text in the environmental justice field.

Bullard's most lasting contribution was the 1983 report on the siting of Houston's municipal disposal sites. He found that six of the eight incinerators, all six of the city landfills, and three or four of the privately owned landfills were located in African American neighborhoods. His findings were used in a class action suit to block the construction of yet another landfill in an African American neighborhood. Although they lost the suit, the action was the first case that used civil rights law to challenge the siting of a waste facility.

Land use is always a part of an environmental assessment. However, justice issues are not necessarily part of these assessments. Most environmental impact assessment handbooks prior to the late 1990s contained little information and few guidelines related to fairness issues in terms of housing and development. They were usually concerned about open space, wetland and farmland preservation, housing density, ratios of single- *versus* multiple-family residences, owner-occupied housing *versus* rental housing, building height, signage and other restrictions, designated land for public facilities like landfills and treatment works, and institutional land uses for religious, health care, police, and fire protection.

When land uses change (usually to become more urbanized), the environmental impacts may be direct or indirect. Examples of direct land-use effects include *eminent domain,* which allows land to be taken with just compensation for the public good. Easements are another direct form of land-use impacts, such as a 100-meter right-of-way for a highway project that converts any existing land use (e.g., farming, housing, or commercial enterprises) to a transportation use. Land-use change may also come about indirectly, such as secondary effects of a project that extend, in time and space, the influence of a project. For example, a wastewater treatment plant and its connected sewer lines will create accessibility that spawns suburban growth.[11] People living in very expensive homes may not even realize that their building lots were once farmland or open space and that had it not been for some expenditure of public funds and the use of public powers such as eminent domain, there would be no subdivision.

Environmentalists are generally concerned about increased population densities, but housing advocates may be concerned that once the land use has been changed, environmental and zoning regulations may work against affordable housing. Even worse, environmental protection can be used as an excuse for some elitist and exclusionary decisions. In the name of environmental protection, certain classes of people are economically restricted from living in certain areas. This problem first appeared in the United States in the 1960s and 1970s in a search for ways to preserve open spaces and green areas. One measure was the minimum lot size. The idea was that rather than having the public sector securing land through easements or outright purchases (i.e., *fee simple*) to preserve open spaces, developers could either set aside open areas or require large lots in order to have their subdivisions approved. Thus, green areas would exist without the requisite costs and operation and maintenance funds entailed by public parks and recreational areas. Such areas have numerous environmental benefits, such as wetland protection, flood management, and aesthetic appeal. However, minimum lot size translates into higher costs for residences. The local rules for large lots that result in less affordable housing is called *exclusionary zoning*. One value (open space and green areas) is pitted against another (affordable housing). In some cases it could be argued that preserving open spaces is simply a tool for excluding people of lesser means or even people of minority races.[12]

CASE STUDY: HABITAT FOR HUMANITY

A recent case reflects the common problem of competing values in land use. The housing advocacy group Habitat for Humanity proposed a development of affordable houses in Chapel Hill, North Carolina, which has one of the most expensive real estate markets in the southeastern United States. Being a college town (home to the University of North Carolina), numerous groups, including churches and student coalitions, are calling for "livable housing."

The cost of housing in this town is well above the state average, so a number of advocates have supported the Habitat model, where potential homeowners invest in their own homes through "sweat equity" and receive voluntary support. But some groups formed in opposition to the plan. In an early meeting, one neighbor stated a desire that the homes be like those in a nearby high-cost subdivision (houses costing much more than even the already expensive town average). She recommended that they be "single-family homes with a nice, friendly, college town look and feel." (The quotes have been changed to protect anonymity, but the meanings are maintained.) Another later said that "From day 1, we have said that that parcel is not suited for a high-density project." That may be the case, but the result of such thinking is, in the end, exclusionary. People are very passionate and protective about their neighborhoods, well beyond concern about property values. This is a form of NIMBY ("not in my backyard"), so common to the environmental engineer, who must balance science and social needs to site unpopular facilities such as landfills and treatment plants.

To serve their clients effectively, engineers must be sensitive to the fact that most of us want to protect the quality of our neighborhoods, but at the same time, engineers and land-use planners must take great care that their ends (environmental protection) are not used as a rationale for unjust means (unfair development practices). Like zoning ordinances and subdivision regulations, environmental laws and policies should not be used as a means to keep lower-socioeconomic groups out of privileged neighborhoods.

How can the engineering profession help people who have historically had no voice or have not been taught how to make it heard? For environmental fairness, everyone potentially being affected needs a voice and a place at the table from the earliest planning stages of a project. The default seems to be changing, but some argue that environmental quality is still being used, knowingly or innocently, to work against fairness. And people who are likely to be exposed to the hazards brought about by land-use decisions need to be aware as options are considered, and well before decisions are made. This principle should be applied not only to land development and civic decisions but to everything engineers do that may have an impact on health, safety, and welfare of the public.

Sometimes we need to remind ourselves of just how far we have come in the past 50 years in battling pollution. We also need to realize that the general perception of environmental quality and the public's expectations have grown substantially in a relatively short time. The revolution in thinking and the public acceptance of strong measures to regulate the actions of private industry has been phenomenal. So what may previously have been considered to be simply the "cost of doing business" (brown haze, smelly urban areas, obviously polluted water, and disposal of pollutants in pits, ponds, lagoons, and by land burial) is now considered to constitute inappropriate and even immoral activities. However, it is not automatic that all private and public entities have gotten that message. Engineers can help continue to raise their client's appreciation of fairness and justice as well as the improvement to the "bottom line" that can result from strong environmental programs.

CASE STUDY: CARVER TERRACE[13]

One of the key characteristics of EJ advocates has been their patience. Another is persistence. In the 1970s the citizens of Carver Terrace, in Texacana, Texas, a predominantly African American community, began to see dark, vile-smelling "gunk" oozing out of their lawns. They could not interest the local authorities in the problem, even when they started to believe that their community was experiencing a higher than usual number of medical problems.

In 1978 the problems at Love Canal made hazardous waste a national issue and problem. A year later, after Congress ordered large chemical companies to identify what chemicals they had disposed of and at what locations, the discovery was made that Koppers Company of Pittsburgh had operated a creosote plant in the area now known as Carver Terrace, and when Koppers closed the plant they bulldozed everything into the ground, including the vats and the pond holding creosote, a known human carcinogen. Because the land was inexpensive, poorer families eagerly bought lots and built homes. About 25,000 people, 85% consisting of racial minorities, lived within four miles of the former creosote plant.

When it became known that the ground contained large quantities of creosote, the U.S. EPA sent a team of hazardous waste experts who nosed around in their "moon suits" and that particular study concluded that there was no problem. The citizens knew better, and soon found out that the EPA had conducted other studies and reports that clearly showed this area to be a candidate for Superfund cleanup, but that these studies were not made available to the citizens of Carver Terrace.

In retrospect, the extent of pollution was large; 45 million gallons of shallow groundwater, along with 2150 cubic yards of soil, was contaminated to a depth of 1 foot. The situation was so bad that the citizens group urged the government to "buy out" the residents of Carver Terrace just as they were now buying out people who lived around Love Canal. They could not see why they were being treated differently from those at Love Canal. The residents pointed out that the only obvious difference between the Love Canal residents and the Carver Terrace people was that the former were mostly European American, whereas Carver Terrace was mostly African American. Eventually, through energetic and determined activism, the government also bought out the residents of Carver Terrace.

It is incorrect to conclude that the only way that environmental injustice occurs is from the profit motive and its driving corporate decisions to site environmentally hazardous facilities where people are less likely to complain. Public decisions have also brought lower socioeconomic communities into environmental harm's way. Although public agencies such as housing authorities and public works administrations do not have a profit motive *per se,* they do need to address budgetary and policy considerations. If open space is cheaper and certain neighborhoods are less likely to complain (or by extension, vote against elected officials), the "default" for unpopular facilities such as landfills and hazardous waste sites may be to locate them in lower-income neighborhoods. Also, elected and appointed officials and bureaucrats may be more likely to site other types of unpopular projects, such as public housing projects, in areas where complaints are less likely to be aired or where land is cheaper.

CASE STUDY: WEST DALLAS LEAD SMELTER[14]

An engineering decision about where to site a facility will affect the lives of people for decades. In 1954, the Dallas, Texas, Housing Authority built a large public housing project on land immediately adjacent to a lead smelter. The project had 3500 living units and became a predominantly African American community. During the 1960s the lead smelter stacks emitted over 200 tons of lead annually into the air. Recycling companies had owned and operated the smelter to recover lead from as many as 10,000 car batteries per day. The lead emissions were associated with blood-lead levels in the housing project's children, and these were 35% higher than the levels in children from comparable areas. Lead is a particularly insidious pollutant because it can result in developmental damage. Study after study in this area showed that the children living near this project were in danger of higher lead levels, but nothing was done for over 20 years. Finally, in the early 1980s, the city brought suit against the lead smelter, and the smelter immediately initiated control measures that reduced its emissions to allowable standards. The smelter also agreed to clean up the contaminated soil around the smelter and to pay compensation to people who had been harmed.

This case illustrates two issues of environmental racism and injustice. First, the housing units should never have been built next to a lead smelter. The short-sighted reason for locating the units there would have been justified on the basis of economics. The land was inexpensive and thus saved the government money. The second issue was the foot dragging by the city in insisting that the smelter clean up the emissions. Once the case had been made, within two years the plant was in compliance. By 2003, blood-lead levels in West Dallas were below the national average. Why did it take 20 years for the city to do the right thing?

Despite the general advances in environmental protection in the United States, the achievements have not been evenly disseminated throughout our history. Like much of the rest of our culture for the past three centuries, environmental science and engineering have not been completely just and fair. The history of environmental contamination has numerous examples where certain segments of society were and are exposed inordinately to chemical hazards. This has been particularly problematic for communities of low-socioeconomic status. For example, the landmark study by the Commission for Racial Justice of United Church of Christ[15] found that the rate of landfill siting and the presence of hazardous waste sites in a community were disproportionately higher in African American communities. Occupational exposures may also be disproportionately skewed in minority populations. For example, Hispanic workers can be exposed to higher concentrations of toxic chemicals where they live and work, in large part due to the nature of their work (e.g., agricultural chemical exposures can be very high when and shortly after fields are sprayed, as shown in Figure 1.2).

In 1992, the U.S. EPA created the Office of Environmental Justice to coordinate the agency's EJ efforts, and in 1994, President Clinton signed Executive Order 12898, "Federal Actions to Address Environmental Justice in Minority and Low-Income Populations." This order directs that federal agencies attend to the environment and human health conditions of minority and low-income communities, and requires that the agencies incorporate EJ into their missions. In particular, EJ principles must be part of the federal agency's day-to-day operation by identifying and addressing "disproportionately high and adverse human health and environmental effects of programs, policies and activities on minority populations and low-income populations."[16]

LEGAL ASPECTS OF ENVIRONMENTAL JUSTICE[17]

The environmental justice legal footing is based on Title VI, § (paragraph) 601 of the Civil Rights Act of 1964, which makes it a crime to discriminate on the basis of race, gender, or ethnic origin. Intentional discrimination is, however, very difficult to prove

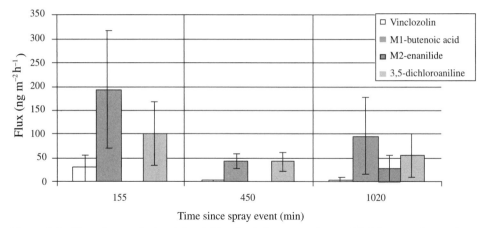

Figure 1.2 Flux of an agricultural fungicide after being sprayed onto soil. These results are from a laboratory chamber study of vinclozolin (5 mL of 2000 mg L^{-1} suspended in water); bars show the time-integrated atmospheric flux of organic compounds from nonsterile North Carolina Piedmont soil (aquic hapludult) with pore water pH 7.5, following a 2.8-mm rain event and soil incorporation. Error bars indicate 95% confidence intervals. The parent compound, vinclozolin [3-(3,5-dichlorophenyl)-5-methyl-5-vinyloxzolidine-2,4-dione], M1 (2-[(3,5-dichlorophenyl)carbamoyl]oxy-2-methyl-3-butenoic acid), and M2 (3′,5′-dichloro-2-hydroxy-2-methylbut-3-enanilide) are all suspected endocrine-disrupting compounds (i.e., they have been shown to affect hormone systems in mammals). This indicates that workers are potentially exposed not only to the parent compound (i.e., the pesticide that is actually applied) but to degradation products as the product is broken down in the soil. [From D. A. Vallero and J. J. Peirce, Transport and Transformation of Vinclozolin from Soil to Air, *Journal of Environmental Engineering,* 128(3):261–268, 2002].

since the defendant would obviously not admit that he or she was biased against a minority. Paragraph 602 of this act, however, makes the point that discrimination can also be proven by actions, know as *de facto* discrimination. That is, if a certain minority population receives fewer benefits from, say, a local government, the only reason for such disparity would reasonably be discrimination. To prove such discrimination, however, substantial effort (read "money") is required, and although Title VI does not explicitly give the right to sue for disparate treatment in court, the courts have upheld the right of private action under §602 for allegation of disparate impact.

Over the years, discrimination has been litigated in numerous cases. For example, in 1983 in *Guardian Association v. Civil Service Commission,* the court found that a written entrance examination required by a police department resulted in disparate results that could be identified by race. African American and Hispanic applicants did not perform as well on the exam, and thus even though the exam was supposed to be neutral, the actual fact of disparate performance demonstrated discrimination.

In 2001, however, the Supreme Court in *Alexander v. Sandoval* ruled that there is no private right of action to enforce disparate impact regulations under Title VI of the Civil Rights Act, thus removing a major judicial tool from individuals and groups who believe that they have been discriminated against. Various circuit court decisions since

Biographical Sketch: Benjamin Chavis, Jr.

Benjamin Chavis, Jr., was born in 1949 in Oxford, North Carolina, and served as a youth coordinator with the Southern Christian Leadership Conference, working in the 1960s with Rev. Martin Luther King, Jr. to desegregate southern schools. When he became an ordained minister, he continued to agitate for racial justice, and got into trouble in Wilmington, North Carolina, where he was convicted of conspiracy and arson. He spent nearly a decade in prison before the charges were thrown out in 1980.

On regaining his freedom, he became the director of the United Church of Christ's Commission for Racial Justice. In 1982 he concluded that the selection of the PCB landfill for Warren County, North Carolina (very near his birthplace) had to be racially motivated. In his view, this poor, predominantly African American county was singled out because its people were unlikely to protest the selection of the disposal site. He called this *environmental racism,* a term he later changed to *environmental justice.*

Teaming with Charles Lee, he wrote the 1987 landmark report "Toxic Wastes and Race in the United States," which documented the uneven distribution of environmentally undesirable land use in African American and other minority communities. They found, for example, that in communities with two or more hazardous waste disposal facilities, the average minority population was more than three times that of communities without such facilities. The report also found that the U.S. EPA took longer to clean up waste sites in poorer areas than in more affluent neighborhoods.

Biographical Sketch: Charles Lee

Charles Lee was Director of Environmental Justice for the United Church of Christ Commission for Racial Justice and worked with Ben Chavis to author the landmark 1987 report "Toxic Wastes and Race in the United States."

Presently, Lee is with the U.S. EPA working on implementation of an executive order promoting environmental justice. He is also a lecturer at the Hunter College School of Health Sciences.

2001 have supported and strengthened this legal concept. As a result, not only do individuals lack standing in court to bring claims of discrimination in violation of the Civil Rights Act, but these claims cannot be enforced through the U.S. Code.

The Sandoval case is interesting and deserves a closer examination. The issue developed in 1990 when the state of Alabama declared English to be the official language of the state. This declaration prompted the Alabama Department of Public Safety to administer its driver's license examinations only in English. Martha Sandoval, a Hispanic American woman, filed a class action suit under §602 of Title VI of the Civil Rights Act against the department and its director, alleging that the policy prevented her, a citizen and taxpayer, from driving a car on roads constructed with federal funds. She won her case, and this was confirmed by the Court of Appeals. But in 2001 the U.S. Supreme Court, in a 5 to 4 vote, held that Congress had not intended the Civil Rights Act to provide private right of action for disparate impact. In other words, Ms. Sandoval may have been discriminated against, but the Court argued that the law did not allow her the right to sue for relief. Justice Antonin Scalia wrote the majority opinion, and argued that a private right of action to enforce federal law has to be created by Congress, and that the Civil Rights Act does not do that.

The majority opinion went further, however, and attacked the text and structure of Title VI, asking if Congress had intended to create private action for any form of discrimination. The ruling suggested that other forms or relief, such as the withholding of funds by an agency, would be legitimate means for enforcing the law, but that private individuals do not have the right to seek relief.

The minority, led by Justice John Paul Stevens, argued that Congress had indeed intended for private right of action. It saw the right of private action as a part of an integrated remedial pattern and that Congress has given agencies broad powers to regulate §602 claims, thus making private action implicit. The minority opinion suggested that future suits might use the U.S. Code to seek relief, specifically 42 U.S.C. §1983.

With the Civil Rights Act, Congress intended to remedy discrimination, seeing this as a national problem. Those opposed to the ruling argue that it seems unlikely that Congress's intent was not to allow private individuals who felt that they had been discriminated against the right to sue for relief and to use only the power of executive agencies to seek such relief. This argument can be viewed as dangerously circular, since often it is the very agencies that practice disparate behavior. One argument is that the ideals of §601 were intended to be carried out by the remedies in §602, and that the intent of Congress was quite clear. The narrow ruling by the majority, however, prevented further court actions.

The effect of the *Sandoval* decision was felt immediately in environmental matters. In *South Camden Citizens v. New Jersey Department of Environmental Protection,* the South Camden Citizens in Action group had filed an injunction, based on Title VI of the Civil Rights Act, to prevent the final approval of a permit to a cement company for discharging its emissions. The essence of the complaint was that the emissions would most severely affect a predominantly minority community that already was the home of 20% of the city's contaminated sites. The District Court of New Jersey had already granted the injunction, basing its decision on §602 of Title VI. But five days after the state court had decided the case, the U.S. Supreme Court rendered the *Sandoval* decision, stating that there is no private right of action to enforce regulations under §602 of Title

VI. The defendant immediately requested a stay of the injunction, and eventually the courts ruled that *Sandoval* had indeed prevented such suits. More important, the Third Circuit Court ruled in the *South Camden* case that 42 U.S.C. §1983 could not be used to enforce federal regulations unless the right had been expressly given to do so. Such language is not in Title VI of the Civil Rights Act.

The effect of *South Camden* was to bar the door from any further legal action based on disparate environmental impact. The only way that environmental justice issues can now be addressed is through the various agencies. When the agencies themselves are the defendants, however, such relief is not possible.

Because the U.S. Supreme Court had not ruled explicitly on the use of 42 U.S.C. §1983 as a vehicle for seeking private relief, there was some hope that this could still be applied to environmental issues. This small hope was eliminated in the case of *Gonzaga v. Doe,* a case having to do with the release of personal information. The plaintiff sued Gonzaga University because information about sexual misconduct had been released to a teacher certification panel, in clear violation of the Federal Education Rights and Privacy Act (FERPA) of 1974. The majority opinion, written by Justice William H. Rehnquist, found that the issue is one of "rights." That is, §1983 can be used to seek relief if rights, and only rights, have been violated. The opinion stated that ". . . Section 1983 provides a remedy only for the deprivation of 'rights, privileges, or immunities secured by the constitution and laws' of the United States. Accordingly, it is the rights, not the broader or vaguer 'benefits' or 'interest' that may be enforced under the authority of that section." The opinion stated that "if Congress wishes to create new rights enforceable under §1983, it must do so in clear and unambiguous terms." The opinion concluded that that the FERPA nondisclosure provision fails to confer such enforceable rights.

The minority opinion stated that even if the standard of rights is used, there is a federal right to have §1983 enforced and that the Court's own rulings had implied the right of action on claims that ". . . reflect a concern, grounded in separation of powers, that Congress rather than the courts control the availability of remedies for violation of statues."

The net effect of the *Gonzaga* ruling was to remove the one remaining possibility of using 42 U.S.C. §1983 for seeking relief in cases of perceived environmental injustice. Despite calls for the U.S. Congress to pass legislation that would supercede *Sandoval,* giving private individuals that explicit right to sue under Title VI of the Civil Rights Act, there has been no such action.

Given the lack of ability to seek relief through the courts, some federal agencies have created their own policies on environmental justice, the most important one being the policy developed by the U.S. EPA. In 1998, in response to Executive Order 12898 from President William Clinton, the agency issued its *Interim Guidance for Investigating Title VI Administrative Complaints Challenging Permits,* and this has become the primary mechanism for analyzing disparate impact complaints. These guidelines do not create any rights enforceable by parties in litigation with the United States, but creates a systematic way of addressing such complaints.

The *Interim Guidance* document specifically lists five actions that must be taken in response to a complaint:

Biographical Sketch: Christopher Stone

Just because an animal, a tree, or even a place cannot hire a lawyer and argue its case in court, does this nonhuman have an absence of standing? Why can't a human being who can argue the case do so on behalf of the animal, tree, or other nonhuman?

This was the question that came before the U.S. Supreme Court in 1967 in a famous case, *Storm King v. Federal Power Commission*. The Federal Power Commission wanted to lease out some federally owned land for a ski area, but the Sierra Club objected, believing that the development would harm the forests. However, the problem was that no single member of the Sierra Club could prove that he or she was being directly harmed by the construction of the ski slopes. The harm to each person would be small, and collectively the harm would be great, but there was no one person who would have standing in the court. This is the problem of the 25-cent debt. The damage to a single person is just not great enough to warrant taking up the court's time.

One of the lawyers on the case was Christopher Stone (born 1937), a professor on the faculty of the University of Southern California law school. Stone, with an undergraduate degree from Harvard and a law degree from Yale, had been active in the Sierra Club and helped them on numerous occasions. In support of the Sierra Club's case, Stone wrote an article cleverly entitled "Should Trees Have Standing?" in which he argued that natural objects have every right to be represented in court if the damage is sufficiently great. The article appeared in the law review and became part of the brief submitted to the Court. Sitting on the Court at that time was Justice William Douglas, who was an avid outdoorsman and was quite sympathetic in cases involving the destruction of natural habitats. Using Stone's arguments, Douglas was able to sway the court in favor of the plaintiffs and to stop development of the ski area. Although it is unlikely that this common law precedent will be widely used in the control of environmental pollution in the future, its successful application in the 1960s fueled and encouraged the environmental movement.

- Identifying the population affected by the facility
- Determining the racial or ethnic composition of the population
- Examining other permitted facilities within the area
- Conducting a disparate analysis
- Determining the significance of the disparity

If, using this procedure, the U.S. EPA believes that an apparently disparate impact exists, it asks the permitted agency to rebut its findings. For example, if a city is intending to build a wastewater treatment plant in an economically depressed neighborhood, and as a result of a complaint, EPA finds that a disparate situation exists, it would ask the city to show why this is not a case of environmental injustice. The city would then have to prove that the benefits accrued by the affected population outweighs the detrimental effects of the facility. Merely demonstrating that the facility has met all of the regulatory requirements is insufficient. This process reverses the onus of responsibility, requiring the agency to prove that the action will benefit the people affected instead of requiring the people to prove that they are being harmed.

The case is examined by the Office of Civil Rights and EPA experts, and if the finding is that the planned facility is such that mitigation is not possible and that the entire project, on balance, will not be beneficial to the affected population, the EPA can then withdraw funding or permits.

There was a great deal of opposition from private parties and from the states to the use of the EPA *Interim Guidelines,* arguing that this process is in conflict with present land-use laws and regulations, that it does not provide standards on which decisions are based, and that the procedure has not been promulgated with the consent or even input from the states. Many saw this as a backdoor way of proving discrimination by disparate impact, a concept specifically struck down by the U.S. Supreme Court.

Others argued that the use of the *Interim Guidelines* resulted in unnecessary delay and litigation in cases where cities and states are attempting to achieve economic development in low-income neighborhoods. The placement of a power plant in a low-income neighborhood, they argue, would create jobs and provide economic improvement to the community. This argument is that the guidelines harm the very people they are intended to protect.

The complaints were loud enough to cause Congress to intervene, and in October 1998 a rider on an appropriations bill forbade the EPA from accepting new Title VI complaints. This drew strong opposition from the Office of Management and Budget, which called the measure inappropriate and antienvironmental. But because of such riders on appropriations bills, the EPA was unable to initiate new studies on disparate environmental impact. Finally, in 2002, the Congress ceased attaching riders to appropriations bills, believing (correctly in retrospect) that the Bush administration would not pursue such policies.

CASE STUDY: SHINTECH HAZARDOUS WASTE FACILITY, ST. JAMES PARISH, LOUISIANA

Before the moratorium on processing complaints went into effect, the EPA accepted a complaint from the Tulane Environmental Law Clinic in cooperation with other environmental groups with regard to a large toxic waste disposal facility that was to be constructed by Shintech, a Japanese firm, in the St. James Parish in Louisiana. This parish is poor and predominantly African American, and is the location of a vast array of industrial plants. The company promised jobs both in the construction and operation of the plant as an enticement to the community. The complaint, however, stated that the emissions from this facility would create a disparate environmental impact on the minority population.

The allegations of disparate impact were supported in part by the fact that 18 toxic waste facilities were located in St. James Parish, and almost a quarter of all the pollutants produced in the state were emitted within a four-mile radius of the parish. The case was accepted by the Office of Civil Rights for review, but they decided not to report their conclusions until the EPA guidelines were published. During this administrative delay, Shintech decided to move the plant to a middle-class neighborhood, thus making the case moot. It should be noted that the new location was advantageous to Shintech since it was close to a Dow Chemical plant, and this allowed the waste to be pumped to the waste treatment facility, saving considerably on the cost of transport. Such decisions point to the complicated nature of environmental justice. Companies optimize on a number of variables. In this case, the cost of pumping may have outweighed any savings from siting the plant in a lower-income neighborhood.

The EPA had accepted a second case for review, and this case did not end in quite the same manner as the Shintech hazardous waste facility.

CASE STUDY: SELECT STEEL CORPORATION RECYCLING PLANT, FLINT, MICHIGAN

In 1998 the Michigan Department of Environmental Quality approved an emissions permit for a steel recycling mini-mill in Flint, Michigan, to be constructed by the Select Steel Corporation. A mini-mill uses an electric arc furnace to melt scrap steel. A local group filed a Title VI complaint asserting discriminatory impact on a minority community. The Office of Civil Rights accepted that case for review and was pressured into quick action by Select Steel's threat to move its plant to Ohio. EPA's delay in insisting on a careful review of this complaint caused significant political pressure. Michigan's Governor Engler criticized the EPA in a press conference, saying in part: "This is about every company that has ever had to deal with the EPA's reckless, ill-defined policy on environmental justice. . . . The EPA is imposing their bureaucratic will over this community and punishing the company with the latest environmental standards, all because of a baseless complaint. . . . The new result is that the EPA is a job killer."[18]

The *Detroit Free Press* relentlessly attacked the EPA, calling it a "rogue agency" and devoting large amounts of news space to the controversy. Whether the public bashing had any effect on EPA is still unknown, but its decision was in favor of the steel company, arguing that all of the permits had been allowed correctly and that no emission regulations would be violated from the emission from the facility. In other words, if there is no standard, the effect of the emissions is not a problem.

The point of environmental justice is not whether or not all the emission guidelines have been met, but rather, whether or not the people affected by this facility are being treated fairly.

Following the decision, Select Steel decided to relocate its plant in Lansing, Michigan, instead of Flint, saying that they no longer wanted to fight with local groups.

The Select Steel case made it quite clear that the *Interim Guidelines* had to be reviewed and modified. If the argument could be made that adverse impact is acceptable when either no standard exists or when all of the emission standards have been met, there would be no argument for disparate impact.

The effect of the Shintech and Select Steel cases demonstrated that the system for determining environmental injustices was not functioning well. The stinging criticism appeared to have an effect on EPA's thinking, and as a result, EPA sought help from its Science Advisory Board, and with considerable input from others, revised the *Interim Guidelines*. The new guidelines were published as the *Draft Revised Guidelines for Investigating Title VI Administrative Complaints Challenging Permits* (*Guidelines for Investigating*) as well as *Draft Title VI Guidance for EPA Assistance Recipients Administering Environmental Permitting Programs* (*Guidance for Recipients*). The objective of the new guidelines was to reduce the chance of Title VI complaints by providing strategies for enhancing public participation in decision making.

The most important part of the first document, the *Guidelines for Investigating*, relates to communication and participation by all concerned parties. The *Guidelines for Recipients* suggests various approaches for analyzing the issues that may lead to Title VI complaints. One of the suggestions in the revised guidelines is the *area-specific approach*, which identifies geographic areas where adverse environmental impact may exist and encourages all interested parties to develop agreements before divisive situations arise. That is, the stakeholders can agree before any development occurs as to what would be acceptable in the specific area. These agreements would then be reviewed by the Office of Civil Rights to make sure that all interested parties have had an opportunity to participate.

The system would work in this way: An economically depressed neighborhood would agree in advance that the placement of a certain industry or municipal facility in the neighborhood would be acceptable and desirable. This plan would then be shelved until the day that such a facility was actually proposed. With the *a priori* agreement in place, there would be no cause for opposition and complaints based on Title VI discrimination.

The *case-by-case* approach allows state and local agencies (recipients of the federal grants) to develop criteria to evaluate permit actions that are likely to raise Title VI concerns. These criteria would then be applied to each case and would dictate whether the complaints had merit. These guidelines are useful for preventing future conflicts, but they do not resolve existing problems.

But even after these guidelines were published, the EPA was still restricted by Congress from pursuing Title VI complaints, and as a result a large backlog of complaints built up. EPA's appropriations bill for fiscal year 2002 did not include this restriction, and the then EPA Administrator Christine Todd Whitman responded by setting up a task force to clear away the backlog.

One way of clearing away the backlog is simply to deny as many of the complaints as possible, something that the task force had been accused of doing. As of June 20, 2003, of the original 136 pending complaints, only 16 had been accepted and three are still under review. The remainder, 86%, were rejected, dismissed, suspended, resolved, or referred to another agency.[19]

ARGUMENTS AGAINST ENVIRONMENTAL JUSTICE

Criticism of the environmental justice movement usually takes three forms:

- Denial that there is such discrimination
- If there is such discrimination, the discrimination is beneficial
- If there is discrimination, and if it is not beneficial, at least it is not racially motivated[20]

Let's consider these arguments in turn.

Argument 1: Environmental Injustice Does Not Occur

This argument questions whether injustice is actually occurring. One argument is that the appearance of disparate treatment is anecdotal and that rigorous studies have not been

---

Biographical Sketch: Gaylord Nelson

Perhaps President Bill Clinton said it best when he presented the highest civilian award, the Presidential Medal of Freedom, to Gaylord Nelson: "As the father of Earth Day, he is the grandfather of all that grew out of that event: the Environmental Protection Agency, the Clean Air Act, the Clean Water Act, and the Safe Drinking Water Act."

Gaylord Nelson, born in Wisconsin in 1916, received his B.A. degree at San Jose State College and his law degree from the University of Wisconsin. He was a state legislator, a two-time governor of Wisconsin, and a U.S. senator, serving a total of 18 years in that capacity. In 1969 he had what has been called one of the most powerful ideas of his time: Earth Day. The national event, during the time of dissent over the Vietnam war, drew over 20 million participants and put into motion the string of legislation that forms the backbone of our environmental law today.

In 1961, while governor of Wisconsin, he created the Outdoor Recreation Acquision Program, funded by a penny-a-pack tax on cigarettes, then acquired a million acres of parkland in Wisconsin. While in the U.S. Senate, he authored the legislation that preserved the 2100-mile-long Appalachian Trail. After leaving the Senate, he served for many years as a consultant to the Wilderness Society. He died in 2005, leaving a legacy of environmental protection and care of the planet.

done. The second approach is to argue that federal agencies such as the U.S. Environmental Protection Agency have not won any cases for complainants, and that a large number of complaints have been dismissed after review. In addition, the decisions of the U.S. Supreme Court and various Appeals Courts suggest that there are no legal grounds for such a case, and hence it does not exist. This is of course a spurious argument. Just because there are no legal precedents for correcting environmental injustice and just because agencies do not pursue with diligence complaints regarding discrimination does not mean that it does not exit.

A second type of denial of the existence of injustice centers on demographics and statistics. First, since it is nearly impossible to agree on an absolute definition of "race," it ought to be impossible to argue for discrimination. If we are somehow able to define race (the argument goes), we have to show that statistically, people of identifiable racial characteristics are being discriminated against.

To illustrate these arguments, let us define a town with four neighborhoods that has the following racial characteristics:

Neighborhood	Percent Minority
A	10
B	5
C	35
D	95

Which of these neighborhoods is considered a minority neighborhood? That is, where would the line be drawn?

A related problem with defining a minority neighborhood is in choosing the size of the land area. The seminal report by the United Church of Christ Commission for Racial Justice, for example, used zip codes to identify locations. This is a blunt tool at best, since so many small communities have only one zip code. The community discussed above, for example, might have a single zip code, and thus the uneven distribution of minorities would never be evident. A wastewater treatment plant located in neighborhood D would not even be on the environmental justice radar screen; and if this town had, say, 45% minority population, it would not show up as a minority community, and any suggestion of environmental discrimination would disappear.

There are numerous variations on the themes of this argument. For example, another argument against environmental justice is that it is not the percentage of people in a neighborhood that matters, but rather the absolute number of people affected. Using our hypothetical community, for example, suppose that neighborhood D in the community above is sparsely populated, and even though it is 95% minority, there are only a few people in that neighborhood.

The point of these arguments is to deny problems with disparate distribution of pollution and the resulting exposure because we cannot conduct well-documented studies to show that the disparacy exists. This classical "head in the sand" approach flies in the face of a precautionary approach. Indeed, if such arguments were universally held, few engineering projects would be implemented for want of 100% known conditions.

A third denial argument is that the undesirable land-use facilities often are not constructed in lower-socioeconomic neighborhoods, but rather, the neighborhoods grow up around such facilities because the land there is affordable. Experience has shown that this point has some validity. The location of airports has often caused major shifts of more affluent, at least more mobile, populations away because of the noise from airplanes. As these shifts may make these neighborhoods more affordable, families of lower socio-economic status replace the previous owners. People with fewer resources are able to afford homes in areas vacated by the more affluent, even though they are aware of the high noise levels. The shifts can even change the land use, changing neighborhoods from predominantly single-family dwelling units to multiple-family apartments and commercial use.

Certainly, airports are not the only type of driver for displacement. Many company towns have changed character over the decades as the workers who had to live near polluting facilities are better able to afford better housing in suburbia and exurbia. In fact, neighborhoods such as those in Gary, Indiana, Sauget, Illinois, and Detroit, Michigan were once vibrant, albeit heavily polluted neighborhoods. They are now almost completely populated by people with little if any affiliation with the adjacent industry.

The need of some citizens to seek more economical housing is not, of course, an excuse for exposing them to higher levels of environmental contaminants. People do not move to less expensive neighborhoods to be nearer contamination and unhealthy conditions. They move there because this is all they can afford.

Finally, some reports claim that all socioeconomic groups resist the siting of undesirable facilities and land use in their neighborhoods, and the final siting of these facilities in lower-socioeconomic-class neighborhoods is as a result of inability or unsophistication in being able to fight off such decisions.[21] The argument goes that sites are initially equally and equitably distributed, and the lower-socioeconomic neighborhoods are not very good at protecting their communities.

Although this may be true, it does not provide justification for disparate treatment. In fact, it is a tacit admission of injustice. The inability to summon sophisticated and expensive resistance to unfair siting practices demonstrates a type of second-order discrimination that results from first-order practices that have led to lower education levels and histories of discrimination in these neighborhoods. If engineers are to take a life-cycle view of their projects (one of our common themes in this book), we have to some extent failed when, after some time, our projects lead to injustices. We have a "social contract" with the public. Breaching of this contract must be avoided.

CASE STUDY: THE ORANGE COUNTY LANDFILL AND UNKEPT PROMISES[22]

Chapel Hill, now a booming community, was once a quaint village hosting the University of North Carolina, the flagship university in the North Carolina higher education system. The town has a storied past: the university being the first state institution to open its doors to students and the town surviving the invasion of Union troops during the Civil War.

Chapel Hill remained a village until the 1960s, when expansion of the university caused a surge in population and pressure on new developments.

During that time, Chapel Hill also was becoming a Mecca for retired people, with its mild climate, great golf courses, beautiful gardens, and of course the advantages of a first-rate university drawing people from the Northeast. The village was becoming a city of over 56,000 people.

During the 1960s progressive era, Chapel Hill organized the first truly integrated school system in North Carolina, carving out the central section of town in a way that essentially integrated all schools. This forward-looking liberal attitude carried through in the election of municipal officers, and it was no wonder that Chapel Hill was the first town in North Carolina to elect an African American as mayor.

Howard Lee was a talented and hard-working mayor who went on to become a state senator. During his tenure as mayor, he had to grapple with intense development pressures that necessitated the organization of many municipal services, including the creation of a bus service.

At that time the town was using a small landfill owned by the university for the disposal of its solid waste, but this landfill was rapidly running out of space and the university wanted to close it, so in 1972 a search began for a new landfill site. Searches then were not nearly as intense as they are today, and the entire process was quite informal. The town council decided that it wanted to buy a piece of land to the north of the town and make this the new landfill. This land seemed like a good choice since it was between Chapel Hill and Hillsborough, the county seat of Orange County, and within a short distance of Chapel Hill. It was also a convenient location for Carrboro, a small community next to Chapel Hill. There were no new housing developments near the proposed landfill site, and it was off a paved road, Eubanks Road, and this would facilitate the transport of refuse to the landfill.

However, a vibrant African American community, the Rogers Road neighborhood, abutted the intended landfill area, and these people expressed their dissatisfaction with the choice of a landfill site and went to Mayor Lee for help. The mayor talked them into accepting the decision, promised them that this would be the only landfill that would be located near their neighborhood, and that if they could endure this affront for 10 years, the finished landfill would be made into a neighborhood park. Most important, they were told that the next landfill for Chapel Hill would be somewhere else and that their area would not become a permanent dumping site. The citizens of the Rogers Road neighborhood grudgingly accepted this deal and promise and then watched as the Orange County Regional Landfill was built near their community.

The site for the landfill was 202 acres, cut into two sections by Eubanks Road, and abutting Duke Forest, a research and recreational facility owned by Duke University. On one side of the site was the Rogers Road neighborhood. The landfill, which had no liner or any other pollution control measures, was opened in 1972. The three communities contributing to the landfill, Chapel Hill, Carrboro, and Hillsborough, along with Orange County, formed a quasi-governmental body called the Landfill Owners Group (LOG) to operate the landfill. The LOG was comprised of elected officials from the four governmental bodies. One of the early actions by this group was to establish a sinking fund that would eventually pay for the expansion of this landfill or a new site when this became necessary.

As the population of Orange County exploded in the 1970s it became quite clear that this landfill would not last very long and that a new landfill would be needed fairly soon. The LOG, using money from tipping fees, purchased a 168-acre tract of land next to the existing landfill, called the Green Tract, with the apparent intent of using it when the original landfill became full, but without actually publicly declaring that this was the intended use for this land.

In the early 1980s it became apparent that a new landfill would be necessary, but by that time the Green Tract was considered to be too small for the next landfill. This would not be a long-term solution, and a need was apparent for a larger site that would accommodate the solid waste needs on a long-term basis. The four governmental agencies asked the LOG to initiate proceedings to develop a new landfill, which could be opened in the mid-1990s.

The LOG set up a landfill selection committee (LSC) to oversee the selection of the new landfill and asked Eddie Mann, a local respected banker and civic-minded citizen, to chair the LSC. The LOG directed the LSC to seek technical help with the selection process, and as a result, Joyce Engineering, a Virginia firm that had assisted other communities in the selection of landfills, was hired to conduct the search.

After a study of Orange County, Joyce Engineering selected 16 locations as potential landfill sites, using criteria established by the LSC such as proximity to cities, airports, and environmentally sensitive areas. One of the 16 sites chosen by Joyce was the Green Tract, which became known as OC-3.

The next step was to hold public hearings and then to cull the list of 16 down to a smaller list for final discussion. As the 16 sites were being considered, each was placed in one of three categories: (1) to be considered further, (2) to be placed in reserve for possible consideration later, or (3) not to be considered further.

Following these hearings, the LSC pared down the original 16 sites to five, one of which was the Green Tract. The LSC did not consider persuasive the argument that the former mayor of Chapel Hill had promised the residents in that neighborhood that future landfills would be located elsewhere. Since Howard Lee, the former mayor of Chapel Hill, did not represent Carrboro, Hillsborough, or Orange County, the well-intentioned promise was not considered binding by the other governmental entities. In addition, although Lee acknowledged making this promise, it was never found on any written document. (This is not uncommon in the southeastern U.S., where oral tradition often holds primacy over written documentation, which in part explains the discrepancy between Northern and Southern histories.) Further, the people who were least able to resist the backdoor expansion of the existing landfill, the Rogers Road neighborhood, were told that the promises made by elected officials were null and void because the new politicians could not be held to these promises. The effect of this argument was to suggest that any promise made by one administration does not need to be kept by another. This is analogous to buying savings bonds from the federal government with no guarantee that they will be redeemed in 10 years since a new administration will be in Washington. Or, in environmental parlance, all political decisions are unsustainable.

One of the problems with the Green Tract was that it was too small to afford a long-term solution, a source of encouragement to the Rogers Road neighborhood. But this was all changed when late in the process and well after the public hearings, Eddie Mann introduced a new site, OC-17. This site abutted the existing landfill and the Rogers Road neighborhood and included a large tract of land in Duke Forest, a section called the Blackwood Mountain region. The introduction of this site and its acceptance by the LSC as a finalist was a case of local politics at their worst.

The opponents of these two tracts, OC-3 (the original Green Tract) and OC-17 (the new Blackwood Mountain area), began to fight the selection process, aided by many Chapel Hillians who saw the inequity in this process. The resisters packed the LSC committee meetings, printed T-shirts ("WE HAVE DONE OUR SHARE"), wrote letters to the newspaper, and fought valiantly to keep the inevitable from happening.

In 1995 the LSC approved the selection of OC-3 and OC-17 as the new landfill but suggested that some form of compensation be made to the citizens in the Rogers Road neighborhood. The decision next went to the LOG for their consideration. The vote in the LOG was 6 to 3 in favor of the selected site. Two of the negative votes were by the representatives from Carrboro. The town of Carrboro would not be directly affected by the location of the landfill in the Eubanks Road area, and thus Carrboro ought to have had a clear selfish motive for choosing this site. But the two Carrboro representatives on the LOG, Mayor Mike Nelson and Alderwoman Jacquelyn Gist, based their negative vote on the promise made by Howard Lee to the Rogers Road neighborhood and announced that they would fight the selection of this site.

Nevertheless, having been approved by the LOG, the decision next went to the four governmental bodies for approval. Chapel Hill, Hillsborough, and Orange County approved the site with little debate. In the meeting of the Chapel Hill Town Council, the previous promise by Mayor Howard Lee was not even brought up. But Mayor Nelson and Alderwoman Gist convinced the Carrboro council to delay the approval until compensation could be worked out *in advance* of the decision, citing the previous broken promises as loss of trust in politicians.

This delay by Carrboro allowed Duke University to marshal its forces and to hire appropriate lawyers and scientists to come to the defense of Duke Forest. The university trustees voted unanimously to fight the siting, and the president of Duke, Nan Keohane, wrote a strong letter to the LOG and the four governmental bodies threatening legal action if the land in Duke Forest was to be taken. Using his knowledge of the area, Jud Edeburn, the manager of Duke Forest, quickly located areas with endangered species and several wetland locations, thus reducing the available acreage for the landfill. A historic African American cemetery was discovered in the forest and placed on the protected National Registry, further reducing the availability of land. But Joyce Engineering found ways to redesign the landfill so as to accommodate these restrictions and still use the major part of the tract for burial of solid waste. Demands for public hearings and more tests did not change the decision, and a year after the vote, OC-17 remained the first choice of the LOG and the three governments. The government of Carrboro was under increasing pressure to withdraw the opposition.

Then, in 1997, Duke University announced that it had deeded the Blackwood Mountain section of Duke Forest to National Aeronautics and Space Administration (NASA) for its use in conducting experiments. The federal government now controlled this land and the fight was over. It took clever legal work, the effective battle fought by the citizens of the Rogers Road neighborhood, and the courage of Carrboro's Mayor Nelson and Alderwoman Gist to stop the landfill from being sited at a location where the people had already done their share, although a just outcome was reached. Unfortunately, this is an example of consequentialist ethics—that is, the ends justify questionable means. This is a common complaint about the legal profession, but it is not uncommon to engineering.

Like many environmental justice problems, the local community on Rogers Road continues to be under the threat of public decisions, even after an apparent compromise is reached.

The most recent threat was in the spring of 2006 when the Orange County Commission was considering the landfill site to be the possible location for a transfer station where trucks unload garbage that is to be taken to another site. A local advocate, Reverend Robert Campbell, stated that the transfer station "is something that we don't want because a transfer station is going to create even more traffic, going to also create more waste along the highway."[24]

Gayle Wilson, the director of the county solid waste management department, said the department would "make it the best we can for the citizens out there. We will meet with them; try to meet their concerns as best as possible."[24] This is likely to be little solace for the citizens. Wilson's choice of words "citizens out there" is quite telling. It is almost unthinkable that the landfill or transfer station would ever be considered in one of Chapel Hill's gated suburban communities, as it is unlikely that these communities would be thought to consist of "citizens out there."

Example:

Know Your Acronyms

The public hearings in the Orange County Landfill Case were classic NIMBY ("not in my backyard") exercises. Two other useful and colorful acronyms have recently emerged in this area: BANANA (build absolutely nothing anywhere near anything) and NOPE (not on planet earth). Language reflects culture so engineers must be prepared, technically and mentally, for conflict on even the most (seemingly) benign project proposals. This perception is a challenge for the engineer who seeks to "improve" the landscape. Roland K. Vosburgh,[23] Director of the Columbia County (New York) Planning Department sums up this challenge as:

> These acronyms cause us to chuckle, but it is no laughing matter if the focus in favor of preservation and *status quo* win in the end. And it has become virtually impossible to develop (read change) a piece of real estate without someone opposing it.

Vosburgh goes on to lament that even environmentally necessary actions, such as reclaiming abandoned industrial sites (known as "brownfields") are avoided because of the unflinching opposition to land use changes. This was manifest in the Chapel Hill landfill siting process. Neighbors who lived around their proposed sites hired lawyers and environmental scientists or were fortunate enough to have lawyers, physicians, and engineers as neighbors, and these representatives tried to persuade the LSC that their site simply was inappropriate. In other cases the members of the LSC themselves had a reason to eliminate a specific site from consideration. Often, the classification of a site into the third (not to be considered further) category was on what appeared to be flimsy evidence. In one case, a member of the LSC who happened to live near a site said that this was nice farmland and that sheep would graze on the hillside. This apparently was given as a sufficient reason for eliminating the site from further consideration. There appeared to be no overt collusion or visible trading of votes, but it became quite clear to observers that the decisions had been made far in advance of the public hearing. The Rogers Road neighborhood (and the Green Tract, which was one of the possible sites being considered) was represented on the LSC by a graduate student who did not live in the neighborhood and who, by her level of participation, seemed to have little interest in the outcome.

Argument 2: If Environmental Injustice Does Occur, It Is Not Bad

The second line of argument advanced against the environmental justice movement is to admit that there might be environmental discrimination but that the discrimination is not harmful. Suppose that the price of land in our imaginary community is as follows:

Neighborhood	Percent Minority	Land Cost ($/acre)
A	10	20,000
B	5	15,000
C	35	10,000
D	95	3,000

Now suppose that a municipal facility such as a solid waste incinerator would need 100 acres. The cost to the municipality would be $2,000,000 if the facility were built in neighborhood A, and $300,000 if it were built in neighborhood D. This is a large savings that would be passed on to the taxpayers in the community. Hence (the argument goes) it is more advantageous to build the plant in neighborhood D because all members of the community, including the residents of neighborhood D, would share in the cost savings by having lower taxes. In addition, the lower property values in neighborhood D caused by the undesirable facility will also represent a savings to the property owners since the property taxes will be lower.

It goes without saying that the greatest savings in taxes would be for those who pay the most, and thus the argument that everyone shares in the economics of using lower-priced land is not persuasive. In addition, lower taxes should not be an excuse for unfair and unequal treatment of all citizens. This also points to the problem of valuation in design decisions. Many values, such as social justice and ecosystem protection, do not readily lend themselves to monetized value. This often means that those benefits and costs that can easily translate into dollars, such as taxation, supplant other values.

Another utilitarian (greatest good) argument for using the lower socioeconomic neighborhood is that people who live there would probably have high unemployment, and a facility located there would provide jobs. These critics point to the influx of less-advantaged people around such urban facilities as incinerators and suggest that the facilities were actually beneficial to the neighborhood. But this is not the point. By disparate distribution of environmental contamination we are still treating unfairly some less advantaged part of our population.

Finally, another argument advanced against environmental justice is to say that whatever the situation might be, the cure is far worse than the disease. For example, the elimination of pollution would be extremely costly, and in the long run, impossible. We could not as a nation set a goal of zero pollution and still hope to pay for other social needs.

This is true, but not germane. In fact, such arguments were posited in the early 1970s against new environmental standards. The issue now is the fair distribution of environmental costs and benefits. The issue then was the overiding need to save the precious ecological resources and to protect public health.

Argument 3: If Environmental Injustice Does Occur and It Is Bad, It Is Not Racially Motivated

The third argument against environmental injustice accepts that there certainly seems to be disparity in the siting of pollution-producing facilities and acknowledges that the less advantaged residents often do not have much choice in the placement of such facilities: that is, that discrimination does occur. But the argument goes, this unfairness is not racially motivated.

Proponents of environmental justice argue, as does philosopher and activist Kristin Shrader-Frechette, that there seems unlikely to be any other reason for such injustice. "If the area closest to a noxious facility tends to have a population of nonwhites rather than whites, then regardless of what zip codes (or any other system of aggregation) reflect, there is likely to be environmental racism."[25]

Racism is a difficult and often contentious issue. It is like any other -ism: It is a belief in something that is taken on faith and cannot be proven. Even further, sometimes rational proof does not change the thinking of one who is a true believer. Racism is the belief that some racially identifiable group is inferior to some other group or does not deserve equal moral or legal protection. This belief then leads to racial discrimination or to actions that are manifested based on this belief, that is, to *de facto* discrimination.

For example, if the belief is that "Orientals" (the title given to Asians throughout much of U.S. history) do not value human life as highly as do Western people (an opinion once famously expressed by General Westmoreland during the Vietnam war), it is a small matter to kill indiscriminately all who have Oriental features, including noncombatants. Drawing such distinctions can be a tool for desensitization. Such dehumanization has been a common tool in warfare throughout history.

In the case of environmental justice, racial discrimination can be defined as ". . . those institutional rules, regulations, and policies of government or corporate decisions that deliberately target certain communities for least desirable land uses, resulting in the disproportionate exposure of toxic and hazardous waste on communities based upon prescribed biological characteristics. Environmental racism is the unequal protection against toxic and hazardous waste exposure and the systematic exclusion of (disadvantaged groups) from decisions affecting their communities."[26]

But saying that this occurs and proving it are two different issues. There seems to be no doubt as to the fact that racially identifiable socioeconomic groups have been unfairly treated in terms of environmental contamination. But is this due to *racial* discrimination? If such discrimination can be shown to be racially motivated, the Civil Rights Act Title VI §601, which makes racial discrimination illegal, can be used to correct the injustice. Proving racial discrimination is, however, quite difficult. What has to happen to prove that racial discrimination has occurred is for the person responsible to admit that he or she intentionally discriminated on the basis of race. This is an admission of guilt and thus unlikely ever to occur.

Also, much of the alleged discrimination is "corporate" (i.e., there is no single person or small group of persons engaging in the acts of discrimination). It is more a manifestation of company or other corporate policies and actions. Finding the "guilty parties" can be a tortuous process.

In fact, some disparate exposure to pollutants may not have been brought about by overt individual or corporate discrimination, but may be the direct result of demographic shifts. For example, at one time in the history of many U.S. cities, working-class whites lived near factories and other facilities known to release contaminants into the air and water. People wanted to live near the plants because transportation was limited. As better transportation systems evolved, along with improved wages as a result of unions and other social movements in the first half of the twentieth century, the people working in the factories were able to move farther away into the suburban or exurban housing developments. This left the remaining neighborhoods adjacent to the pollution sources less desirable and consequently, available at relatively low prices. As a result, lower-socioeconomic-status families moved into these areas. Thus, an entirely new demographic group evolved (i.e., people who received no wages or other monetary benefits of the factories received most of the exposure to the pollutants being released). Although this is not an example of direct discrimination, it does reflect an overall injustice: People are put at risk without any contravening benefit except a cheap place to live. In a sense, they are tenants or owners of "last resort."

Discussion: Engineering Justice Systematically

One of the key lessons learned as we crossed the threshold into this new millennium is that of the need to provide sustainable solutions (See Chapter 6). In environmental matters, no project can be viewed as truly independent. It affects and is affected by other projects. In order to ensure that the public health and welfare are protected, engineers must make certain to view each design as a part of a system. We can take a lesson from thermodynamics, which requires that we account for every input and output of mass or energy within a control volume. Every element of that system is related in some way to every other element. Although the elements may be independent, they are always interrelated to every other element in the system. This systematic reality has been recognized in the political arena as well. For example, Executive Order 12898 calls upon the agencies of the federal government to determine up front whether their projects and decisions will burden any groups unfairly.

However, the track record for a systematic view is not so good in the relatively brief history of environmental science. One would hope that the entire project life, from design through use to decommissioning (i.e., the "life cycle") would be part of every important environmental decision. And, by extension, one of the lessons learned for environmental justice is to think about long-term impacts on society. Unfortunately, there are a number of cases that indicate our approach is all too often that of reaction and "retrofitting," much like the decisions of convenience in the 1970s regarding the best way to remove pollutants found in car exhaust that were contributing to smog. At that time, decision makers in the government and in automobile companies were assessing the optimal means of reducing hydrocarbons that were being released from mobile sources (i.e., cars, trucks, buses, and trains). The options boiled down to whether to redesign and retool internal combustion engines to improve efficiency and consequently reduce emissions or to find a way to retrofit or "add on" a product without making major changes to the engines.

The large U.S. automobile manufactures decided to retain the basic engine designs, merely adding a catalytic converter to the exhaust system. A catalytic converter uses metal catalyst pellets (e.g., platinum) to oxidize hydrocarbons to water and CO_2 and convert carbon monoxide, CO, to CO_2 (see Figure 1.3.) This could be likened to putting "Band-aids" on inefficient engines, but the automotive engineers insisted that this would be the most cost-effective way to reduce emissions. These same engineers were shocked and embarrassed when Japanese automobile manufacturers, most notably Honda, chose the second alternative, to produce highly efficient engines that did not require catalytic converters. The Japanese engineers decided that the reduction of emissions was an essential and integral part of engine design. American engineers, on the other hand, decided to use a bolt-on device to solve the problem without attacking the fundamental source of the emissions.[27]

A case can be made that this is what is often the approach taken to address environmental injustices. At the federal level, each department and

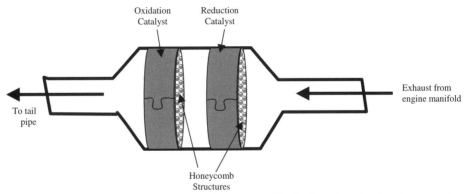

Figure 1.3   Design components of a catalytic converter. The hydrocarbons in the exhaust are first catalytically reduced (e.g., the addition of hydrogen atoms), followed by catalyzed oxidation (e.g., to carbon dioxide and water). If untreated, the exhaust's volatile organic compounds will react photochemically in the atmosphere to form ozone and other air pollutants.

agency has a unique mission defined by enabling legislation. These agencies are evaluated by the American people and their elected representatives on how well the missions are accomplished. That is one of the purposes of congressional oversight, for example. So when a new initiative comes along, agencies are more likely to see it as an ancillary objective, or worse, as an obstacle in the way of the "real" business of achieving their mission. This was a common occurrence in the early days of the National Environmental Policy Act (NEPA), which called upon federal agencies to rethink their missions regarding the environment. Like the EJ executive order, NEPA defined a new "ethic" in how the federal government does business, making environmental policy a leading priority. Specifically, NEPA created the environmental impact statement (EIS), which was required for any major federal action with potential impacts on the environmental quality.[28] But agencies often resisted the new policy. The frustration of environmental advocates was that the agencies were simply "making environmental silk purses from the bureaucratic mission-oriented sow ears."[29] Can this be said about the governmental and legal response to past environmental injustices today?

WHAT ENGINEERS CAN DO

Environmental injustice may seem intractable, but progress is being made. It is a problem that we are not going to solve in this book, although we do hope to give a few pointers on how to recognize and deal with injustice. The facts are that environmental inequality exists and that often it is the minority populations and lower socioeconomic groups in our country who bear the brunt of the pollution. We may help to solve some of these problems if the engineering community is increasingly aware of its influence on preventing injustice. As such, we point out a few things along the way that the individual professional can do to avoid inadvertently becoming a party to injustice and to take positive steps in one's profession to be empathic to all clients, not just those who procure our services directly.

The problems highlighted above seem to be a blend of legal, moral, and technical factors with one common outcome (i.e., injustice). But engineers are trained to be technical experts only. Yes, we practice in a milieu of law, politics, and social sciences, but our forte is within the realm of the physical sciences and engineering principles.

The contemporary engineering profession is demanding that we be better equipped technically and technologically as well as in the social and human sciences. This calls for a systematic approach to engineering education and practice, which is consistent with elements defined by the National Academy of Engineering to be included in guiding strategies for the engineer of the future: Applying engineering processes to define and to solve problems using scientific, technical, and professional knowledge bases requires:

- Engaging engineers and other professionals in team-based problem solving
- Using technical tools
- Interacting with clients and managers to achieve goals
- Setting boundary conditions from economic, political, ethical, and social constraints to define the range of engineering solutions and to establish interactions with the public[30]

The remainder of this book navigates through both of the engineer's worlds, the technical and the social. It does so without excuses. If an equation, reaction, or chemical description is pertinent to a discussion, it is included. If a topic were watered down to make a point about a social issue, it would lose its import and impact. The only way to be a just engineer is to know one's business and to apply that business in a manner sensitive to contemporary social needs. Thus this book is both technical and presents the lessons with an eye toward the socially important issues surrounding each engineering core competency.

CASE STUDY: THE WARREN COUNTY PCB LANDFILL REVISITED

As noted above, the Warren County PCB landfill was constructed in 1982 to contain soil that was contaminated by the illegal spraying of oil containing PCBs from over 340 km of highway shoulders. The landfill received soil contaminated with over 100,000 liters of oil from 14 North Carolina counties.

The landfill was located on a 142-acre tract about three miles south of the town of Warrenton and held about 60,000 tons of contaminated soil collected solely from the contaminated roadsides. The U.S. EPA permitted the landfill under the Toxic Substances Control Act, which is the controlling federal regulation for PCBs. The state owns approximately 19 acres of the tract, and Warren County owns the remaining acreage surrounding the state's property. The containment area of the landfill cell occupied approximate 3.8 acres enclosed by a fence. The landfill surface dimension was approximately 100 meters by 100 meters with a depth of approximately 8 meters of contaminated soil at the center. The landfill was equipped with both polyvinyl chloride and clay caps and liners, with a dual leachate collection system. The landfill was never operated as a commercial facility.

In 1994, a state-appointed working group, consisting of members of the community and representatives from the state, began an in-depth assessment of the landfill and a study of the feasibility of detoxification. Tests using landfill soil and several treatment technologies were conducted. In 1998, the working group selected base-catalyzed decomposition (BCD) as the most appropriate technology (see Figure 1.4). Approximately $1.6 million in state funds had been spent by this time. In 1999 the working group fulfilled its mission and was re-formed into a community advisory board. In the BCD process, PCBs are separated from the soil using thermal desorption. Once separated, the PCBs are collected as a liquid for treatment by the BCD process. BCD is a nonincineration chemical dechlorination process that transforms PCBs, dioxins, and furans into nontoxic compounds. In the process,

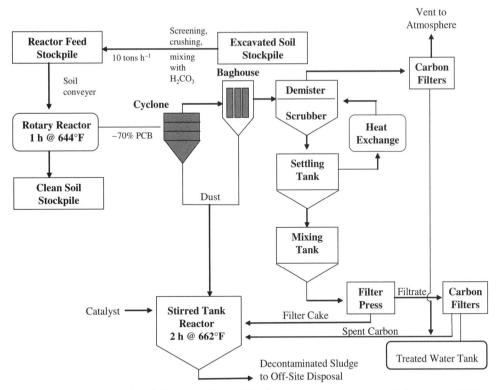

Figure 1.4 Base-catalyzed decomposition. This is the process recommended to treat PCB-contaminated soil stored in Warren County, North Carolina. (From Federal Remediation Technologies Roundtable, *Screening Matrix and Reference Guide,* 4th ed., FRTR, Washington, DC, 2002.)

chlorine atoms are chemically removed from the PCB, and dioxin and furan molecules and replaced with hydrogen atoms. This converts the compounds to biphenyls, which are nonhazardous. Treated soil is returned to the landfill and the organics from the BCD process are recycled as a fuel or disposed off-site as nonhazardous waste.

The cleanup target of 200 parts per billion (ppb) was established by the working group for the landfill site and was made a statutory requirement by the North Carolina General Assembly. The EPA cleanup level for high-occupancy usage is 1 part per million (ppm). EPA's examples of high-occupancy areas include residences, schools, and day care centers. The plan is an example of a conservative and precautionary design, since these areas are likely to have greater exposures than those at a landfill, which limits contact and access, and because the cleanup target is five times lower than the EPA requirement.[31] The removal of PCBs from the soil will eliminate further regulation of the site and permit unrestricted future use.

A public bid opening was held on December 22, 2000 for the site detoxification contract. The IT Group, with a bid of $13.5 million, was the low bidder. Existing funds were sufficient to fund phase I. A contract was established into with the IT Group, and a notice to proceed was issued on March 12, 2001. Site preparation work was completed in December 2001. Work included the construction of concrete pads and a steel shelter for the processing area, the extension of county water, an upgrade of electrical utilities, and the establishment of sediment and erosion control measures.

The treatment equipment was delivered in May 2002. An open house was held onsite the next month so that community members could view the site and equipment before startup. Initial tests with contaminated soil started at the end of August 2002. The EPA demonstration test was performed in January 2003. An interim operations permit was granted in March based on the demonstration test results. Soil treatment was completed in October 2003. A total of 81,600 tons of material was treated from the landfill site. The treated materials included the original contaminated roadside soil and soil adjacent to the roadside material in the landfill that had been cross-contaminated. The original plan, which specified using the BCD process to destroy the PCBs after thermal desorption and separate them from the soil, was overdesigned. With only limited data available to estimate the quantity of liquid PCBs that would be collected, conservative estimates were used to design the BCD reactor. In practice, however, the quantity of PCBs recovered as liquid was much less than anticipated. Thus, the BCD reactor tanks were too large to be used for the three-run demonstration test required under TSCA to approve the BCD process. As an alternative, one tankload of liquid containing PCBs was shipped to an EPA-permitted facility for destruction by incineration. Most of the equipment was decontaminated and demobilized from the site by the end of 2003. Site restoration will be complete once vegetation has become established. The total cost of the project was $17.1 million.

REFERENCES AND NOTES

1 R. Nieburhr, *The Children of Light and the Children of Darkness,* Charles Scribner's Sons, New York, 1944, foreword.

2. Martin Luther King, Letter from Birmingham Jail, in *Why We Can't Wait,* HarperCollins, New York, 1963.

3. Presidential Executive Order 12898, Federal Actions to Address Environmental Justice in Minority Populations and Low-Income Populations, February 11, 1994.

4. For example, this is the fourth canon of the American Society of Civil Engineers' *Code of Ethics,* ASCE, Washington, DC, adopted in 1914 and most recently amended November 10, 1996. This canon reads: "Engineers shall act in professional matters for each employer or client as faithful agents or trustees, and shall avoid conflicts of interest."

5. Even this is a challenge for environmental justice communities, since certain sectors of society are less likely to visit hospitals or otherwise receive early health care attention. This is not only a problem of assessment but can lead to more serious, long-term problems compared to those of the general population.

6. W. Burke, W. D. Atkins, M. Gwinn, A. Guttmacher, J. Haddow, J. Lau, G. Palomaki, N. Press, C. S. Richards, L. Wideroff, and G. L. Wiesner, Genetic Test Evaluation: Information Needs of Clinicians, Policy Makers, and the Public, *American Journal of Epidemiology,* 156:311–318, 2002.

7. Institute of Medicine, *Toward Environmental Justice: Research, Education, and Health Policy Needs,* National Academies Press, Washington, DC, 1999.

8. The principal source for the facts and timelines associated with the Warren County PCB landfill is the North Carolina Department of Environment and Natural Resources, Warren County PCB Landfill Fact Sheet, *http://www.wastenotnc.org/WarrenCo_Fact_Sheet.htm,* accessed June 28, 2005.

9. A *substituent* is an atom or group of atoms that replace the hydrogen atom on an organic compound. In this case, the substituent is chlorine (Cl), but numerous other elements can replace hydrogen. After replacement, the physical and chemical properties can change significantly. For example, halogens such as Cl frequently raise the flash point and boiling point and make compounds less water soluble. Unfortunately, the replacements may render a rather nontoxic compound carcinogenic or increase the toxicity of an already toxic compound.

10. Both Ward and Burns were convicted and did jail time for their crime.

11. B. B. Marriott, Land Use and Development, in *Environmental Impact Assessment: A Practical Guide,* McGraw-Hill, New York, 1997.

12. See M. Ritzdorf, Locked Out of Paradise: Contemporary Exclusionary Zoning, the Supreme Court, and African Americans, 1970 to the Present, in *Urban Planning and the African American Community: In the Shadows,* J. M. Thomas and M. Ritzdorf (Eds.), SaGE Publications, Thousand Oaks, CA, 1997.

13. From K. Shrader-Frechette, *Environmental Justice: Creating Equality, Reclaiming Democracy,* Oxford University Press, New York, 2002.

14. From D. E. Newton, *Environmental Justice,* Oxford University Press, New York, 1996.

15. Commission for Racial Justice, United Church of Christ, *Toxic Wastes and Race in the United States,* UCC, Cleveland, OH, 1987.

16. Presidential Executive Order 12898, note 3.

17. Much of the materials for the next two sections can be found in *Not in My Backyard: Executive Order 12898 and Title VI as Tools for Achieving Environmental Justice,* U.S. Commission on Civil Rights, Washington DC, October 2003.

18. http://www.great-lakes.net/lists/enviro-mich/1998-09/msg00016.html, accessed June 29, 2005.

19. Around the Nation: Congressman and 120 Arrested at PCB Protest *The New York Times,* September 27, 1982, p. A16.

20. Shrader-Frechette, note 13.

21. R. Rosen, Who Gets Polluted? The Movement for Environmental Justice, in *Taking Sides,* T. D. Goldfarb (Ed.), McGraw-Hill, New York, 1997.

22. This case is based on the report by S. Azar, The Proposed Eubanks Road Landfill: The Ramifications of a Broken Promise, Duke University, Durham, NC, 1998.

23. R. K. Vosburgh, *Economics, Change, and the Law of Unintended Consequences,* Pace Law School, New York, February 19, 2003.

24. E. Coakley, "Landfill Transfer Station Opposed," *The Herald-Sun,* Durham, NC, March 19, 2006.

25. Shrader-Frechette, note 13.

26. B. Bryant, (Ed.), *Environmental Justice: Issues, Policies, and Solutions,* Island Press, Washington, DC, 1995.

27. Often, there are blessings in disguise in environmental protection. In this case, the fact that the auto makers vied for catalytic converters led to the need for oil companies to provide gasoline without lead compounds. Up to the 1970s, gasoline commonly contained lead-based, antiknock, octane-enhancing compounds because this allowed for less expensive refining. A common engineering economics concept is that of the law of diminishing returns, which in this case meant that the costs of refining to get just a bit more caloric efficiency (octane boosting) become far more expensive with each incremental increase. Thus the companies would refine to a certain octane rating and then add compounds, in this case tetraethyllead, to the fuel. However, since the lead has affinity for the metal pellets (catalysts) in the catalytic converter, after a few tanks the converters were rendered ineffective (i.e., the pellets were coated with lead, which is not a catalyst). This meant that "no lead" gasoline had to be provided. As a result, studies began to show marked decreases in the concentrations of lead in the air and soil near roadways, which in turn lead to an all-out lead ban in most of North America in the coming decades. This is an example of doing the right thing for the wrong reasons.

28. Section 102 of NEPA, 42 U.S.C. §4321 et seq., Public Law 91-190, 83 Stat. 852.

29. Although we have been unable to locate the citation, this is very close to a mid-1970s quote heard by Vallero from Timothy Kubiak, who at the time was an EIS reviewer for the U.S. EPA and who had studied NEPA under Lynton K. Caldwell, a strong advocate for a national environmental policy, at the University of Indiana.

30. National Academy of Engineering, *Educating the Engineer of 2020: Adapting Engineering Education to the New Century,* National Academies Press, Washington, DC, 2005.

31. Similar protective approaches have frequently been used in emergency response and remedial efforts, such as those that followed the attacks on the World Trade Center towers. For example, the risk assessments assumed long-term exposures (e.g., 30 years) to contaminants released by the fire and fugitive dust emissions, even though the exposures were significantly shorter.

2

Moral Justification for Environmental Justice

Engineers are a practical lot. We apply the sciences with the intent of changing things for the better. We see a problem or a blank slate and envision something new, something better. At times, this practicality pushes us toward a mode of going with what works and not thinking too deeply about the theoretical underpinnings of what we do on a day-to-day basis. We are reminded to "think outside the box" when confronted with seemingly intractable problems. This is not by accident, but is the result of our academic and professional training.

We have been well prepared for our fields, for some of us beginning long ago, by first grasping the mathematics and physical principles of science. We are often put off by philosophy and its ilk, but these disciplines really can be valuable to us. Most of us do not consider the theory behind physical principles to do our jobs, but we have considered these concepts along the way as part of our academic preparation. By analogy, an understanding of environmental justice must also be steeped in an appreciation, a moral justification, of what we do. The moral principles and canons espoused in our codes of ethics are practical manifestations of deeper moral and philosophical justifications, in much the same way that our designs and calculations are rooted in mathematical and scientific foundations. Thus, it is worthwhile to consider the moral rationale for our daily practice in what makes us not only competent, but also moral, professionals.

Justice requires reason, but is seldom obviously rational. Reason must be informed by practical experience and a set of values. For engineers, these values are to a limited extent codified in our standards of practice (i.e., codes of ethics). But—here is the challenge. Justice cannot be formulaic. We cannot plug in some values, set certain initial and boundary conditions, and expect some proven and general principle to yield a "Navier–Stokes"-like result. If the principles of environmental justice or the fair treatment of all people were universally upheld by everyone at all times, there would be no need to justify it. For example, we do not need to justify that pain is bad. Pain is simply bad, and that's all there is to it.[1] Yes, there are times when pain is necessary, such as may be true for a visit to the dentist, but we put up with it because it is for our long-term benefit. Pain is still something to be avoided if possible.

The fair treatment of all persons is not in that category, however. What exactly is *fairness,* and why ought we to be fair to others? The concept of fairness and justice needs to be clarified and solid arguments advanced if we are to convince others that these are worthy goals. This matters not only as a large engineering issue, but as a way that the individual engineer conducts business. It is not simply a philosophical or theoretical concept, but is part of the engineer's tool kit. No project is complete unless matters of justice are incorporated. Thus, just as any project must include a good design, a reasonable approach for building and implementing the design, and a means for ensuring that the design criteria have been met, so should that project be designed to be fair and just.

FAIRNESS AND DISTRIBUTIVE JUSTICE

Part of the difficulty in defining justice is that justice and injustice are not often distinguished by the "what" so much as the "how." For example, the act of "taking" is morally neutral (i.e., neither good nor bad), depending on the conditions of the taking. Taking, in fact, has been in the news recently, as the U.S. Supreme Court ruled in July 2005 that certain private concerns may use eminent domain ostensibly to take private property for the public good. Some would say that this is good, since the private enterprise is improving things (e.g., enhancing the local tax base). This is a utilitarian perspective; that is, the proponents perceive a greater good, with the end (larger tax revenues) achieved by defined means (the taking of private property that yields much lower tax revenues). Others consider such taking to be immoral and a violation of the intent of the U.S. Constitution, since such powers are granted only to public entities, and the *public good* is a strictly defined term. Also, they see the takings as an encroachment or even an outright assault on individual freedoms. At a more basic level, most faith traditions consider "stealing" to be an immoral form of taking.

As a moral concept, fairness is a relatively new idea and is thought to be at some higher level than most basic moral rules, such as rules against lying, stealing, and the like. Fairness, the equal application of morality to all people, is a much more sophisticated concept. However, fairness has been built into many value systems throughout recorded history. For example, during the time of the Roman Empire, tax collectors were local citizens in the remote provinces who were required to collect a certain sum from citizens. In addition, they were allowed to collect monies beyond what was due to the empire as personal commissions. Thus, the tax collectors were despised by the local people because they were seen as disloyal and because their methods of collection were deemed unfair.[2] This is one of the first examples of professional ethics, or more correctly, of public dissatisfaction with the ethics of a profession. It is also an example of how justice is defined by the "how" *versus* the "what" in a matter. Many reasonable persons at the time may not have begrudged the local tax collectors rightful wages. The injustice consists of the inflated amounts taken as well as the extortion and the tactics used in the gain. Fast-forward to contemporary times and there is similar discomfort with unfairness, such as insider trading in the stock market, exorbitant interest rates, price gouging, corporate cheating, excessive corporate salaries, "big box" department stores replacing local downtown businesses, political chicanery, and even "legacy" college admissions. Fundamentally, these are perceived as unfair practices (the "big guy" exploiting the "little guy").

The idea of fairness as a moral vehicle for individual and professional ethics, however, was not adequately explored until John Rawls wrote his hugely influential book, *A Theory of Justice* (first published in 1958), in which he proposed that justice is fairness.[3]

For Rawls, justice emerges when there is a fair compromise among members of a true community. If people are fairly situated and have the liberty to move and better their position by their own industry, justice results when they agree on a mutually beneficial arrangement. Fairness is the right ordering of distributed goods or bads, and fair persons are those who, when they control distributive processes, make those processes fair. This is a circular definition, of course, defining fairness as the fair distribution of goods, so let's try another approach.

The concept of distribution is familiar to the engineer. So, for a moment, let us think of justice as a commodity that can and should be distributed fairly throughout society (i.e., distributive justice). Allocating things of value that are limited in supply relative to demand has various dimensions. Justice in this regard depends on what is being allocated, such as wealth and opportunities. For example, in discussions on fair taxation, one often hears arguments concerning the meaning of a "fair redistribution of wealth." Other variables include the diversity of the subjects of distribution: for example, people in general, people of a certain country or national origin, citizenship status, socioeconomic status, or even people *versus* other components of ecosystems. The basis of *how* goods *should* be distributed also varies. For example, some philosophies call for equal distribution to every member of society (known as *strict egalitarianism*), others for the characteristics of people comprising a population (varying by age, handicaps, or historical biases), and still others based purely on market forces (such as *strict libertarianism*).[4]

Whereas the idea of fairness is tied to many ethical principles, such as justice, reciprocity, and impartiality, the word *fair* can have other meanings as well. For example, there is the problem of the *free rider,* a person who uses the contributions of others in society to better his or her position but does not participate in the cost of the society. A person who does not pay taxes for religious reasons still uses the roads and public services, for which others pay. We would deem such actions "unfair" since that person would be taking social goods without contributing to the social welfare.[5]

Another meaning of *fair* is the receipt of good or bad events beyond the control of society. For example, a person whose trailer is destroyed by a tornado while other trailers in the vicinity are spared would call this "unfair," although there is nothing unfair (in moral terms) about a random event of nature. However, if the random occurrence is followed by a willful act, such as increasing the costs of needed supplies following a natural disaster, commonly called *gouging,* such an act *would* be considered unfair. A corollary type of unfairness would be an engineer's decision to provide substandard services to public clients, such as the design of a public housing or school project, simply because the opportunity presents itself.

A popular use of the word *fair* relates to how events beyond the control of society treat the person. For example, a person might get a debilitating disease such as multiple sclerosis, a neurological illness that strikes only young people. Although it is a tragedy for that person and his or her family and friends, contracting multiple sclerosis is not a case of unfairness. It is a sad event, but it is not unfair. On the other hand, if human suffering is caused by premeditated human actions, such as decisions to release toxic pollutants into the environment, thereby increasing the risk of human illness, such decisions *would* constitute unfairness.

Thus, we are looking for a connotation of fairness that distinguishes such unfortunate confluences of events (such as the genetic expression of chronic diseases) from those where human decisions have caused unfairness or have not accounted properly for certain groups and which have led to adverse consequences. The definition we want to propose is that fairness occurs with the honoring of appropriate and just claims. Another way of saying it is that fairness is a process where the legitimate claims of each person are respected.[6]

The ancient Greeks considered fairness to exist when equals were treated equally and unequals were treated unequally. That is, fairness occurs when identically situated people are treated identically. When there are no significant differences among various people, they all ought to be treated equally.[7]

The problem of course comes in the definition of *significant*. What characteristics are "sufficiently significant" to allow for disparate treatment?[8]

| **Example:** | **Significant Difference** |

Two people, a man and a woman, apply for two identical jobs at a private company. The man is offered a salary of $40,000 and the woman is offered a salary of $30,000. Is this difference in pay morally right or fair. If not, what makes it immoral?

Since the jobs are identical, there has to be a justification for the difference in pay. Since gender is not a significant difference, this appears to be a clear case of discrimination and unfairness.

But let's complicate the example. Suppose that the job was to unload trucks, and the strength and stamina of the worker made a difference in productivity. Would the company be justified in hiring a person who will be physically able to perform the needed task? The company might argue that a man would probably be able to perform the job as required, whereas a woman would not. But this is stereotyping. Perhaps a woman can prove that she is able to perform as well as a man. If in that case her salary is still lower than the salary for the man, this would be a case of unfairness. The problem is that most private companies prejudge the ability of a person to do a job based on stereotypes and would not give a woman the opportunity to show what she can do.

In most cases, gender or race or country of origin is not a "significant" difference that allows disparate treatment; Discrimination on the basis of such difference is patently unfair.

Let us complicate the example even further. What if a job were only offered to males or to women over the age of 50? Is this unfair? If the job entails exposure to dangerous levels of chemicals known to be *teratogens* (those that cause birth defects), is it fair to allow women of child-bearing age to work there? Should *all* women in this group be prohibited?

The equal treatment of equals is also one of the conundrums of *affirmative action*. Does fairness demand retribution for past wrongs committed to an identifiable social group? When is fairness the same as *equity* (equal treatment) such as equal housing and employment opportunites, and when does fairness require a more affirmative approach to repair past and ongoing injustices (e.g., lingering effects of generations of uneven educational achievements, union membership, and career opportunites due to institutional, intentional, and even sanctioned biases). Equal opportunity seems to imply "equals treated equally," whereas affirmative action calls for some effort to treat "unequals unequally." One way to resolve this might be to define fairness as a lack of envy, when no participant envies the lot of any other. This is not, however, necessarily fair, since the claims of some people might be exaggerated.

Example:

Fair Distribution

A farmer is retiring and wants to distribute his farm of 300 acres among his three sons. What is a fair way of distributing the land?

If the sons are equal in all significant (there's that word again!) ways, the farmer would divide his farm into three 100-acre plots. But suppose that one son claims to be a better farmer than the other two and insists that this ought to result in his having a larger share of the 300 acres. A second son might need 120 acres because he wants to sell the land for a new airport, and thus stakes his claim for the larger lot. A third might say that since he has more children than the first two, he needs a larger share because eventually, he will have to subdivide his plot among more offspring.

Are any of these claims significant enough to change the initial distribution of 100 acres each? It would be unlikely that a disinterested arbitration board would respect any of these claims, and thus the different claims should not result in a division different from the 100 : 100 : 100 distribution. Each of the three sons might go away unhappy, but the process has nevertheless resulted in a "fair" division of the goods.

The units used to divide some scarce resource are also important. In the example above, the units are acres of land. But not all land is the same, and some of the 100-acre plots might have water, others trees, and others valuable minerals, and a truly fair distribution would then take all such variables into account. If these can be expressed in a common denominator such as dollars, a fair division is at least theoretically possible. On the other hand, some land might have special meaning or memories, and this value cannot be included in terms of dollars, and since these valuables are probably viewed differently by each son, fair distribution is possible.

The injustices done to Native Americans in moving them from their ancestral lands to reservations were an example of using the wrong units for compensation. The land areas given in compensation were supposedly equal to those taken by the government, but the loss of sacred lands was devastating to nations such as the Cherokees. Even if the land area was equal, this was in no way a just or fair process.

Western culture, especially engineers, like to quantify, but some valuables defy such calculation. This is a common challenge at town meetings and hearings. The facts and figures may be wasted on many in attendance simply because the river, lake, building, neighborhood, or school is more than its physical dimensions. It has subjective and abstract meaning and value that can easily be missed in an environmental impact assessment or actuarial report.

Another problem with an envy-free approach to fairness is that it depends on each person having a similar personality. Suppose that of the three farmer's sons in the example above, one is not very astute in business, and the other two brothers convince him that he should take only 60 acres, leaving 240 acres to be divided between the other two brothers. The naïve brother does not object and the deal is consummated. It is an envy-free division. But we recognize that such a division is eminently unfair to the less astute brother.

We have to conclude that defining fairness as a lack of envy thus does not seem to be useful; and at its worst, it can be a tool for unfair distributions. After all, there is no shortage of those who live by the maxim "Never give a sucker an even break."[9] Unfortunately, there is no shortage of those who would take advantage of another's ignorance, naiveté, and sense of fair play.

Perhaps we can get some help from other professions in trying to define fairness. One means of determining fairness in the legal profession is the *reasonable person standard.* A fair distribution of goods occurs when an objective outsider, taking into account all the claims of the participants, renders a decision that would be agreed to by most rational, impartial people as being equitable to all, regardless of each individual claim. In common law, the reasonable person standard is a "legal fiction" since there really is not such a person. But this is not necessarily a bad thing because it provides a means to analyze a situation that is evoking strong emotions for and against a decision. By creating a hypothetical person whose view is based solely on reason provides a means of looking at the situation in a less biased way. *Bias* is another of those terms with a distinct engineering meaning, that is, it is a systematic error. Thus, the reasonable person standard helps to recalibrate our sense of fairness in the same way that we calibrate our scientific apparatus against a known (i.e., rational) standard. We would expect an arbitration board to apply such a rational approach to determining fairness.

Another way of describing fairness is to define what we mean by its opposite: unfairness. Rescher[10] identifies three types of claims of unfairness that might be valid:

- Inequity
- Favoritism
- Nonuniformity

Inequity   Giving people goods not in proportion to their claim is an inequity. The opposite would be *equity,* a condition where people's shares are proportional to their just and appropriate claims.

For example, suppose that a business goes broke and creditors are lining up for their share. Say that the business has $100, but three creditors each are owed $50, $100, and $250. An equitable distribution of the available funds would pay each one 25 cents on the dollar, so the three claimants would get $12.50, $25, and $62.50 each. Of course, the claims have to be proven to be just claims.

Favoritism   Some conditions that have nothing to do with the issue at hand (e.g., one's relations or one's religion) ought to have nothing to do with the situation or claim. The opposite would be *impartiality,* the even-handed distribution of goods without favoritism.

In the bankruptcy example above, suppose that the executor decides to pay out $50, $50, and $0 to the three creditors because the first creditor is a local merchant and the last is an international bank, believing that it is more important to support the local merchant than some far-off impersonal bank. This distribution is unfair since the type of business ought not to be germane to the distribution of the available funds.

Nonuniformity   "Equal treatment under the law" means that the law is to be applied to all people regardless of their status or wealth. The opposite is *uniformity,* the uniform application of the rules.

Suppose a dinnertime rule in a family is that all vegetables have to be eaten before dessert is served. Some children will not always kill off the last pea, hoping to get away with the small transgression. If the parents allow this for all of their children save one, the one who is held to the strict rule can rightfully claim to have been treated unfairly. The rule was not evenly applied.

However, returning to the Greek definition, fairness is not equalitarianism: that is, the treatment of all people equally. To function, society occasionally has to impose unequal treatment of some. For example, the military draft was patently inequalitarian. Only some people were to be drafted, others were not. Those drafted may have ended up in harm's way, and certainly would lose time out of their lives. But it is not possible to send partial persons into the army. If the need is for 100,000 soldiers out of an eligible population of 10,000,000, everyone has a 1% chance of being drafted. The key here is that the draft, the process by which the 100,000 will be chosen, has to be fair. Everyone ought to have an equal chance of being drafted unless they are able to show some significant reason why they should be exempt. If you recall the Vietnam era, exemptions such as college deferments and conscientious objection, were the stuff of controversy and moral debates.

One of the principles of our society is that all persons are to be treated equally under the law. But this does not mean unqualifiedly equal. Some identifiable groups of people such as professionals are treated differently under the law. All professional pharmacists, for example, are allowed to dispense drugs, whereas this activity is illegal for the nonprofessional. All people in the category "pharmacists" are then being treated differently from other people. Unfairness occurs when a pharmacist, because of some irrelevant differences such as gender, religion, or shoe size, is not allowed to dispense drugs. Similarly, although we want to treat all people the same when they have committed a crime, this is seldom done. For the same crime, a first offender might receive a different sentence from that of a repeat offender, and most would agree that that is "fair."

Equality before the state is also important, in that goods distributed by the state (and goods taken by the state) are not equal but are equitable. The progressive income tax requires rich people to pay a larger percentage of earnings than poor people on a per person basis, and welfare recipients need to show that they are destitute before they can receive assistance. The important objective of fairness is that each person be treated equitably within the process. So a rich woman ought not to have to pay more taxes than a rich man, all other things being equal.

Perhaps the best definition of fairness that is useful in our discussions of environmental justice is to say that fairness is treating each person the same according to democratically accepted and agreed-on rules, and whenever these rules result in unequal treatment, there has to be a good and acceptable reason for the inequality.[11]

What makes unfair treatment immoral? Such treatment becomes immoral when the claims of individuals are not respected. For example, is it unfair to exclude women from the Rotary Club? The male members might claim that this is a private club and they have a right to exclude whomever they wish. There is, after all, a vote of the membership for admission of new members. But the Rotary Club is not purely a social club because it is made up primarily of business people, who often discuss and transact business during the meetings. To exclude women from such interaction is unfair and immoral. On the other hand, is it fair for a women's golf team to exclude male members? One argument is that since golf is partly dependent on strength and on average men are stronger than women, allowing men to compete on the same level with women would destroy the integrity of the women's golf game, and the exclusion of men is neither unfair nor immoral. Unfairness in this case could result from applying standards that do not help achieve the "actual ends" (e.g., golf playing) but some other hidden end (e.g., men-only golf clubs).

At this writing this is a strongly debated issue for professional golf and its prestigious Masters tournament (which is based at an exclusive golf club in Georgia). Arguably, this points to a more important aspect of fairness: systematic exclusion. For example, it has only been a few decades since job selection criteria included questions about one's intentions to have children, whether one owned an automobile, or whether one can lift 50 pounds. It is, of course, no one's business how a person gets to work (private automobile, bike, public transportation, or a Jetsons' flying car!), and thus an irrelevant question can only result in unfair treatment. A general rule is that if the job that needs to be performed is not affected by the answer given to a question in an interview, such a question should not have been asked in the first place.

A fair division of goods may not be democratically popular. Consider a country with two primary religious sects, one with 40% of the population and the other with 60%. An election is held, and the majority of people (the 60%) decide to prevent any and all goods from going to the minority 40%. This is obviously unfair, even though the result has been arrived at democratically.

The siting of undesirable facilities such as a landfill is just such a problem. If fairness is to be decided by the majority, the wealthier and more powerful members of a community would choose to site a landfill at one location and then hide the decision behind a democratic vote, saying that this is legitimate and fair. Such a selection of a landfill site may have been legitimate and democratic, but it would have been patently unfair if the claims of the people who get the landfill in their backyard have not been respected.

Fairness also has a time component; that is, like other aspects of engineering, justice is constrained in time and space. The example of siting the facility unfairly is a spatial injustice (i.e., where we put the landfill determines the injustice). Sometimes, it is not so much what or where, but when an action takes place that determines its fairness. In the words of the British statesman William E. Gladstone (1809–1898): "Justice delayed is justice denied." Excluding people from key decision points or waiting to involve them until enough momentum for the project has been gathered is a manner of injustice. Not accounting sufficiently for future generations (e.g., nuclear waste dumps, mine subsidence, or future declines in property values) is arguably an unfair practice.

Finally, a different connotation of the word *fair* is to denote that something is neither very good nor very bad. For example, in the old adjective grading systems, a grade of C was often described as fair. This was to let the guardians of the student know that although the student was not failing, he or she was not doing an excellent job in the course. At first blush, this connotation may be seen as very similar to "equal opportunity." Rather, fair in this use is actually a utilitarian or statistical concept. A fair decision from a utilitarian viewpoint is one that provides the greatest amount of benefit for the greatest number of people. So if a benefit-to-cost ratio (B/C) is calculated from different segments of society, the option that provides the best overall B/C value for the largest segment of society would be chosen. Such a perspective is akin to the statistical concept of a normal distribution (see Figure 2.1). In other words, if we assume that the benefits and costs (or risks) in a given situation are normally distributed within a population, we would be able to select the fairest option as the one where most of the population receives, on average, the largest benefits *versus* costs or risks. The normal distribution translates into the very well off *and* the very poor receiving the least benefits and most of the population receiving the lion's share of benefits. This can occur, for example, in economics, where the very poor receive a stipend from the government or low-paying jobs and the very rich are taxed at an increasingly high marginal rate (i.e., a "progressive" tax), but the majority of the population receives the most goods and services from the government. The highest B/C ratios are those near the statistical measures of central tendency: the mean, median, and mode.

The normal distribution analogy appears plausible until some of the other aspects of fairness are considered. In fact, this can be one of the most unfair ways to decide on environmental issues, since it places an undue and disparate burden on a few groups; and

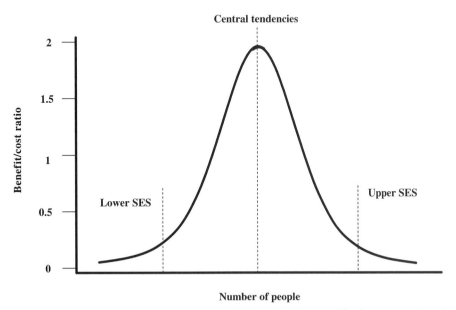

Figure 2.1 Conceptual model for selecting a "fair" option from a utilitarian perspective when the benefits are normally distributed.

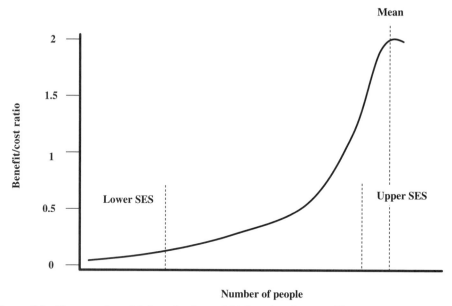

Figure 2.2   Conceptual model for selecting a "fair" option from a utilitarian perspective when the benefits are skewed in favor of the higher socioeconomic strata.

these are often the ones least likely to be heard in terms of pointing out costs and risks. This is known as the *tyranny of the majority*.[12] Often, the curve is not normally distributed but is skewed in favor of the higher socioeconomic strata (see Figure 2.2). Sometimes, it is doubly unfair because the people assuming most of the costs and risks are those that receive the fewest benefits from this particular decision.

These curves demonstrate the importance of the concept of "harm" in fairness. No group should have to bear an "unfair" amount of costs and risks. This is why John Stuart Mill added the *harm principle* to utilitarianism and why John Rawls argues that one must empathize with the weakest members of society. Rawls argues that the only fair way to make a moral decision is to eschew personal knowledge about the situation that can tempt a person to select principles of justice that will allow them an unfair advantage. This is known as the *veil of ignorance,* but it is really a way to implement Mill's harm principle. So fairness also involves more than utility and more than a good *B/C* ratio; it requires virtue.

Discussion: Harm and the Hippocratic Oath

The Hippocratic Oath for physicians is an example of a precautionary principle when it states: "First do no harm. . . ." The traditional text of Hippocrates of Cos (ca. 460–377 B.C.) is:

> I swear by Apollo the physician, by *Æsculapius, Hygeia,* and *Panacea,* and I take to witness all the gods, all the goddesses, to keep according to my ability and my judgment, the following Oath. To consider dear to me as my parents

him who taught me this art; to live in common with him and if necessary to share my goods with him; to look upon his children as my own brothers, to teach them this art if they so desire without fee or written promise; to impart to my sons and the sons of the master who taught me and the disciples who have enrolled themselves and have agreed to the rules of the profession, but to these alone the precepts and the instruction. I will prescribe regimens for the good of my patients according to my ability and my judgment and never do harm to anyone. To please no one will I prescribe a deadly drug nor give advice which may cause his death. Nor will I give a woman a pessary to pro- cure abortion. But I will preserve the purity of my life and my art. I will not cut for stone, even for patients in whom the disease is manifest; I will leave this operation to be performed by practitioners, specialists in this art. In every house where I come I will enter only for the good of my patients, keeping my- self far from all intentional ill-doing and all seduction and especially from the pleasures of love with women or with men, be they free or slaves. All that may come to my knowledge in the exercise of my profession or in daily commerce with men, which ought not to be spread abroad, I will keep secret and will never reveal. If I keep this oath faithfully, may I enjoy my life and practice my art, respected by all men and in all times; but if I swerve from it or violate it, may the reverse be my lot.

Many contemporary physicians continue to take a reworded oath based on the Hippocratic Oath. The oath actually includes a number of precautionary el- ements instructive to the engineer. The closest parallel is the first engineering canon: "Hold paramount the health, safety, and welfare of the public," another example of a precautionary principle. It is a call to empathize with those who may be affected by our decisions and actions.

Weaknesses in this connotation of fairness can be demonstrated using an ethical analytical tool: a line drawing.[13] Ethicists use a number of tools to analyze cases for ethical content, but since the engineering profession makes much use of graphical tools, line drawing is a popular technique to analyze engineering cases. This technique is most useful when there is little disagreement on what the moral principles are but when there is no consensus about how to apply them. The approach calls for a need to compare several well-understood cases for which there is general agreement about right and wrong and to show the relative location of the case being analyzed. Two of the cases are extreme cases of right and wrong, respectively. That is, the positive paradigm (PP) is very close to being unambiguously moral and the negative paradigm (NP) unambiguously immoral:

	NP	Our Case	PP	
Negative feature 1		×		Positive feature 1
Negative feature 2		×		Positive feature 2
Negative feature 3			×	Positive feature 4
Negative feature n	×			Positive feature n

Next, our case (T) is put on a scale showing the positive paradigm and the negative paradigm, as well as other cases that are generally agreed to be less positive than PP but more positive than NP. This shows the relative position of our case T:

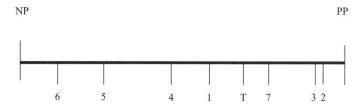

This gives us a sense that our case is more positive than negative, but still short of being unambiguously positive. Does this not imply that if we take this action we are being fair? In fact, two other actual, comparable cases (2 and 3) are much more morally acceptable. This may indicate that we should consider taking an approach similar to these if the decision has not yet been made. Consider the following example, adapted from Fledderman[14]:

Example:

Disposal of a Slightly Hazardous Waste

A company is trying to decide how and to what extent it should dispose of a "slightly hazardous" waste. The company's current waste stream contains about 5 parts per million (5 ppm) of contaminant A. The state environmental department allows up to 10 ppm of contaminant A in effluent drained into its sanitary sewers. The company has no reason to suspect that at 5 ppm any health effects would result. In fact, most consumers would not detect the presence of contaminant A until relatively high concentrations (e.g., 100 ppm). The city's wastewater treatment plant discharges to a stream that runs into a lake that is used as a drinking water supply. So is it ethical and fair for the company to dispose of slightly hazardous waste directly into the city sewers?

The positive paradigm in this case is that the company would do what is possible to enhance the health of people using the lake as a drinking water source. The negative paradigm is an action that causes the drinking water to be unhealthy. Here are some hypothetical options the company may have pursued:

1. Company dumps contaminant A at the regulatory limit (10 ppm). No harm, but unusual taste is detected by a few sensitive consumers.

2. The company lowers its concentration of contaminant A so that it is effectively removed by the town's existing drinking water system.

3. Company discharges contaminant A into the sewer at 5 ppm or below, but ensures that it is effectively removed by the town with new equipment bought by the company and donated to the town.

4. Contaminant A can be removed by equipment paid for by taxpayers.

5. Seldom, but occasionally, contaminant A concentrations in water will make people feel sick, but the feeling lasts only an hour.

6. Contaminant A passes through untreated (i.e., 5 ppm), into the steam, where it builds up and causes sensitive people to become acutely ill, but only for a week and with no long-term harm.

7. Equipment is installed at the company that reduces the loading of contaminant A to 1 ppm.

Drawing each case and our proposed case (T) gives us a relative location with respect to the negative and positive paradigms:

Cases 2 and 3 are clearly the most ethical, since the amount of contaminant A that reaches the public is kept well below the regulatory limit and any health threshold. Case 7 is also close to the positive paradigm, since it is well below the regulatory limit, but even at these levels some sensitive people (e.g., newborn, the immunocompromised, the elderly) could experience effects. Cases 5 and 6 are less ethical because they resemble the negative paradigm (i.e., actions that make the water less safe to drink). The key, though, is that in the middle of the diagram (case 4), the burden of the problem caused by the company is shifted from the private company to the public. This is not the fairest option by any means.

Although being right of center means that this case is closer to the most moral than to the most immoral approach, other factors must be considered, such as feasibility and public acceptance. The location on the line indicates that being fair is different from receiving a grade of C. Fairness implies that we need to search for options that move us closer to the positive paradigm (i.e., the ideal). As we migrate toward options in the negative direction, we give up a modicum of fairness. This is the nature of balancing benefits and costs, but the engineer must be fully aware that these balances are taking place. So, like risk assessment, professional judgment in selecting the fairest designs and projects must account for trade-offs (e.g., cost-effectiveness *versus* fairness, security *versus* liberty, and short-term needs versus long-term effects).

Care should always be taken when trying to apply objective and quantitative tools to concepts such as ethics and justice. Social sciences and philosophical principles are often highly subjective. Although the natural sciences and engineering strive for objectivity, they, too, must deal with subjectivity from time to time.

Discussion: Physical Science Is Not without Subjectivity[15]

Engineers and physical scientists can become frustrated with the social sciences and their uncertainties that result from the inherent difficulty of controlling variables when studying human populations. These uncertainties are manifested in larger errors and greater variability of conditions than those associated with the physical sciences. However, this should not lead one to conclude that there is no subjectivity in the physical sciences. Consider fluid dynamics and, more specifically, viscosity. A recent study[16] has shown marked variability in the applications of one of the basic concepts of fluids: the critical Reynolds number (R_c), which describes whether a flow is laminar, turbulent, or in a transitional state. R_c is defined differently for flow within a circular pipe and flow in an open channel, so the number changes for different systems. Engineering textbooks vary in the way that R_c is defined, leading to various and inconsistent interpretations of whether flow is turbulent or laminar, which is a very important distinction in fluid mechanics.

Some of the subjectivity is the result of the steady march of inquiry and the iterative nature of science. As we learn more, key scientific concepts are refined. Some are even abandoned, but their remnants remain in the lay literature (e.g., watching the sun "rise") or even in scientific circles (e.g., "nature abhors a vacuum"). Frequently, the changes are subtle. The value of R_c was brought home again during a recent seminar held by a Duke engineering graduate student regarding research being directed by Zbigniew Kabala. They were considering how, at a small or micro scale, the geometry of a conduit can have profound effects on whether flow is laminar or turbulent. In fact, engineers generally expect a flow between laminar and turbulent conditions (i.e., the critical flow region to have Reynolds numbers greater than 2000 and less than 4000). Critical flow may also be defined as a flow with velocity = 0 at the walls and twice the average velocity at the center of the conduit (laminar) and a flow with no relationship to the proximity of the wall, due to mixing (turbulent). The student pointed out that at very low Reynolds numbers, in small conduits, flows behaved more turbulently than would be expected in larger systems. In fact, the visual demonstration of the flow using dye showed that the size, and especially the shape, of the pockets lateral to the flow changed the critical range substantially, even to the point where a finite amount of the fluid remained in the pockets (adhering to the walls) well after the remainder of the flow had moved downstream. In other words, when clear water was sent through the conduit, some of the blue dye remained out of the streamlines. Is this why, even though at the meso or macro scale, soil or other unconsolidated material may still contain measurable concentrations of contaminants, even if they have relatively high aqueous solubility? Sometimes, subjective judgment beyond the number is needed to describe a system: in this case, whether R_c represents a turbulent, laminar, or critical system,

This discussion points to the fact that a certain amount of subjectivity can present itself in all sciences, and we must take care not to be condescending to our colleagues in the social sciences and the humanities. Two books by Sheldon Rampton and John Stauber provide ample cases of how the public has been manipulated by experts: S. Rampton and J. Stauber, *Trust Us, We're Experts: How Industry Manipulates Science and Gambles with Your Future*, Penguin Putnam, New York, 2001; and S. Rampton and J. Stauber, *Toxic Sludge Is Good for You*, Common Courage Press, Monroe, ME, 1995.

These books may be better classified as muckraking than as scholastic endeavors, but they do point out some of the ways that *spin* is used to justify decisions. As such, the books provide cautionary tales to engineers on how they may, unwittingly, be parties to deception. The lesson for environmental justice is that the perception of uncertainty can be amplified when engineers communicate with diverse audiences. We may well know that the physical principles hold and that we are applying them appropriately in a particular project, but certain segments of society may perceive that we are not being completely honest with them. This is further exacerbated in neighborhoods that have traditionally been excluded from decision making or with whom the track record of "experts" has been tainted with prevarications and unjustifiable "bills of clean health."

At the risk of stating the obvious, good risk communication is *not* an invitation to compromise the quality of science or to introduce pseudoscientific methods or junk science to keep everyone happy. To the contrary, it is a reminder that we must be just as open and honest about what we do not know as we are about what we know. On more than a few occasions, we have witnessed engineers who believe in a project so strongly that they become advocates to the point that they begin to compromise the actual scientific rationale for the project.

Some of this is the result of *sunken costs,* costs that are so committed and so far down the road that rethinking a project's design and approach is not a workable option. Another factor is the "us and them" problem, wherein the engineer begins to see those who complain, less as clients and more as obstacles that must be overcome. Whatever the reasons, an unwillingness to examine and reexamine a project in terms of its scientific credibility is dangerous.

It does not have to be a fellow engineer who points out scientific weaknesses. For example, some years back, Vallero addressed a group of elementary students in a small town in Missouri. After a brief overview of pollution, including a discussion of the depletion of the ozone layer, Vallero took questions from the students. One precocious student asked a fairly straightforward question: If we have too much ozone down here [in the troposphere] and we are losing it up there [in the stratosphere], why don't we just build a system to move the ozone from here to there? Vallero's first reaction was to "school" the student in the ways of science, especially that it was the activities that were causing the two problems and that they were very different from each other. However, the kids weren't buying it, making Vallero's arguments increasing defensive. At some point, he stopped himself and realized that he really wasn't listening and that he hadn't in fact considered the student's idea. After a very pregnant pause, he admitted as much and said he would need to think about it further. Indeed, Vallero has thought about it often since. The "system" may not be what the fourth grader had in mind (e.g., tubes stretching into the sky), but since the atmosphere is a system, stratospheric–tropospheric exchanges must be considered. If they are part of the problem, they may well be part of the solution. The student is probably out in the real world now. We can only hope that he has become an engineer and may be able to put his thoughts into action!

VIRTUE AND EMPATHY

We argue that being fair and advancing the cause of justice is morally admirable. People who devote their lives to doing the right thing are said to behave *virtuously*.

If one reads the classical works of Aristotle, Aquinas, Kant, et al., the case is made for life being a mix of virtues and vices available to humans. *Virtue* can be defined as the power to do good or a habit of doing good. In fact, one of Aristotle's most memorable lines is that "excellence is habit." So if we do good, we are more likely, according to Aristotle, to keep doing good. Conversely, vice is the power and habit of doing evil. The subjectivity or relational nature of good and evil, however, leads to some discomfort in scientific circles, where we place great import on certainty and consistency of definition. Aristotle tried to clarify the dichotomy of good and evil by devising lists of virtues and vices, which amount to a taxonomy of good and evil. One of the many achievements of Aristotle was his keen insight as to the similarities of various types of living things. He categorized organisms into two kingdoms, plants and animals. Others no doubt made such observations, but Aristotle documented them. He formalized and systematized this

---

Biographical Sketch: Pietro Angelo Secchi

In 1865, the Pope decided that he wanted to test the quality of the water in the Mediterranean Sea and sent the commander of the papal navy to investigate. Pietro Angelo Secchi (1818–1878), the Vatican astronomer at that time, was asked to come up with a way to accomplish the task of measuring water quality. Secchi devised a white iron disk that is lowered over the side of a boat and the depth at which the disk is no longer visible is noted. The deeper the depth, the better the light penetration, and the clearer the water. On April 20, 1865, what became known as the *Secchi disk* was first lowered over the side of the papal steam yacht *l'Immaculata Conczione*. The idea was so simple and worked so well that the Secchi disk was soon adopted by water quality scientists all over the world.

Pietro Secchi was almost famous for another reason as well. He was the first person to use photography to study solar bodies, and his pictures of Mars revealed lines that might have looked like canals. He did not suggest that these were artificial canals, but the imagination of science fiction writers took over and the myth of canals and civilization on Mars was born. Fascination with the possibility of life on Mars continues to this day.

taxonomy. Such a taxonomic perspective also found its way into Aristotle's moral philosophy.

We will not all agree on which of the virtues and vices are best or even whether something (e.g., loyalty) is a virtue or a vice, but one concept does seem to come to the fore in most major religions and moral philosophies: *empathy*. Putting oneself in another's situation is a good metric for virtuous acts. The Golden Rule is at the heart of Immanuel Kant's *categorical imperative*. With apologies to Kant, here is a simplified way to describe the categorical imperative: When deciding whether to act in a certain way, ask if your action (or inaction) will make for a better world if all others in your situation acted in the same way. This is an argument in several environmental areas, including recycling, midnight dumping, sustainable development, and selecting low-toxicity source materials in manufacturing. An individual action's virtue or vice is seen in a comprehensive manner. It is not whether one should pour a few milligrams of a toxic substance down the drain—it is whether everyone with this amount of toxic substance should do so. The overall stewardship of the environment may cause one to rethink an action (as has been the case for decades now). A corollary to this concept is what our colleague Elizabeth Kiss of Duke's Kenan Center for Ethics calls the "six o'clock news" imperative. That is, when deciding whether or not an action is ethical, consider how your friends and family would feel if they heard about all of its details on tonight's TV news. That may cause one to consider more fully the possible externalities and consequences of one's decision!

The concept of empathy is central to environmental justice. Justice is the virtue that enables us to give others what is due them as our fellow human beings. This means that we must not only avoid hurting others by our actions but that we ought to safeguard the rights of others in what we do and what we leave undone.

The categorical imperative is emblematic of empathy. Kant uses this maxim to underpin duty ethics (called *deontology*) with empathetic scrutiny. However, empathy is not the exclusive domain of duty ethics. In teleological ethics, empathy is one of the palliative approaches to dealing with the problem of "ends justifying the means." Other philosophers also incorporated the empathic viewpoint into their frameworks. In fact, Mill's utilitarianism's axiom of "greatest good for the greatest number of people" is moderated by his *harm principle,* which, at its heart, is empathetic. That is, even though an act can be good for the majority, it may still be unethical if it causes undue harm to even one person. Empathy also comes into play in contractarianism, as articulated by Thomas Hobbes as social contract theory. For example, Rawls has moderated the social contract with the veil of ignorance as a way to consider the perspective of the weakest—one might say most disenfranchised—members of society. Finally, the rationalist frameworks incorporate empathy into all ethical decisions when they ask the guiding question: What is going on here? In other words, what benefit or harm, based on reason, can I expect from actions brought about by the decision I am about to make? One calculus of this harm or benefit is to be empathetic to all others, particularly the weakest members of society, those with little or no "voice."

The word *empathy* has an interesting beginning. It comes from the German word *einfühlung,* which means the ability to project oneself into a work of art, such as a painting. Psychologists at the beginning of the twentieth century searched for a word that meant the projection of oneself into another person, and chose the German word, translated into English as *empathy*. The concept itself was known, such as the Native Americans' admonition to walk in another's moccasins, but it needed a construction. The early meaning of empathy was thus the ability to project oneself into another person, to imitate the emotions of that person by physical actions. For example, watching someone prick a finger would result in a visible winching on the part of the observer because the observer would know how this feels. Some observers actually feel the pain, similar to the pain of the person having the finger pricked, although often not as intensely.

From that notion of empathy it was natural to move to more cognitive role-taking, imagining the other person's thoughts and motives. From here, empathy began to be thought of as the response that a person has for another's situation. Psychologists and educators, especially Jean Piaget,[17] began to believe that empathy develops throughout childhood, beginning with the child's first notion of others who might be suffering personal stress. The child's growing cognitive sense eventually allows him or her to experience the stress in others. Because people are social animals, this understanding of the stress in others, according to the psychologists, eventually leads to true compassion for others.

A problem with this notion of empathy development is that some experiments have shown that the state of mind of a person is very important to that person's ability to empathize. Small gifts or compliments apparently significantly increase the likelihood that a person will show empathy toward third parties. A person in a good mood tends to be more understanding of others. If this is true, empathy is (at least partly) independent of the object of the empathy, and empathy becomes a characteristic of the person.[18]

The psychologist Charles Morris defines empathy as[19] "the arousal of an emotion in an observer that is a vicarious response to the other person's situation. . . .Empathy depends not only on one's ability to identify someone else's emotions but also on one's capacity to put oneself in the other person' place and to experience an appropriate emotional response. Just as sensitivity to non-verbal cues increases with age, so does empathy: The cognitive and perceptual abilities required for empathy develop as a child matures."

Such a definition of empathy seems to be widely accepted in the moral psychology field. But there are serious problems with this definition. First, we have no way of knowing if the emotion triggered in the observer is an accurate representation of the stress in the subject. We presume that a pinprick would be felt in a similar way because we have had this done to us and we know what if feels like. But what about the stress caused by a broken promise? How can an observer know that he or she is on the same wavelength as the subject when the stress is emotional?[20]

If a subject says that she is sad, the observer would know what it is like to be sad, and would share in the sadness. That is, the observer would empathize with the subject's sadness and be able to tell the subject what is being felt. But is the observer really feeling what the subject is feeling? There is no way to define or measure "sadness," and thus there is no way to prove that the observer is actually feeling the same sadness that the subject is feeling.[21] An existentialist might say that this is true for everything, even physical realities, but that is beyond the scope of this discussion.

The second problem relates to nonhuman animals. Psychologists have studied empathy exclusively as a human–human interaction, yet many nonhuman animals can exhibit empathy. Witness the actions of a dog when its master is sick. You can read the caring and sympathy and hopefulness in the dog's eyes.[22]

Humans also have strong emotional feelings toward nonhuman animals. The easiest to understand in these terms is the empathy we feel when animals are in pain. We do not know for sure that they are in pain, of course, since they cannot tell us, but they act in ways similar to the way that humans behave when they are in pain, and there is every reason to believe that they feel pain in the same way. Anatomical studies on animals confirm that many of their nervous systems do not differ substantially from those of humans, and thus there is every indication that they feel pain.

More problematical are the lower animals and plants. There is some evidence that trees respond physiologically when they are damaged, but this is far from certain. The response may not be pain at all but some other biochemical messaging or sensation (if we can even suggest that trees have sensations). Yet many of us are loathe to cut down a tree, believing that the tree ought to be respected for what it is, a center of life. This idea was best articulated by Albert Schweitzer in his discussions on the "reverence for life," the idea that all life is sacred.

Empathy toward the nonhuman world cannot be based solely on sentience. Something else is going on. When a person does not want to cut down a tree because of caring for the tree, this is certainly some form of empathy, but it does not come close to the definitions used by the psychologists.

The third problem with this definition of empathy is that there is a huge disconnect between *empathy* and *sympathy*. If an observer watches a subject getting a finger pricked, the observer may know exactly what it feels like, having had a similar experience in the past. So there is great empathy. But there might be little sympathy for the subject. The observer might actually be glad that the subject is being hurt, or it might be funny to the observer to watch the subject suffer.

Years ago on the popular television show *Saturday Night Live* there was an occasional bit where a clay figure, Mr. Bill, suffered all manner of horrible disasters and ended up being cut, mangled, crumbled, and squashed. Watching this may have elicited some empathy on the part of the observers, but there certainly was no sympathy for the destruction of the little clay man. Its destruction was meant to be funny.

We could argue that a lack of sympathy might indicate that there must also be a lack of empathy. How is it possible for someone to empathize with another person getting a finger pricked but think it to be humorous? Perhaps there has been no empathy there at all. Or perhaps we have conditioned ourselves to laugh at others when they get hurt as a defense mechanisms (e.g., "whistling in the dark") to somehow separate the violence from our own experience. Or, we have learned from and have become desensitized by video games to destroy others without regret.

ENGINEERING AND FAIRNESS

Empathy is not a moral value in the same way that loyalty, truthfulness, or honesty are moral values. We can each choose to tell the truth or to lie in any particular circumstance, and a moral person will tell the truth (unless there is an overwhelming reason not to,

Biographical Sketch: Albert Schweitzer

 Albert Schweitzer (1875–1965) was born in Alsace, and following in the footsteps of his father and grandfather, entered into theological studies in 1893 at the University of Strasbourg, where he obtained a doctorate in philosophy in 1899 writing a dissertation on religious philosophy. He began preaching at St. Nicholas Church in Strasbourg in 1899 and served in various high-ranking administrative posts. In 1906 he published *The Quest of the Historical Jesus,* a book on which rests much of his fame as a theological scholar.

Schweitzer had a parallel career as an organist. He had begun his studies in music at an early age and performed in his father's church when he was 9 years old. He eventually became an internationally known interpreter of the organ works of Johann Sebastian Bach. From his professional engagements he earned funds for his education, particularly his later medical schooling.

He decided to embark on a third career, as a physician, and to go to Africa as a medical missionary. After obtaining his M.D. at Strasbourg in 1913, he founded his hospital at Lambaréné in French Equatorial Africa. In 1917, however, the war intervened and he and his wife spent 1917 in a French internment camp as prisoners of war. Returning to Europe after the war, Schweitzer spent the next six years preaching in his old church, and giving lectures and concerts to raise money for the hospital.

Schweitzer returned to Lambaréné in 1924 and except for relatively short periods of time, spent the remainder of his life there. With the funds earned from his own royalties and personal appearance fees and with those donated from all parts of the world, he expanded the hospital to 70 buildings, which by the early 1960s could take care of over 500 patients in residence at one time.

On one of his trips up the Congo to his hospital, Schweitzer saw a group of hippopotamuses along the shore and had a sudden inspiration for a new philosophical concept that he called *reverence for life,* which has had wide influence in Western environmental thought. His idea was that all life is sacred and that we should hold it in awe and reverence. Schweitzer would not harm any animal, and at night in the jungle would not have a candle for fear that a moth would fly into it. He agreed that we needed to eat to survive, but he argued that this should be at the lowest level of harm as possible, including not eating any meat.

Schweitzer's contribution is not so much that he established a set of rules for others to follow, but that he articulated by word and example a new way of living—of having respect for the least of nature's creatures.

He was awarded the Nobel Peace Prize in 1953.

such as to save a life). But it is not possible to choose to have or not to have empathy. One either has empathy or one does not. One either cares for those in need, or one does not.

Because we believe that empathy is worthwhile, and respect and admire people who have empathy, we tend to assign moral worth to this characteristic and we believe that people with empathy are virtuous. On the other hand, we do not condemn those who do not have empathy. For example, people who contribute to various relief organizations such as CARE and Oxfam do so because they have empathy for those in need, but many people choose not to contribute. They lack empathy for others in need in this instance, but this does not make them bad people. They simply choose not to contribute.

Can engineers not have empathy and still do good engineering? That is to say, is empathy necessary for good engineering? Certainly on a personal level, engineers are human and they read the same newspapers and watch the same TV news as everyone else, and thus their lack of empathy ought not to be any more or less criticized than the lack of empathy by anyone else. But the truth is that the responsibility of professional engineers is supererogatory to everyday ethics. Engineering ethics is a different layer on top of everyday common morality, and engineers share many responsibilities not required of nonengineers. By virtue of their training and skills, engineers serve others and have certain responsibilities that relate to their place in society. The oft-quoted first canon in many codes of engineering ethics,

> The engineer shall hold paramount the health, safety, and welfare of the public.

is very clear. It states that the engineer has responsibility to the *public,* not to a segment of the public that he or she likes or gets along with, or the segment that employs the engineer, or the segment that has power and money. The engineer is responsible to the public. Full stop. And in doing so, the engineer must help that segment of the public least able to look out for themselves. There is a *noblesse oblige* in engineering, the responsibility of the "nobles" to care for the less fortunate.

Thus, to be an effective and "good" engineer requires that we be able to put ourselves in the place of those who have given us their trust. The implications for environmental justice are that it has been much easier to export "canned" answers and solutions to problems from our vested viewpoints. This view must span time and space. What will the community look like in 10 years if the project is implemented? What happens if some of the optimistic assumptions are not realized? The neighbors will be left with the consequences. It is much better, but much more difficult, to see the problem from the perspective of those with the least power to change things. We are empowered as professionals to be agents of change. So, as agents of change and environmental justice, engineers must strive to hold paramount the health, safety, and welfare of all of the public, we must be competent and we must be fair.

REFERENCES AND NOTES

1. All right, we can hear the physiologists and clinical psychologists, as well as more than a few theologians, vigorously protesting such statements. Indeed, pain is an inherent defense mech-

anism. Without it, animals would repeat many harmful behaviors. The point here is that the response itself is not pleasant, and if given the choice, most of us would choose not to have pain. Granted, the psychopathologists may also be stirred by this contention, noting the masochists, who allegedly enjoy pain. This is probably a learned activity, overcoming the innate avoidance response. It does indicate that even the seemingly universal assumptions have their exceptions.

2. One early example of distributive justice is the advice of John the Baptist to the tax collectors and soldiers who asked how to follow the precepts he had laid out. Luke 3:10–14 states:

   "What should we do then?" the crowd asked.
   John answered, "The man with two tunics should share with him who has none, and the one who has food should do the same." Tax collectors also came to be baptized.
   "Teacher," they asked, "what should we do?"
   "Don't collect any more than you are required to," he told them.
   Then some soldiers asked him, "And what should we do?"
   He replied, "Don't extort money and don't accuse people falsely—be content with your pay."

3. J. Rawls, *A Theory of Justice* (1971), Belknap Press Reprint, Cambridge, MA, 1999.

4. Interestingly, John Rawls (mentioned earlier) is considered to be a contractarian, as are the libertarians, since both subscribe to a form of social contract theory as posited by Thomas Hobbes. Rawls modulated strict contractarianism by adding the veil of ignorance as a protection for weaker members of society.

5. The concept of the free rider shows up in economics on some supply–demand curves. In fact, in a pure supply–demand relationship, the free rider would not exist. That is, since the supply of all goods and services is provided according to their demand (more demand, the greater the cost), no person would receive any goods or services without payment.

6. N. Rescher, *Fairness: Theory and Practice of Distributive Justice,* Transaction Publishers, New Brunswick, CT, 2002.

7. The concept of treating equals equally and unequals unequally shows up in theological and religious discourses. For example, the late Harry Werner, a Jesuit missionary who lived among Native Americans on reservations, noted that the concepts of borrowing and lending are empathic. For example, Werner noted persons who asked to borrow $10 but never returned the money. Werner observed this behavior regularly and was informed by the locals that it makes little sense for someone who has little money to return it to those who have much. Werner had a car and lived in a decent abode, so until the people had such stations in life, there was no moral obligation to repay the money.

   Another instance of treating unequals unequally is the concept of usury. This is mentioned in the Bible, for example, but draws seemingly little attention in Western society. While the contemporary definition is more akin to that of *gouging,* the term *usury* originally meant the charge of *any* interest on a loan. The practice is prohibited in Islam. In the strictest sense, Jews are prohibited from charging interest on loans to other Jews. St. Thomas Aquinas considered the act of charging interest to be immoral since it charges doubly [i.e., both the item (money) borrowed and the use of the thing]. The lender charges for the loan by insisting that the loan be repaid. Thus, the repayment is the charge for the loan. Any further charge is a charge for using the loan. To Aquinas, charging interest on a loan is likened to selling a person a bottle of wine and adding another charge if the person actually drinks the wine! Of course, this is moral reasoning and not business acumen. That is, lending can be a moral good, but this goodness is diminished if interest is required.

8. The term *significant* also has a statistical meaning, related to the probability that an observed result is due to chance alone. An experimental finding is statistically significant if there is a probability of less than some percentage (e.g., 1%) that the difference observed would occur by chance alone (a p-value of less than 0.01). Significance is an expression of the probability of a hypothesis being true, given the data. But how does this help us get a handle on fairness?

Often, since engineering involves numerous elements of uncertainty, engineers are willing to accept significance levels much less restrictive that those of our colleagues in the basic sciences. Physicists, for instance, may require a p-value several orders of magnitude more restrictive (e.g., $p < 0.000001$). In this instance the physicist will not accept a hypothesis if there is more than a 1 in a million probability that the outcome is due to chance. On the other hand, our colleagues in the social sciences and medicine, who deal with people with all their uncertainties, may be quite happy to accept a 95% confidence (or $p < 0.05$). If nothing else, this shows that people are difficult to predict and to study.

9. This is the title of a 1941 film by the comedian W. C. Fields. It is similar to the sentiments in the phrase "there's a sucker born every minute." This phrase, which in its simplest connotation means that people are easy to manipulate, has been erroneously attributed to the circus entrepreneur P. T. Barnum. In fact, the quote was made by a competitor of his, David Hannum. In the late 1860s, Hannum and Barnum were both bidding on what they thought was a petrified giant that had been "found" (it was actually planted by a hoaxer). When Barnum allegedly had his own giant made, Hannum was quoted as lamenting that these people were being fooled, not knowing that he was the original brunt of the hoax. This account was shared by R. J. Brown online at http://www.historybuff.com/library/refbarnum.html, accessed July 13, 2005.

10. Rescher, note 6.

11. There is, of course, the problem of the majority in a democracy choosing to act immorally. Racial discrimination in the South was for years supported by the majority, but this did not make it morally right or fair to African Americans. We have to assume in this definition that the decisions of fairness are based on defensible moral principles by the popular majority.

12. The phrase was coined by Alexis de Tocqueville and considered at some length by John Stuart Mills. In a democracy, the majority is very powerful. It can influence and even control an entire population, as was done in Germany, where elected officials, including Hitler, gained and abused power. This was much on the mind of the framers of the U.S. Constitution: that the duly elected do not become the tyrants and that those with little power not be crushed. That is why the U.S. is fundamentally a representative democracy and *not* a pure democracy.

13. C. B. Fleddermann, *Engineering Ethics,* Prentice Hall, Upper Saddle River, NJ, 1999.

14. Fleddermann, note 13.

15. S. A. Lowe, Omission of Critical Reynolds Number for Open Channel Flow in Many Textbooks, *Journal of Professional Issues in Engineering Education and Practice,* 129:58–59, 2003.

16. Lowe, note 15.

17. J. Piaget, *The Moral Judgment of the Child,* Free Press, New York, 1965.

18. S. Vaknin, *Malignant Self Love: Narcissism Revisited,* Lidija Rangelovska Narcissus Publications, Shopje, Macedonia, 2005.

19. C. G. Morris, *Psychology: An Introduction,* 9th ed., Prentice Hall, Upper Saddle River, NJ, 1996.

20. This is one of the problems with B. F. Skinner's brand of behaviorism, as articulated in *Beyond Freedom and Dignity* (Hackett Publishing, Indianapolis, IN, 1971). Certainly, we act out on what we have learned, and learning is an aggregate of our responses to stimuli. However, human emotions and empathy are much more than this. Empathy is a very high form of social and personal development. So although one might be able to "train" an ant or a bee to respond to light stimuli, or a pigeon to "play Ping-Pong" (as Skinner did), even these lower animals have overriding social complexities. At the heart of humanity are freedom and dignity, despite what some behaviorists tell us.

21. Vaknin, note 18.

22. The concept may be innate and extended to other animals, such as elephants' sensing "awe" for their ancestral graveyards.

3

Justice and the Engineering Profession

The engineering profession is rooted in the builders of antiquity, and these builders used their common sense to construct amazing edifices. There was no engineering theory, and thus all building was by experience and intuition.

The first division of the profession occurred in the sixteenth century, when engineers began to combine practice with theory in the construction of military (both offensive and defensive) facilities. Architecture and mechanics were both important in the construction of forts and other facilities, and military engineers who designed and placed artillery had the highest level of learning and education in ballistics and hydraulics. Ancient military engineers sometimes even enlisted the support of theorists, such as the King of Syracuse's employment of the Greek mathematician Archimedes to build weaponry and to calculate trajectories to defend against the assaults of Roman armies (212 B.C.).[1]

The builders of nonmilitary facilities were artists first and engineers second, and for them form was far more important than function. These artist-engineers became what today we call *architects*. The builders who relied increasingly on the theory of mechanics for constructing facilities became the engineers, and their work was almost exclusively for military purposes.

The term *ingenia* was first used to describe the technical aspects of defensive and offensive warfare. In seventeenth-century France, the *Génie* officers in the military were also asked, during peacetime, to work on nonmilitary projects, and became known as *Génie civil*.

Military engineers were long a mainstay of any successful civilization, and this was true as well in the new United States of America. The first engineers in the colonies were educated in Europe, but by the late eighteenth century the military academy at West Point, New York, was turning out engineers who had successful careers first in the military and then as civilian engineers. The notable first jobs that engineers undertook in peacetime were in the construction of roads. The National Road, or Cumberland Pike, begun in 1787, allowed the movement of pioneers west and solidified the country by providing communication and transportation. In 1821, Congress directed the U.S. Army's Corps of Engineers to begin surveying routes for canals and roads, and asked that the work be conducted by both "military engineers and civil engineers." William Wisely, former director of the American Society of Civil Engineers (ASCE) and an engineering historian, believed that this may have been the first time the word *civil* was used in the

United States to differentiate officially between engineers whose primary responsibility was military and those whose primary responsibility was civil works.[2]

The importance of West Point as a cradle of American engineers is proven by the statistic that during the period 1802–1829, of the total of 572 graduates, 49 of them had become chief engineers on major civilian projects.[3] West Point was the single source of educated engineers in the United States until the forerunner of Norwich University, and then Rensselaer Polytechnic Institute (RPI), began educating engineers whose primary responsibility was civil construction. The first civil engineering degree was granted by RPI in 1835, and one of the best known early graduates of the RPI program was Washington Roebling, who with his father and wife, was responsible for the design and oversight of the construction of the Brooklyn Bridge.

American civil engineers first organized in Boston as the Boston Society of Civil Engineers, and in 1852 engineers in New York created the ASCE. Civil engineering has at its heart betterment of the human condition. In fact, the contemporary motto of the ASCE is "building a better world." Many prominent engineers have worked tirelessly not only for the construction of public facilities, but also for the public good that comes from such activities.

There have been major divisions within the profession of engineering. First, engineers split from architects because the engineers emphasized the application of mathematics and mechanics in their work. Second, civil engineers split from military engineers when they recognized that their efforts were not related militarily but were intended for peaceful use. The third split is between engineers who, adopting the model of the educated robot, blindly do the bidding of clients or employers, and those engineers who recognize that they have a professional responsibility to society. We call the latter engineers *just engineers.*

Biographical Sketch: John Smeaton

The first person to identify himself publicly as a "civil" engineer (as opposed to a military engineer or just an engineer) was John Smeaton. Born in 1724 in England, he showed early promise as a builder of machines and structures, and being finally allowed to leave his training in law, he became the most respected engineer in England in the eighteenth century. He went on to build the first successful lighthouse on Eddystone Reef, south of Plymouth, which lay directly in the path of a shipping channel and had been responsible for the destruction of many ships. Later, he constructed bridges in Perth, Coldstream, and Hexham, and he designed a new water supply for the Edinburgh and the Forth and Clyde canal, which allowed passage from ocean to ocean. Recognizing that his works were not of a military nature, Smeaton began to sign his name using the title "civil engineer." He and his engineering colleagues used to meet at the Queens Head Tavern in London, and this group became the nucleus of the Institution of Civil Engineers, which was chartered in 1818.[4]

Biographical Sketch: Vauban

 The notion of the engineer's responsibility to society is an old one. For example, one of the greatest *Génie* officers in seventeenth-century France was the Marquis de Vauban, who was first a soldier, participating in numerous battles, and then constructed some of the most enduring forts in the seventeenth century. He introduced the idea of describing his works in words and drawings so that other builders could use these to duplicate the facilities. But Vauban is most famous for his last work, *The King's Tithe*, which is mostly an economics text for civil engineers in which he argues for the betterment of living conditions for the working people.[5] Included in this pamphlet was a review of the finances available to build such forts, and Vauban concluded that the brunt of the costs were being borne by the working people. The nobility apparently was not asked to contribute at all, and this seemed quite unfair to the engineer. Understandably, this did not sit well with the politically powerful and the court cronies, and Vauban was in essence "fired" from his job as the king's engineer. He spent his last year in virtual exile.

Vauban was an engineer who recognized that his job was to tell the truth regardless of who might be offended, and he paid the price for his honesty. A century after his death, Napoleon recognized Vauban's genius and dedication to France and had him reburied in the Panthéon in Paris.

ENGINEERING AND SOCIETY

If you don't lie to me, I won't lie to you, and we will both be better off. It is good and useful therefore to agree not to lie. Under certain circumstances, we punish people who lie by fines or even by sending them to jail. Socrates (469–399 B.C.) expounded a type of social contract, as recorded in Plato's *Crito*. Basically, Socrates argued that he could not ethically avoid the death penalty because it was part of the contract that he and other citizens had with the laws of Athens. These laws, he argued, gave order to society: provided for the marriage of his parents, his own civil upbringing, and even his chance to be a scholar. Thomas Hobbes (1588–1679) is given credit for articulating modern social contract theory in his masterpiece, *Leviathan* (1651). Hobbes argued that without a social contract, human beings would revert to a primitive, highly selfish environment, which he referred to as the *state of nature*. Social norms are the prescription against this state. So it is in our own self-interests to enter into such a contract. Also, Hobbes considered humans to be rational and reasonable, wherein human reasoning is likened to "Scouts, and Spies, to range abroad, and find the way to the things Desired" (*Leviathan*). Thus, *contractarianism* is a form of teleology (*telos,* Greek for "far") where ethical content of an act is based on what results from that act.

The same type of contract exists between the professional and the public. Although it is not directly reciprocal, there is an understanding of behavior that when adhered to

Biographical Sketch: Tadeusz Kościuszko

 Originally from Poland, military engineer Tadeusz Kościuszko (often Anglicized as Thaddeus Kosciuszco) came to America and presented himself to the Continental Congress in August of 1776. He was granted the rank of Colonel of Engineers and he served with the revolutionary forces until the war ended seven years later.

Distinguishing himself in Yorktown and in New York, he is most remembered for his efforts in designing the fortifications and for fighting with the troops at the Battle of Saratoga, where American soldiers withstood British troops fighting under Gen. John Burgoyne for nearly a month before capturing the British as they retreated. He was then assigned the task of building the fortifications at West Point, which later became the United States Military Academy.

In 1783 he received the commendation he so richly deserved. He was made a citizen of the United States, promoted to Brigadier General by Congress and was admitted to The Society of the Cincinnati, one of only three foreigners allowed to join.

Kościuszko never received a salary for all of his efforts on behalf of the United States and performed his duties out of love for freedom and liberty. George Washington described him as "a gentlemen of science and merit." President Thomas Jefferson wrote to him, "From one man we can have but one life, and you gave us the valuable and active part of yours, and we are now enjoying and improving its effects."

After the war in 1784, Kościuszko returned to Poland to provide support to the growing reform movement, which ultimately led to the signing of the May Constitution of Poland, the world's second constitution after the Constitution of the United States.

will benefit all. Part of the contact for professional engineers is that they agree to hold paramount the health, safety, and welfare of the public. This is an example of a professional duty. Socrates spoke of duty ethics or *deontology* (*deon,* Greek for obligation) when he defined morality as "how we ought to live." The deontological decision school of thought was more fully articulated later by Immanuel Kant (1724–1804) in *Groundwork of the Metaphysics of Morals* (1785). He argued that ethical decisions are made on the basis of what is "right" rather than on the consequences of the decision. Kant argued that an act is either ethical or unethical if, when it is universalized, it makes for a better world. This is the *categorical imperative,* which instructs one to "act only according to that maxim whereby you can at the same time will that it should become a universal law." In other words, engineers and other professionals must behave in a way that we would want all engineers to behave. The public is, of course, to decide what is in its

Biographical Sketch: Peter Palchinsky

A direct spiritual descendant of Vauban in the twentieth century was the great Russian engineer Peter Palchinsky. Palchinsky's most perplexing problem was that he took seriously the idea that engineers should hold paramount the health, safety, and welfare of the public.

Born to poor parents in 1875 in central Russia, he worked his way through school, graduating with a degree in mining engineering from the prestigious institute at St. Petersburg. On leaving school, he was given the job of investigating why the production of coal at the Czar's mines in the east was so low. What he found was that the miners were living and working under appalling conditions and with no concern for occupational safety. His report, at first applauded by the leaders in St. Petersburg, got Palchinsky into trouble once the implications of what he was suggesting were realized. Providing better working conditions for the miners would have caused great upheaval in other parts of the frail economy. Palchinsky was soon arrested for speaking the truth, and spent some time in jail. He escaped to the West and developed a distinguished career as an engineer designing harbors and port facilities, always recognizing that these facilities were large systems that included the need for worker protection and comfort.

But he missed Russia, and after being pardoned by the Czar, in 1913 he went back. When the Bolsheviks came to power in 1917, Palchinsky was arrested once again as a collaborator and spent more time in jail. His honesty and skills prevailed, however, and he eventually became a well-respected engineer, founding a journal and an institute to study mining engineering. But he made the mistake of not toeing the party line, and was highly critical of top-down planning, the emphasis on huge projects that neglected secondary consequences, and especially the poor living conditions of the workers. This last concern was not popular with the leaders, and when Stalin decided to get rid of anyone who might challenge his rule, he had Palchinsky arrested and shot.

His wife, Nina (shown with Peter in the photograph), learned of her husband's fate from a small news article describing the crimes against the revolution supposedly committed by Peter Palchinsky. As the wife of a criminal, she was also suspect, and was eventually sent off to the death camps so graphically described by Alexander Solzinitzin.

In *The Ghost of the Executed Engineer*, an exceptional book describing the conditions of engineering in the former Soviet Union, Loren Graham argues that the destruction of the engineering profession in Soviet Russia led directly to the downfall of the USSR. Eliminating the engineers abolished the one group of educated persons who would speak out against the incredibly stupid plans concocted by state planners.[6]

For more on Peter Palchinsky, see Loren Graham's *The Ghost of the Executed Engineer*, Harvard University Press, Cambridge, MA, 1993.

best interest, and the social contract between the engineer and the public requires that the public decide how the engineer is to fulfill that obligation. The evolution of the engineering profession shows how integrated engineering has always been dictated by the needs of society.

The moral responsibilities of engineers are spelled out in various engineering codes of ethics. As we noted in Chapter 2, the first canon of most engineering codes of ethics states: *The engineer shall hold paramount the health, safety, and welfare of the public.* Engineers make a distinction between *should* and *shall* in the code of ethics and in the writing of engineering specifications. If the word *should* is used, the understanding is that the requirement is voluntary and the specific item is not enforceable. If the word

---

Biographical Sketch: Ellen Swallow Richards

 Ellen Swallow (1842–1911) was born in an age when the opportunities for women in higher learning were severly limited, but this did not prevent her in 1868 from graduating from Vassar College, where she studied chemistry and biology. Because many believed that women could not get a college education without injuring their health, pioneers such as Ellen Swallow had to do more than men to prove herself—and she did. At one point she wrote home to her parents: "The only trouble here is that they won't let us study enough."

Upon graduation, she got herself admitted to MIT as an "experiment" and was allowed to work under the watchful eye of the only other woman on the staff. Her ample intelligence, even demeanor, and gentle helpfulness soon won over the faculty, and she was the first woman to be awarded a bachelor's degree from MIT. She had continued to study at Vassar, and in the same year that she received her B.S. degree from MIT, she received a Master of Science degree from Vassar. A few years later she married an MIT colleague, Robert Richards.

Ellen Swallow Richards was never paid a salary by MIT but earned her own income through teaching and consulting projects. The students who went though her laboratory and learned analytical techniques of water chemistry reads like the Who's Who of early public health engineering. She herself became an expert in the chemical and biological quality of water and in the late 1880s conducted the first thorough stream survey for the state of Massachusetts, a survey that became a template for all stream quality projects for other states.

She believed in including environmental questions in decision making and found the public's apathy toward environmental destruction intolerable. Her world view and personal ethical outlook no doubt influenced a generation of public health professionals. At one point, she wrote: "One of the most difficult lessons to learn is that our tolerance of evil conditions is not proof that the conditions are not evil."

shall is used in the language of the code, however, the item is enforceable by the engineering society.

The second important word is *paramount*. This word, also, is unambiguous. The engineer must place the health, safety, and welfare of the public above all other considerations, such as cost and aesthetics. A pertinent question at this juncture is whether or not there is any wiggle room in the word *paramount*. Under what circumstances would engineers decide *not* to hold paramount the health, safety, and welfare of the public? Imagine a public hearing on a controversial project where the engineer is asked if the health, safety, and welfare of the public was deemed to be paramount. The engineer might respond that no, it was not, for some overriding reason. What might such a reason be, and would such a reason ever be morally justified?

Engineers are required to consider the public welfare because they have special skills on which the public depends. Because a cross section of the public cannot monitor and evaluate the work of engineers, the public allows engineers to self-regulate and expects certain assets from the profession, such as honesty, truthfulness, and a commitment to public service. There is, in short, an implied contract between professional engineering and the public.

Perhaps the best illustration of the responsibility of engineers to society is the DC-10 story. The main characters of this story are Dan Applegate and other talented aerospace engineers who were placed in a difficult ethical situation. But situations do not dictate ethics, and engineering ethical decisions always require that the health and safety be paramount in *every* situation.

CASE STUDY: THE DC-10

The DC-3, manufactured by Douglas Aircraft, is perhaps the most storied of the cargo/passenger airships. Designed to provide transport during World War II, it served with distinction in all theaters, dropping parachute battalions during the Normandy invasion, ferrying supplies in Burma, and even providing the backbone for the Berlin Air Lift after the war. During and immediately after the war, this aircraft lifted Douglas Aircraft to preeminence in the manufacture of cargo/passenger aircraft. After the war, building on the success of the DC-3, Douglas introduced the DC-4, a four-engine straight-wing plane with a front nosewheel, and soon followed with the similarly propeller-driven DC-5, DC-6, and DC-7. But in their enthusiasm to manufacture propeller-driven aircraft, Douglas made a strategic error. They did not anticipate the move to jet engines for commercial aircraft and allowed first deHavilland in Britain, and then Boeing in Seattle, to gain a foothold in the passenger airplane market. DeHavilland's ill-fated Comet kept falling out of the skies due to metal fatigue, consequently leaving the field to Boeing, which introduced the four-engine 707, the most widely used long-distance aircraft in the world, soon followed by the immensely popular three-engine medium-range 727.

Douglas scrambled to catch up by introducing the DC-9, a small twin-engine craft that soon found competition from the Boeing 737. The rush to catch up with Boeing caused Douglas to experience financial difficulties. In trouble for funds, in 1967, the McDonnell Corporation, a manufacturer of military aircraft, purchased Douglas Aircraft and the company became known as McDonnell Douglas.

Now having the necessary capital, the new McDonnell Douglas Corporation decided to challenge Boeing in the jumbo jet department. Boeing was ready to introduce the wide-bodied 747, and since this was a unique airplane, McDonnell Douglas decided to build the DC-10 as competition.[7] The same idea occurred to Lockheed Aircraft, which began work on the L1011, a craft remarkably similar to the McDonnell Douglas DC-10. Since Boeing had such a lead in the jumbo market, the race between Lockheed and McDonnell Douglas was for number two. There would be no number three.

The development costs for an airplane such as the DC-10 are immense, and each corporation in effect was "betting the company" on the success of the airplanes. Although no information is available on just how much this development costs, we do know that after building a number of L1011s, Lockheed decided that they would never make money in the jumbo jet market and ceased to manufacture the L1011, taking a $2.5 billion loss.

The airplane as conceived by McDonnell Douglas engineers had two engines on the wings and one engine under the tail assembly. As with all large jetliners, the airplane fuselage was divided into the passenger compartment and the cargo compartment, with the two separated by the passenger compartment floor. This floor was made of open trusses, thus providing an ideal conduit for running control lines and electric cables from the cockpit to the rear of the airplane. Three hydraulic systems, each independent of the other for the sake of redundancy, were designed and all three lines ran through the passenger compartment floor. If any two of the hydraulic systems should fail (a very low probability, so thought the engineers), the third system would still allow the pilot to fly the airplane. The fuselage of the DC-10 is shown in Figure 3.1. Note the passenger compartment above the cargo compartment and the placement of the control lines between the two compartments. The rear cargo door is also shown.

New airplanes go through a thorough series of tests to prove their airworthiness, beginning with ground tests to simulate flight. Because the Federal Aviation Administration (FAA) does not have the necessary technical personnel to conduct these tests, the company engineers are in effect deputized to conduct the airworthiness tests themselves, a clear conflict of interest. Such self-testing and reporting is not uncommon in many regulatory situations, however, and depend on periodic auditing by oversight agencies. For example, a great deal of environmental testing (e.g., pollutants leaving a company's stack or outfall structure) is tested by the companies themselves and reported to state or federal agencies. If in periodic visits the agencies find something amiss, the penalties can be quite severe, including civil fines and criminal charges leading to jail time. Most of us are familiar with this self-reporting system when we file our 1040 forms with the Internal Revenue Service.

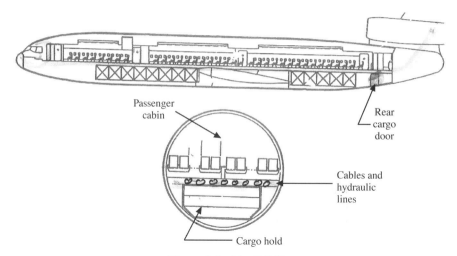

Figure 3.1 The DC-10.

In one of the DC-10 ground tests, as the fuselage was being pressurized to simulate flight, the rear cargo door flew open, causing rapid depressurization in the cargo compartment. Since the passenger compartment was still pressurized, the sudden decompression caused the floor to collapse into the cargo compartment, severing all hydraulic and electric lines. The engineers obviously realized that if this had occurred in flight, the airplane would not have been flyable and the craft would have been lost. The investigation of this incident centered on the design of the rear cargo door, not on the placement of control lines in the passenger compartment floor. The McDonnell Douglas engineers concluded that the depressurization occurred because the door was not closed properly and that it was unlikely that this would happen again. They decided that the door was adequate but that some small modifications were needed and asked Consolidated Vultee Aircraft Corporation (Convair), the subcontractor, to do the redesign.

Doors in the passenger compartment are known as *plug doors* because they are larger than the door opening, so that the pressure in the compartment cannot blow them out. Higher pressure will just force the doors to be sealed more tightly. In the cargo compartment, however, where flight attendants cannot open the door from the inside, the doors have to be opened from the outside and cannot be plug doors.

The cargo doors on the DC-10 open out and are hinged at the top. The locking mechanism is a series of hooks on the door that fit over spools attached to the body of the airplane. An electric motor (the actuator) drives a shaft that moves the hooks over the spools. A locking mechanism is supposed to assure that the shaft has gone "over center," thus transferring the pressure from the compartment to the hooks and spools. Higher pressure would simply shut the door even tighter. A manual handle on the outside of the door, when pushed shut and flush with the door, causes a lock pin to freeze the locking mechanism in the "over center" position. When the lock pins are in place, a warning light in the cockpit goes off, assuring the pilot that the door is locked properly. In summary, the door is closed by turning on the activator to move the hooks over the spools, and shutting the handle on the outside of the plane, causing the lock pins to lock the hooks in place.[8] A simplified sketch of the locking mechanism as originally designed for the rear cargo door is shown in Figure 3.2.

The problem with this design, however, is that if some object interferes with the door being shut correctly or if there is a misalignment, the actuator will not be able to push the hooks all the way over the spools. Because the manual handle will not be able to push the lock pins into position, the door should not be able to be locked because the lock pins would not be able to move to the shut position. But if the manual handle is pulled very hard, it is

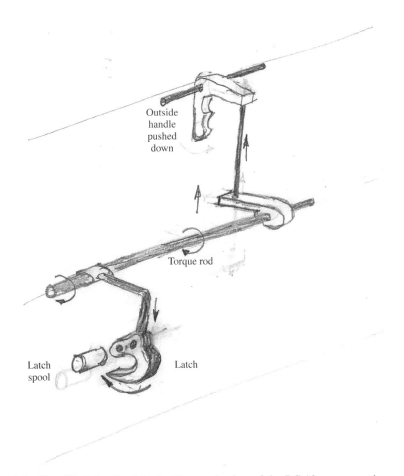

Figure 3.2 Simplified sketch of the locking mechanism of the DC-10 rear cargo door.

possible to bend the rods and stow the handle flush with the door in the shut position. The warning light in the cockpit goes off even though the door is not closed properly.

The DC-10 eventually passed all its tests and was certified as airworthy. Deliveries to customers began, with American Airlines being the largest customer. In June 1972, a lightly loaded American Airlines DC-10 that had taken off from Detroit experienced a sudden decompression while flying over Windsor, Ontario, at 12,000 feet, with the loss of two of the three hydraulic systems. The rear cargo door flew off, collapsing the passenger compartment floor. The floor did not totally collapse, however, due to the small number of passengers. The pilots, using the remaining backup hydraulic system and steering the plane by modulating the thrust of the engines, were able to return safely to the airport.

An investigation by the National Transportation Safety Board concluded that the loss of the rear cargo door caused the catastrophic decompression and ordered further modifications to the airplane to prevent similar accidents. The baggage handler responsible for shutting the rear cargo door admitted that he had difficulty shutting the cargo door and had used his knee to gain leverage before he was able to force the handle down. The National Transportation board concluded that the design of the rear cargo door represented a serious safety problem and that the DC-10 ought not to have been certified as airworthy. The board recommended grounding all of the DC-10s then in service.

Based on this conclusion, the Federal Aviation Administration should have issued an "airworthiness directive," which would have grounded all DC-10s while the modifications were being made. Such a move would have been highly detrimental to McDonnell Douglas and might have caused the company to follow Lockheed in abandoning the project. The administrator of the FAA and McDonnell Douglas management instead reached a "gentleman's agreement" to get the problem fixed as soon as possible, using the mechanism of the service bulletin to alert all airlines to get the aircraft modified. All doors were to have small view holes through which it was possible to see if the locking mechanism was in place. A decal with the instructions for looking through the holes was to be attached to the door. Records indicate that two years later not all of the existing aircraft had been so modified probably because a service bulletin is simply an advisory that can be ignored by the airlines.

After the Windsor incident, Dan Applegate, the chief product engineer at Convair, in charge of designing the cargo door, sent a remarkable memorandum to his superiors, warning Convair management that in his opinion the specifications for the rear cargo door had been changed to the point where the door was now unsafe. He believed that the design of the airplane was such that the next loss of the rear cargo door on a fully loaded DC-10 could result in total collapse of the passenger compartment floor and loss of the airplane.

The two-page memorandum, written in a matter-of-fact engineering style, discusses the problem of the catastrophic decompression and the collapse of the passenger compartment floor, resulting in the loss of all hydraulic systems, and concludes: ". . . once this inherent weakness was demonstrated by the July 1970 [ground] test failure, [McDonnell Douglas] did not take immediate steps to correct it. It seems inevitable that, in the twenty years ahead of us, DC-10 cargo doors will come open and I would expect this to usually result in the loss of the airplane. . . . It is recommended that overtures be made at the highest management level to persuade Douglas to immediately make a decision to incorporate changes in the DC-10 which will correct the fundamental cabin floor catastrophic failure mode."[9]

Convair management discussed Applegate's memorandum, and he was told that if they went to McDonnell Douglas with these concerns, Convair would open itself up to criticism and possible liability. Convair management believed that since the FAA certified the aircraft, Convair should not get involved. In effect, Applegate was told to shut up, which he did.

In March 1974, a Turkish Airlines fully loaded DC-10 took off from Orly Airport in Paris. At 12,000 feet the craft experienced a catastrophic decompression and 346 people lost their lives. Most of the wreckage was strewn over a wide area and none of the bodies were identified, except for six victims who were all found near the rear cargo door, six miles from the rest of the wreckage. The Turkish baggage handlers remembered having difficulty shutting the door and did not look through the sight hole. They did not know to look, since the directions were written in English, which they could not read.[10]

Such disasters always demand a sacrifice, and the investigation focused on Dan Applegate. This is without question unfair. Applegate was, by all accounts, an excellent engineer. (He has since retired.) But by studying this incident we can learn something about engineering. In the face of his management's refusal to follow his recommendations, Applegate chose to do nothing. It is this decision that is so interesting and deserves analysis.

Dan Applegate is a stereotype of a technically excellent engineer thrust into a decision-making role where personal and professional ethics are tested. Applegate was probably ill prepared to make the kind of decision required of him. His education was in physics and chemistry, followed by mechanics and structures, thermodynamics, materials, circuits, and other engineering topics. He probably took few, if any, social science courses and probably was never introduced to concepts of ethical reasoning. In his time, engineerng education was, and to a great degree still is, thought of as applied natural science.

All of this applied natural science was useful and necessary for Dan Applegate, but in his role as chief engineer, these skills were no longer the only ones he needed. They were not even the most critical. In Applegate's position as the chief engineer for Convair, he was not doing applied natural science at all, but what he was practicing was applied social science.

Consider Applegate's problems:

1. *Finance and Economics.* He states several times in his famous memorandum how important to Convair the continuation of subcontracts from McDonnell Douglas are and what financial effect discontinuance of these contracts would have. In his capacity as chief engineer he had to consider the financial health of his company.

2. *Political science.* He and Convair and McDonnell Douglas were involved in a complex arrangement with the federal government whereby the actions of the FAA significantly affected the welfare of his company. The decisions of the FAA, governed in some part by political motives, were crucial to his company.

3. *Management.* He had to understand the actions of McDonnell Douglas engineers. He also needed to understand why his own management was refusing to pass on his warning to McDonnell Douglas. He needed to know what a manager does and how these people operate.

And most important:

4. *Ethics.* Dan Applegate probably had never even thought about how value-laden decisions are made. He probably never heard of whistle-blowing, and he never considered any action other than the steps that he took up to his tacit acquiescence.

Dan Applegate was faithful to his employer. Loyalty is morally defensible if the object of the loyalty is engaging in moral actions. But modern Western cultural norms established at Nuremburg during the Nazi war crimes trials consider blind loyalty to be morally unacceptable. "Following orders" is not honorable behavior if the action leads to immoral results. But Dan Applegate no doubt firmly believed that his job and his loyalty to Convair were both honorable and moral, but given the nature of the ethical problem, ought he to have taken some other action? Achieving no satisfaction to his concerns within the company, he would have had the option of going public, an act known as *whistle-blowing*.

Ethicists writing on whistle-blowing make a distinction between *permissive* and *obligatory* whistle-blowing.[11] The first test in a situation such as that faced by Dan Applegate would be to ask if he was permitted to go outside the company to air his concerns. Usually, this permission is valid if the potential harm is significant and if the whistle-blower has exhausted all avenues within the organization. The risk of harm in this case was great, and by sending the memorandum to top management, Applegate had exhausted his options within the company. He thus had "permission" to take the next step—to go outside the company.

Would he have been obligated to do so? Some ethicists argue that the "obligation" test is based on two conditions:

1. The action will not result in great harm to oneself.
2. The action has a reasonable chance of being successful.

According to these conditions, if the effect of the action will result in great harm to the whistle-blower or if the action has little chance of being successful, the action is ethically unwarranted and there is no obligation to blow the whistle. As the most extreme example, suppose that a German railroad employee during World War II realizes that he is switching trains full of people who will be killed at a concentration camp.[12] Clearly, there is no chance whatever of success if he chooses to be a whistle-blower, and there is a high probability of his own death. This is not the same as saying that the German railroad employee was acting ethically in switching the trains, only that whistle-blowing in this case would not be an effective means to address the immorality. However, if there were a system in place that gave the railroad employee a reasonable chance of calling attention to this immoral act without himself getting killed, whistle-blowing would have been obligatory. Others argue that this is a gross and dangerous oversimplification and that doing the right thing is not so conditional. In fact, the moral obligation is to do what is right, no matter the likelihood of success.

In the case of Dan Applegate, if he had gone public with his concerns, he would certainly have lost his job. Although this is not trivial, it pales in comparison to the loss of 300 lives. A skillful engineer can get other jobs, and his fame might even have resulted in enhanced income. In our litigious society, going public with his concerns would have had an immediate effect in making McDonnell Douglas redesign the rear cargo door. Thus, Applegate was, by most ethical tests, obligated to tell someone outside the company, in order to get the door and floor redesigned.

There is, of course, the possibility that such a disclosure would have resulted in sufficient adverse publicity that McDonnell Douglas would have abandoned the DC-10 altogether and thousands of people would have lost their jobs. Although to most people and most readers of the first canon of engineering ethics, job loss is not morally tantamount to loss of life, the obligation is thus not always so clear-cut, and it is somewhat more difficult to blame Dan Applegate out-of-hand. But what seems to be clear is that Dan Applegate was ill prepared to make the decision. His education had probably been a rigorous series of basic and applied physical science courses. Ironically, there is no doubt that technological excellence helped him advance in this profession to a position where these skills were no longer adequate to make the decision he had to make. Had he had a stronger education in the social sciences (including ethics) it is possible that he would have not allowed his memorandum to die once the management had rejected his concerns. He might have thought more deeply about his responsibilities both as an engineer and as a human being, and prevented the disaster that followed.

What does a cargo door on an airplane have to do with environmental justice? The DC-10 is one of a series of well-known engineering ethics case studies, and if we concentrate only on these well-reasoned situations we would be tempted to believe that such ethical conflicts are few and far between for engineers. We might conclude that we need not worry about understanding ethics, justice, and the other social values, since the chances of having to make a value-laden decision are remote. The fact is that most stories of ethical decision making in engineering are seldom made public. Consider the following two examples of ethical decisions by engineers who had to make difficult engineering decisions.

The first engineer started his own consulting engineering firm. A developer was converting a large apartment building into luxury apartments and wanted this engineer

Biographical Sketch: Hardy Cross

In the United States, the sense that engineering involves working for the public good is a long and honored tradition. Hardy Cross (1885–1959), one of the engineering giants of the early twentieth century and a name familiar to most civil engineers, understood that engineering was not just bricks and mortar. In his memoir entitled *Engineers and Ivory Towers,* he wrote: "Many engineering problems are as closely allied to social problems as they are to pure science."[13]

to do the structures and utilities drawings for the conversion. The engineer asked what was to become of low- and middle-income families presently living in the building. The developer had no idea, nor did he see that as being his concern. This was now his building and he wanted to convert it to luxury condominiums. The engineer thought about it a long time and finally decided that he could not bring himself to be party to the displacement of these people from their homes. The building has since been converted and the engineer did not do the work.

The second engineer is also in private practice. His firm had been working with a local developer who had greatly overextended himself and was in deep financial trouble. One day the developer asked the engineer to meet him at the bank for a conference. It turned out that the developer had found some investors who were willing to place $250,000 in the development company. As they were sitting there transferring the money to the developer, the engineer knew that the financial situation of the developer was grave and that he owed millions of dollars and was without doubt going into bankruptcy. The $250,000 was a drop in the bucket and would not make any difference to the success of the development corporation. But his code of ethics required him to keep quiet and watch the investors throw their savings away.

In both of these cases the engineers were out of their technical fields. No part of the engineering code of ethics would help the engineer who refused a job because it would dislocate families. He had to draw on his informal experiences in ethical reasoning, human sociology, and political science. The engineer who could not prevent the loss of investors' funds because of his commitment to be a faithful representative of his client had to have skills in communication and professional ethics. In these cases, the decisions made by the engineers were nontechnical decisions and the skills they had to use were skills based not on natural science but on the social sciences.

Interestingly, both of these cases involve decisions about development. We may think that land is a "blank slate" or that buildings are simply three-dimensional structures ready to be built, changed, or demolished as means to our engineering ends. In fact, they are human enterprises that will affect people's lives directly. A notorious example of an engineering failure because the designers lacked skills in social sciences is the Pruitt–Igoe public housing development in St. Louis.

CASE STUDY: PRUITT–IGOE PUBLIC HOUSING

The Pruitt–Igoe housing development in St. Louis, Missouri, was a modernist monument, emblematic of advances in fair housing and progress in the war on poverty. Regrettably, Pruitt–Igoe has become an icon of failure of imagination, especially imagination that accounts properly for the human condition.

Although we think of such public projects in terms of housing, they also often represent elements of environmental justice. Contemporary understanding of environmental quality is often associated with physical, chemical, and biological contaminants, but in the formative years of the environmental movement, aesthetics and other "quality of life" considerations were essential parts of environmental quality. Most environmental impact statements have addressed cultural and social factors in determining whether a federal project would have a significant effect on the environment. These factors have included historic preservation, economics, psychology (e.g., perception of open space, green areas, and crowding), aesthetics, urban renewal, and the *land ethic*. In his famous essays, posthumously published as *A Sand County Almanac,* Aldo Leopold argued for a holistic approach: "A thing is right when it tends to preserve the integrity, stability and beauty of the biotic community. It is wrong when it tends otherwise."[14]

The land ethic was widely disseminated about a decade after the Pruitt–Igoe project was built, so the designers did not benefit from the insights of Leopold and his contemporaries. However, the problems that led to the premature demolition of this costly housing experiment may have been anticipated intuitively if the designers had taken the time to understand what people expected. Then we must ask who was to blame. There is plenty of culpability to go around. Some blame the inability of the modern architectural style to create livable environments for people living in poverty, largely because they "are not the nuanced and sophisticated 'readers' of architectural space that the educated architects were."[15] This is a telling observation and an important lesson for engineers. We need to make sure that the use and operation of whatever is designed is sufficiently understood by those living with it.

Other sources of failure have been proposed. Design incompatibility was almost inevitable for high-rise buildings and for families with children. However, most large cities have large populations of families with children living in such environments. In fact, population density was not the problem since St. Louis had successful luxury town homes not too far from Pruitt–Igoe. Another identified culprit was the generalized discrimination and segregation of the era. Actually, when originally inhabited, the Pruit section was for blacks and Igoe was for whites.

Costs always become a factor. The building contractors' bids were increased to a level where the project construction costs in St. Louis exceeded the national average by 60%. The response to the local housing authority's refusal to raise unit cost ceilings to accomodate the elevated bids was to reduce room sizes, eliminate amenities, and raise densities.[16] As originally

Figure 3.3 Demolition of Pruitt–Igoe housing development buildings in St. Louis, Missouri. The development was completed in 1956 and began to be torn down in 1972. (From O. Newman, *Creating Defensible Space,* U.S. Department of Housing and Urban Development, Washington, DC, 1996.

designed, the buildings were to become "vertical neighborhoods" with nearby playgrounds, open-air hallways, porches, laundries, and storage areas. The compromises eliminated these features; and some of the amenities removal led to dangerous situations. Elevators were undersized and stopped only every third floor and lighting in the stairwells was inadequate. So, another lesson must be to know the difference between desirable and essential design elements. Human elements essential to a vibrant community were eliminated without much accommodation.[17]

Finally, the project was mismatched to the expectations of the people who would live there, many of whom came from single-family residences. They were moved to a very large, imposing project with 2800 units and almost 11,000 people living there. This quadrupled the size of the next-largest project at the time.

When the failure of the project became overwhelmingly clear, the only reasonable decision was to demolish it, and the spectacular implosion shown in Figure 3.3 became a lesson in failure for planners, architects, and engineers. In the designer Minoru Yamasaki's own words: "I never thought people were that destructive. As an architect, I doubt if I would think about it now. I suppose we should have quit the job. It's a job I wish I hadn't done."[18]

As both the Pruitt–Igoe public housing development and the ill-fated DC-10 case studies illustrate, engineering is not only applied natural sciences, but many engineers,

Biographical Sketch: Aldo Leopold

Aldo Leopold (1887–1948) was born in Burlington, Iowa, and spent his younger years roaming the woods near his home. He went east to go to school, first Lawrenceville School in New Jersey and then Yale, where he received a graduate degree in forestry. In 1909 he joined the U.S. Forest Service, where he worked to preserve wilderness areas. He joined the Forest Products Laboratory in Madison, Wisconsin, but soon became anxious to be back in the woods. His knowledge of timber and game management soon made him an acknowledged expert, and he was offered a faculty position at the University of Wisconsin. He bought a poor farm in a county where the topsoil had been badly eroded and started to husband it back to health. He died fighting a forest fire on a neighbor's farm. His writings, especially his book *The Sand County Almanac,* had enormous influence on the environmental movement, and he is acknowledged as the founder of the environmental ethics movement of the twentieth century.

especially when they advance to leadership positions in engineering, find themselves in professional situations where the social sciences, particularly ethics, would be the most valuable set of skills that would dictate their success as engineers. Teaching our students first to recognize and then to think through ethical problems is like providing a viewing port in the professional cargo door to see that the ethical mechanism is properly locked and "over center." We often overlook "teachable moments": For example, we repeatedly miss opportunities to relate engineering and social science lessons from even the most life- and society-changing events, such as the fall of the World Trade Center towers.[19]

Thinking of engineering as "applied social science" redefines engineering from a profession that builds things to a profession that helps people. The extension of this conclusion should encourage educators to reevaluate what it is we teach our engineering students. We believe that all engineers should include in their educational quiver at least some arrows that will help them make the difficult ethical and social decisions faced by all professional engineers.

ENGINEERING AND POLITICS

We may not like to admit it, but even scientists and engineers are influenced by things other than the natural laws of science. In fact, everything we do is to some extent a social construct. Anne Fausto-Sterling said it well: "Science is a human activity inseparable from the societal atmosphere of its time and place. Scientists, therefore, are influenced—consciously or unconsciously—by the political needs and urgencies of their society."[20] Fausto-Sterling, a biology professor at Brown University, is widely recognized for her research in gender roles in scientific research, but her arguments about the influence of politics and social norming in science reminds us that engineering is also influ-

Biographical Sketch: Karl Imhoff

Engineering excellence is measured by how effective a design is in meeting the needs of society. Occasionally, a device or system of process is so successful that it serves the public good for generations after its invention. One such device is the Imhoff tank.

One of the most influential wastewater engineers ever was the German engineer Karl Imhoff (1876–1965), who was responsible for training many of the leading American sanitary engineers, such as Gordon Maskew Fair at Harvard. Imhoff's most lasting contribution is an unassuming little treatment system now called the *Imhoff tank.*

The Imhoff tank incorporates primary settling and sludge digestion all in one tank. The sludge solids settle to the bottom, into a cone-shaped hopper, where they begin to digest. The gases they produce will bubble up, but the really clever part of the tank is how the deflectors force the gas bubbles to the side, thus allowing the solids to continue to settle unhindered by the rising gas bubbles. The sludge is drawn out periodically and placed on a sand drying bed. The tank has no moving parts and is able to achieve at least 40% solids and BOD (biochemical oxygen demand) removal. Many Imhoff tanks are still in use, even in the United States. They just have not worn out, and in small communities where effluents from the tanks can be pumped to holding ponds prior to discharge, the system continues to work beautifully—100 years later!

Although the use of Imhoff tanks for large wastewater treatment plants would be quite inefficient, the tank is perfect for small plants serving a limited number of people where money and operator skill are both in short supply. It is appropriate for that use, and there is little doubt that Imhoff had this principle in mind when he came up with the idea. As technologies are transferred to less developed nations, the Imhoff tank stands as an example of a feasible, sustainable, and "low-tech" solution to environmental problems.

enced by the political environments in which we work. Politics informs our decisions as to what is and what is not a high-priority area of scientific research and engineering practice. Such priority setting has disadvantaged certain groups during much of the history of Western civilization. For example, in North America and Western Europe up to relatively recently, most health research has been conducted on adult white males; further, occupational and environmental epidemiological studies frequently have limited this attention to healthy adult white males (i.e., the "healthy worker syndrome"). This practice has at times led to inappropriate and erroneous extrapolations to women, children, the elderly, minority groups, and people with special health conditions (e.g., immunocompromised, asthmatic, and disabled persons).

Even the choice of what to study is influenced by social factors generally and politics specifically. Further, preconceived and preconditioned attitudes will affect how data are

Biographical Sketch: Edwin Chadwick

In the 1840s, Edwin Chadwick (1800–1890) launched the "great sanitary awakening," arguing that filth was detrimental and that a healthy populace would be of higher value to England than would a sick one. He had many schemes for cleaning up the city, one of which was to construct small-diameter sanitary sewers to carry away wastewater, but this did not endear himself to engineers who wanted to build large sewers. A damaging confrontation between Chadwick, a lawyer, and the engineers ensued, with the engineers insisting that their hydraulic calculations were correct and that Chadwick's sewers would be plugged up, collapse, or otherwise be inadequate. The engineers wanted to build large-diameter egg-shaped brick sewers for both wastewater and stormwater that would be large enough to allow human access for cleaning and repair. These sewers were, however, anywhere from three to six times more expensive than Chadwick's vitrified clay conduits, and their high cost prevented their construction in most areas of London. Eventually, the answer was a compromise, with pipes used for collecting (small-diameter) sewers and the interceptors constructed of brick.

Chadwick stands as a champion for the practical. Like Imhoff, his approach has much to teach today's engineers about sustainable design.

interpreted. Even the most objective data of the highest quality are subject to bending and shaping by people. That is, in the process of turning data into information, we change its meaning. This has to be done. At its most basic level, changing raw data into information requires some type of quality check, often called *data verification and validation.* Verification is concerned with how correct and complete data are according to some set of standards. It does not concern itself *per se* with whether the data are "true." Validation goes one step further to see whether the data collected are accurate, complete, and meet specified criteria. The U.S. Environmental Protection Agency defines *data validation* as ". . . an analyte- and sample-specific process that extends the evaluation of data beyond method, procedural, or contractual compliance (i.e., data verification) to determine the analytical quality of a specific data set."[21] This is often done using verification and validation software. For example, one basic check of water quality data is that pH values always range between 0 and 14.

Discussion: Data Validation and the Negative pH Value

Computers are idiots. They will do whatever our software codes tell them. For example, a common data validation check is to instruct validation software to look for physically impossible values, such as negative areas and volumes (e.g., −20 acres or −10 liters). Another instruction is to look for unrealistic values, such as negative pH values or pH values greater than 14. The pH value is

(*continued on page 82*)

Biographical Sketch: George E. Waring, Jr.

 George E. Waring Jr. (1833–1898) was educated at College Hill, Poughkeepsie, and then studied agriculture with a private tutor. In 1857 he was appointed agricultural and drainage engineer of Central Park in New York City, and then received a commission in the U.S. Army, eventually rising to the rank of colonel in the cavalry, and was involved in numerous Civil War battles. After the war he settled in Newport, Rhode Island, where he became the manager of a large farm, but his knowledge of drainage led him to become a full-time engineering consultant.

After the destruction and deprivation of the Civil War, many southern cities were poor and unsanitary. In Memphis, Tennessee, the death rates from communicable diseases were so high that this problem caught the attention of the nation and a commission was formed to study the health problems in the city. One of the recommendations was to construct a sewerage system that would be limited to household wastewater. Instead of a combined system that would carry both stormwater and human waste, Waring proposed a system that would allow the stormwater to run off in channels while the household sewage was collected in small-diameter sewers with flush tanks. His plan, which he based on the small-diameter sewers first promoted by Edwin Chadwick in London, was to be only one-tenth the cost of constructing a combined system. After much political fighting, the city decided to adopt Waring's plan, and a system was constructed consisting of 6-inch vitrified clay pipes with flush tanks leading from homes into increasingly larger collecting sewers. The main argument advanced by Waring was that these sewers were necessary to enhance public health. Waring believed in the since disproven *miasma theory* of disease, that people became ill because they came into contact with sewer gas, and thus the totally buried and tight system was supposed to reduce the incidence of cholera, typhoid, and other such diseases by not allowing the miasma to waft into the community.

After the sewers had been constructed, the incidence of communicable diseases dropped markedly and Waring claimed the system to be a success. But most people considered the system a failure because of operational problems. The small lines from households often clogged and had to be cleaned with "snakes," and when the collecting lines clogged, streets had to be dug up to unblock the sewers. Eventually, manholes were constructed, which if factored into the original cost, would have increased the price of the system significantly.

Nevertheless, the controversy as to whether a city should build separate sewers instead of combined sewers raged for decades, with most engineers favoring combined systems since they were less expensive to build than two separate systems. At that time there were no wastewater treatment plants, and all water—stormwater and wastewater—went to the same convenient place, such as the Mississippi River in the case of Memphis. Only when the polluted waterways became a national concern did the cities recognize that Edwin Chadwick and George Waring had been right all along.

Incidentally, one of the most difficult water pollution challenges that has carried over into the 21st century is that of *combined sewer overflows*. Many big city public works and water engineers likely wish that Waring's and Chadwick's calls for separate systems had been more closely heeded.

Biographical Sketch: Allen Hazen

Allen Hazen has had an immeasurable effect on the sanitary engineering profession. Starting with this MIT degree in 1888, he became a chemist at the new experiment station set up by the Massachusetts State Board of Health. This eventually evolved into the world-famous Lawrence Experiment Station that was responsible for much of our early understanding of water and wastewater treatment. Hazen and his colleagues found, for example, that rapid sand filtration would work well in removing bacteria from water, and were able to show that such filters could compete with slow sand filters that required ten times more land area, allowing cities such as Cincinnati and Louisville to install rapid sand filters. Hazen also showed that tall columns filled with rocks could treat wastewater and that the treatment was due to microorganisms, proving the effectiveness of what became known as trickling filters.

In 1895 Hazen left the Experiment Station to form a series of engineering consulting firms (including what today is Malcolm Pirnie), but he kept involved in laboratory work. One of his enduring contributions is what all students studying environmental engineering learn as the Hazen-Williams equation for calculating pressure loss. Co-developed with Gardner Williams, the equation allowed, for the first time, for the roughness of the pipe to be taken into account in calculating the loss of pressure in a pipe over distance.

In 1910 Hazen was a key player in what became known as the "Pittsburgh case." A physician with the State Board of Health was pushing for the City of Pittsburgh to treat its wastewater prior to discharge into the Ohio River, where the water was used for drinking supply by the small towns. The river was highly contaminated, and it seemed reasonable that the waste ought to be treated prior to discharge. But at the time cities did not treat their own wastes, and the large volume of water in the rivers diluted it to where most smaller downstream communities could treat the water for their own needs. Logically, waterborne disease could be reduced if the city treated its wastewater prior to discharge, and the State began to demand that Pittsburgh do so. The City called in the most eminent sanitary engineers of the time, including Allen Hazen, who argued that it was not the City's responsibility to treat its wastewater, but rather it was up to the downstream communities to do a better job of treating the water withdrawn from the river. Their argument was that other cities upriver would still be dumping waste into the river, so compelling Pittsburgh to specifically treat its waste would not eliminate the need for downstream communities to continue to treat the water they pulled from the river. Hazen and his heavyweights prevailed with the argument that dilution of the waste was all that was required, and it was not until 1958 that Pittsburgh had its own wastewater treatment plant. In hindsight the position taken by Hazen and his colleagues was faulty in that it did not address the primary public health issue of cities polluting otherwise clean waterways and forcing others to deal with the pollution they created. However, it was commonly argued at the time that rivers were for waste disposal, a view illustrated perfectly by N.S. Sprague, superintendent of the Pittsburgh Bureau of Construction, who observed in his summary of the Hazen-Whipple report that "Rivers are the natural and logical drains and are formed for the purpose of carrying the wastes to the sea."

(continued from page 79)
one of the few very complicated concepts that has gained currency in most areas of engineering and medicine and is grasped widely by the general public. They may know that their shampoo is "pH balanced" and even that this means that it is chemically neutral, with a pH value near 7. Elementary students are taught that the range of pH is between 0 and 14. However, to a theoretical chemist, the range is not so limited.

Water not only exists as molecular water (H_2O) but also includes hydrogen (H^+) and hydroxide (OH^-) ions:

$$H_2O \leftrightarrow H^+ + OH^- \tag{3.1}$$

The negative logarithm of the molar concentration of hydrogen ions [i.e., $[H^+]$ in a solution (usually, water in the environmental sciences)], is referred to as pH. This convention is used because the actual number of ions is extremely small. Thus, pH is defined as

$$pH = -\log_{10} [H^+] = \log_{10} ([H^+]^{-1}) \tag{3.2}$$

The brackets refer to the molar concentrations of chemicals, and in this case it is the ionic concentration in moles of hydrogen ions per liter. The reciprocal relationship of molar concentrations and pH means that the more hydrogen ions in solution, the lower the pH value.

Similarly, the negative logarithm of the molar concentration of hydroxide ions (i.e., $[OH^-]$) in a solution is pOH:

$$pOH = -\log_{10} [OH^-] = \log_{10} ([OH^-]^{-1}) \tag{3.3}$$

The relationship between pH and pOH is constant under equilibrium conditions:

$$K = [H^+] [OH^-] = 10^{-14} \tag{3.4}$$

When expressed as a negative log it becomes obvious that the pH and pOH scales are reciprocal and that they both range from 0 to 14. Thus, pH 7 must be neutral (just as many hydrogen ions as hydroxide ions). The log relationship means that for each factor pH unit change there is a factor of 10 change in the molar concentration of hydrogen ions. Thus, a pH 2 solution has 100,000 times more hydrogen ions than those in neutral water (pH 7), or $[H^+] = 10^{12}$ *versus* $[H^+] = 10^7$, respectively.

However, for very strong acids, the molarity of hydrogen ions greater than 1 yields a negative value of pH. For example, a 12 molar (i.e., 12 M) HCl solution has a theoretical pH of $-\log(12) = -1.1$. So for highly acidic, extremely hazardous conditions (such as the occasional mine drainage system), we may measure negative pH values in environmental situations.

We have never observed these situations in engineering practice. In fact, we have never seen a pH near zero. Even the strongest acids do not completely dissociate when they are dissolved in very high concentrations. In the 12 M HCl solution, a portion of the hydrogen will remain bound to the chlorine, so the pH will be higher than what is predicted from the acid molarity. In addition, the number of water molecules is dwarfed by the number of acid molecules, so the influence of the H^+ ions (the effective H^+ concentration, i.e., the activity) is highly elevated above the actual concentration. So although we commonly show pH to be the negative log of the molar concentration of hydrogen ions, it is actually the negative log of the hydrogen ion activity (i.e., pH $= -\log a_H^+$).

Even if one sees a negative pH, it is very likely to be a measurement error. At very high pH, using existing equipment (glass pH electrode), we regularly see a large positive measurement bias (measured pH $>>$ true pH), which is very difficult to correct but is commonly known as the *acid error*. By extension, a pH value above 14 is theoretically possible for highly caustic solutions (strong bases at very high concentrations). However, such conditions are seldom, if ever, found in the environment.

Thus, in engineering practice, it is usually prudent to be highly suspicious of any reported pH value less than zero or greater than 14; and it is reasonable to assume that these values are artifacts of either improper data logging or sampling error. However, unlike impossible physical conditions (e.g., -20 hectares of land), there are rare situations where the unexpected and highly improbable are in fact true. Thus, the protocol for removing the data points and how to treat the gaps they leave must be detailed in the quality assurance plan for a project. Even if our computers are idiots, that is no excuse for us to be.

Dropping erroneous values makes scientific sense, but what is done about the deletion? If we just leave it blank, we may not really be representing the water body's quality properly. If we put a value in (i.e., "impute"), such as a neutral pH 7, we have changed the representativeness of the data. Even a more sophisticated method, such as interpolating a data point between the two nearest neighbors' values, is not necessarily good. For example, we might miss an important but highly localized "hot spot" of pollution. So even at this very early step of manipulating data, bias (i.e., systematic error) can be introduced. Such unknowns can affect the selection of sites and other environmental decision making.

One clear and recent example of the range of perception of what, on the surface, appears to be a well-defined and objective scientific issue is that of arsenic in drinking water. The range of perspectives has to do with the characteristics of arsenic itself and the manner of how public health standards are developed.

CASE STUDY: ARSENIC AND THE POLITICS OF SCIENTIFIC TRUTH

Arsenic is actually a metalloid (i.e., it is a lot like a metal, but it does have some nonmetallic qualities). It shares, for example, some properties with phosphorus and nitrogen (all group VA elements in the periodic table). For general environmental and toxicity purposes, however, it is usually lumped in with the heavy metals. The principal reasons for this are that it is generally removed from water and soil with technologies that work for metals (such as precipitation/coprecipitation techniques), its toxicity and bioaccumulation behavior is similar to that of metals, and it is often found in nature and in contaminated sites along with metals. In fact, it is the second most commonly found contaminant in hazardous waste sites in the United States (see Figure 3.4). Arsenic has been used in industrial products and processes,

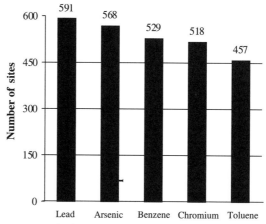

Figure 3.4   Five most commonly found contaminants at high-priority waste sites in the United States (National Priority Listing sites). [From U.S. Environmental Protection Agency, *Proven Alternatives for Aboveground Treatment of Arsenic in Groundwater,* Engineering Forum Issue Paper, EPA/542/S-02/002 (Revised), U.S. EPA, Washington, DC, 2002; www.epa.gov/tio/tsp.]

including wood preservatives, paints, dyes, metals, pharmaceuticals, pesticides, herbicides, soaps, and semiconductors, but since it is also a rather commonly occurring element, it is found in natural backgrounds in rocks, soils, and sediment. The range of potential sources makes dealing with arsenic complicated. For example, some water supplies happen to be in areas where arsenic and metals are found in relatively high concentrations because of leaching from surrounding rocks and soils.

Consider the position of a municipal engineer trying to adhere to federal and state drinking water standards [known as *maximum contaminant levels* (MCLs)], who must also rely on wells that receive water from arsenic-laden rock formations. It is not difficult to remove large concentrations of chemicals from water, but it becomes increasingly difficult and expensive, as the required concentrations decrease. For example, it is a general rule that if it costs $1 per gallon to remove 90% of a contaminant, it will require another $1 to remove 99% of it, and another dollar to remove 99.9% of the contaminant. Thus, the cost of removal as the required concentration approaches zero is exponential.

For metals and arsenic, the available technologies are more limited than for organic contaminants. For example, many organic contaminants (especially those that are not chlorinated) can be treated thermally, where they are broken down into harmless elemental constituents (i.e., carbon, hydrogen, and oxygen). Since arsenic is an element, this is not possible. All we can design for is moving arsenic from one place to another, where people are less likely to be exposed to it. Like heavy metals, arsenic's mobility and toxicity are determined by its oxidation state, or valence. As^{3+}, for example, is up to 10 times more water soluble and is more toxic to humans than when it is reduced to As^{5+}. Arsenic in some valence states is much less likely to move in the environment or to cause health problems than in others. However, once a person is exposed to the arsenic, metabolic processes can change these less toxic forms back to highly toxic forms.

Exposure to any form of arsenic is bad. Engineers need to know the forms (valence states) of the arsenic to optimize treatment and removal, but health scientists are often concerned about total arsenic exposures. The physical and chemical properties of arsenic are complex, but protecting people from the exposures to arsenic is even more complicated. All three branches of the federal government become involved. Congress has passed numerous laws addressing arsenic exposure, such as the Safe Drinking Water Act, which requires that the executive branch (in this case, the EPA) establish a standard (an MCL) for contaminants in drinking water. The actual concentration allowed is based on scientific evidence, professional judgment, and ample margin of safety (commensurate with uncertainties, *and there are always uncertainties!*). The courts become involved when there is disagreement as to whether the law is being upheld and whether the standards are sufficient. For local water supplies (e.g., towns), this can translate into hundreds or even thousands of plaintiffs (i.e., people living in the town being sued).

While everyone agrees that arsenic is toxic, they cannot agree on where to draw the line on allowable exposures. Recently, the MCL was lowered from 50 μg L^{-1} to 10 μg L^{-1}. This meant that water supplies just meeting the old standard would have to remove five times more arsenic. The town engineer may know that the present equipment at the plant would have to be replaced or upgraded, but the way such information is shared can affect what people perceive. For example, the town engineer may quote Robert Goyer, Chair of the National Research Council, Subcommittee to Update the 1999 Arsenic in Drinking Water Report, from his 2001 testimony before Congress: ". . . chronic exposure to arsenic is associated with an increase incidence of bladder and lung cancer at arsenic concentrations below the current MCL. This conclusion was strengthened by new epidemiological studies."[22] However, after delving a bit further, the town engineer may have found that the National Research Council also said in 1999 that "no human studies of sufficient statistical power or scope have examined whether consumption of arsenic in drinking water at the current maximum contaminant level . . . results in an increased incidence of cancer or noncancer effects."[23]

Had the science changed that much in the two years between the 1999 report and Goyer's testimony? Had new studies or better interpretations of those studies led to the change? Or is it simply a matter of whose perspective carries the day? The National Research Council is a highly respected science organization. The committee members are at the top of their fields, but they come from different organizations and often differ as to how data and information should be interpreted. Although their sources are the same epidemiological studies and models, it is not uncommon for subcommittee members to log minority opinions, based on differences in professional judgment.

What complicates controversies such as the acceptable level of arsenic in water is that groups with strong and divergent ideologies, such as the Sierra Club *versus* the Heritage Foundation, will buttress their positions based on political differences. Pity the engineer who has to tell the town council at a public meeting that they will have to spend money for improved arsenic removal. The engineer will inevitably be asked to justify the request. Although the correct answer is that the MCL is set by the U.S. EPA and is now mandated, politics will influence the perception of the decision makers. Although engineers are prone to emphasize science and professional ethics, they need to listen for the third factor, politics, as well. And the town engineer must listen to both the nonscientific and scientific influences.

For countries with sufficient financial and technical means and infrastructures, the arsenic debate represents a trade-off of values. It gets into some very complicated and controversial issues, such as the costs of preventing one cancer. Some have argued that if you include all costs of cleaning up hazardous waste sites, certain sites would amount to billions of dollars to prevent a single cancer. Obviously, that is worth it if the cancer is your own or that of someone you care about, but what if it is an anonymous, statistical person? Is there a threshold when something is just too costly? If so, are we not defining that point as the value of one human life? This is an important matter for those writing health and environmental regulations.

CASE STUDY: ASBESTOS IN AUSTRALIA
Australia's National Occupational Health and Safety Commission is responsible for developing regulations to protect workers from asbestos exposures. In so doing, the commission must consider scientific and economic information. However, the assumptions that are used can greatly affect the expected costs. In other words, the goal is to reduce the number of future deaths and diseases, such as mesothelioma and asbestosis (which derive from a virulent form of asbestos, chrysotile fibers). This exposure occurs when products containing chrysotile are imported, manufactured, and processed. Regulators must choose from several alternatives based on safety, health, and cost-effectiveness. The cost differences can be dramatic. In this instance, the Australian commission chose from three alternatives (see Tables 3.1 and 3.2):

1. Maintaining the *status quo* (base case)
2. Legislative prohibition or ban
3. Reduction in the national exposure standard

The commission recommended the second option, the legislative ban, because of the lack of sufficient information on the safety of alternative ma-

Table 3.1 Comparison of Quantifiable Cost Impacts of Proposed Phase-out of Chrysotile in Australian Products[a]

Item	Phase-out option assumptions	Present value over 40 years at 8%
Savings in death and illness		
Exposure standard	1.0 fibers/mL	
Number of persons exposed	22,300	
Value of human life	$1.5 million	
Cost of lung cancer + mesothelioma	$667,000 × 1.05	$24,187,596
Savings in business compliance costs		
Savings in OHS controls	Waste disposal and medical exams only	$29,511,511
	Present value benefits	$53,699,107
Increase in costs to business		
Increased cost of substitutes for small businesses	20% brakes, 17% gaskets	($ 6,014,403)
Capital and recurrent costs to large businesses	$8.3 million year 1, $1,098,900 per annum	($20,789,143)
	Present value costs	($26,803,546)
	Net result	$26,895,561

Source: Commonwealth of Australia, National Occupational Health and Safety Commission, *Regulatory Impact Statement of the Proposed Phase Out of Chrysotile Asbestos,* NOHSC, Canberra, Australia, 2004.

[a]Based on the national exposure standard of 1.0 fiber per milliliter of air, the maximum number of exposed employees, the lower figure used for the value of human life, and a 5% annual cost for mesothelioma, a cancer associated with exposure to asbestos.

terials, the cost of compliance compared to net benefits, and if and when chrysotile products are prohibited, the expected exemptions needed when suitable substitute materials are not available or in areas of competing values and national interests, such as defense.

Among other matters, the scenario analysis indicates that the net overall benefit of option 2 diminishes as the phase-out period extends. It would appear that were the phase-out period adopted to approach 10 years, the costs to business would outweigh the offsetting benefits to business and workers.

As complicated as this probably appears to most engineers and other technical professionals, think of how it appears to the general public. They do not have to be paranoid to fear that we might be "putting one over on them!"

Table 3.2   Summary of Scenarios Demonstrating Sensitivity of the Costs and Benefits Analysis to Key Factors

Key assumption	Net present value (NPV) over 40 years at 8% discount rate[a]			Discussion
	Scenario 1	Scenario 2	Scenario 3	
Time frame for phase-out[b]	$26,895,561 3 years	$17,486,930 5 years	−$2,327,666 10 years	Highly sensitive to changes in phase-out period. The shorter the period, the higher the NPV. Longer phase-outs continue costs associated with illness and other business costs, which lower the overall NPV[c]
	Scenario 4	Scenario 5		
Workers exposed	$26,895,561 22,300 workers	$13,880,157 10,300 workers		Highly sensitive to numbers of workers exposed. Halving the estimated number of workers exposed still results in a positive NPV.
	Scenario 6	Scenario 7		
Compliance cost savings	$26,895,561 Selected cost savings	−$2,615,951 No cost savings		If there are no savings in costs of complying, the NPV becomes slightly negative.
	Scenario 8	Scenario 9		
Substitutes and cost convergence	$26,895,561 Costs converge	−$46,507,961 Cost do not converge		Highly sensitive to cost convergence for substitutes. Highly negative NPV if no convergence over next 40 years.

Source: Commonwealth of Australia, National Occupational Health and Safety Commission, 2004, *Regulatory Impact Statement of the Proposed Phase Out of Chrysotile Asbestos,* NOHSC, Canberra, Australia, 2004.

[a] Health outcomes have been quantified and expressed as savings in the potential costs of death and illness over 40 years at a discount rate of 8%.

[b] The findings show the proposal to be highly sensitive to changes in underlying assumptions, but it has not been possible to fully quantify the current cost to a community of illnesses, such as asbestosis and other malignancies arising from chrysotile exposure. Hence, the net present value is not a complete quantification of all quantitative impacts and should be used as a guide to decision making only.

[c] The NPV reflects net benefits derived from the following annual (unless otherwise stated) cash flows:

- Benefits each year from savings in health and illness ($2.194 million) from the reduction in business costs of waste disposal ($1.179 million) and from the reduction in business costs of a medical exam ($0.64 million).
- Benefits every three years from a reduction in business costs of a medical exam ($2.56 million).
- Fewer costs imposed on small businesses each year during the phase-out period only ($6.66 million).
- Fewer costs imposed on large businesses each year ($1.098 million) and in year 1 only from investment in new production equipment ($8.3 million).

Truth can be a casualty of politics. One of the vexing, yet refreshing characteristics of engineers is their idealism. Engineers are forward thinking and have been selected, usually self-selected, as "can do" types. They see a problem or an opportunity, and think about ways to solve or achieve it. We characterize these qualities as vexing because we have observed the quandaries of engineers who engage in policy or politics. They sometimes even seem naïve about whether truth can be anything other than what the laws of nature demand. Newton's laws don't lie! We are reminded of an engineer who was interviewed September 12, 2001 about the collapse of the World Trade Center towers. After the engineer had shared some of the technical information and given an excellent explanation of the factors that led to the towers' collapse, the reporter commented that the team of terrorists that planned the attack must have included an engineer. The engineer was visibly shaken by the assertion and made the comment that he hoped that was not the case because "engineering is a helping profession."[24] Politics is an example of how different perspectives (or the more unkind term, *spin*) can be put on even very technical issues.

ENGINEERING AND COMMUNICATION

Listening is an active enterprise and is a necessary component of communication, which is analogous to the electrical engineering phenomenon of the signal-to-noise ratio (S/N) in a transceiver. S/N is a measure of the signal strength compared to background noise. In electrical engineering, our signal is the electrical or electromagnetic energy we are trying to send from on place to another. Conversely, noise is any energy that degrades the quality of a signal. In other words, we want as much of the energy to go to the signal as possible. As shown in Figure 3.5, in perfect communication, the message intended by the sender is exactly what is collected by the receiver. In other words, $S/N = \infty$, because $N = 0$. This is the goal of any technical communication, but this is seldom, if ever, the

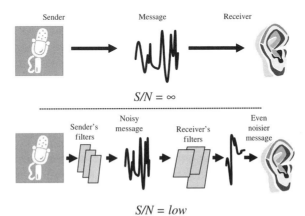

Figure 3.5 Transceiver analogy for communications, consisting of three main components: the sender, the message, and the receiver. The distortion (noise) that decreases the S/N is caused by filtering at either end of the message.

case. There is always some noise. A message that differs from what was meant to be sent (i.e., is "noisy") because of problems anywhere in the transceiver system. For starters, each person has a unique set of perspectives, contexts, and biases. We can liken these as "filters" through which our intended and received message must pass. Since both the sender and the receiver are people, each has a unique set of filters. So even if the message were perfect, the filters will distort it (i.e., add noise). The actual filters being used depend on the type of message being conveyed. In purely technical communications, the effect of cultural nuances should be minimal compared to most other forms of communications. Translating highly technical reports written in Spanish or another non-English language might be much easier and more straightforward than translating literature and poetry.

One of the skills that engineers have to develop is listening to their clients or to those who have a stake in a project. It is easy to ignore people, especially if they are perceived to be uneducated or uncultured. Assuming that their contribution cannot matter much and erecting cultural barriers can hinder listening and communications between engineers and those they serve. But engineers ignore the experience of the "locals" at their own risk, as demonstrated by the following case.

CASE STUDY: "THE DEVIL'S WATER" IN BANGLADESH

In Bangladesh in the 1990s, elevated levels of arsenic in drinking water had become epidemic. As many as 77 million of the 125 million Bangladeshi people are being exposed to elevated concentrations of arsenic in their drinking water, already resulting in about 100,000 related, debilitating skin lesions (see Figure 3.6), with chronic diseases expected to increase with time.[25] Sad to say, an engineering solution to another problem has played a major role in exacerbating the arsenic problem. Surface water sources, especially standing ponds, in Bangladesh have historically contained significant microbial pathogens, causing acute gastrointestinal disease in infants and children. To address this problem, the United Nations Children's Fund (UNICEF) began working with Bangladesh's Department of Public Health Engineering in the 1970s to fabricate and install tube wells in an attempt to give an alternative and safer source of water (i.e., groundwater). *Tube wells* consist of a series of 5-cm-diameter tubes inserted into the ground at depths of usually less than 200 m. Metal hand pumps at the top of each tube are used to extract water.

The engineering solution appeared to be a straightforward application of the physical sciences. It included many elements of a sustainable design, such as not depending on electricity and simple operation and maintenance. However, when the tube wells were installed, the water was not tested for arsenic. This was despite the fact that local people had originally protested the use of groundwater in some locations as "the devil's water." Was it possible that the indigenous folklore was rooted in intergenerational information about possible contamination that would have been valuable for the foreign engineers to know? Is it also possible that the educational, cultural, and technical differences contributed to poor listening by the engineers?

Figure 3.6 Skin lesions resulting from arsenic exposure in Bangladesh. [From A. H. Smith, E. O. Lingas, and M. Rahman, Contamination of Drinking-Water by Arsenic in Bangladesh, *Bulletin of the World Health Organization,* 78(9):1093–1103, 2000. Photo credit: World Health Organization.]

ENGINEERING AND ETHICS

Philosophers have been telling us for a long time how they believe we should behave. The diversity of the advice, however, may lead us to assume that everything is up for grabs and that morality is "situational." Actually, if one studies the counsel regarding what is right and wrong from the ancients to the most contemporary conceptions of moral philosophy, there is a great deal of commonality of thought. Surely, philosophers and teachers of philosophy at the university level frequently subscribe to one classical theory or another for the most part, but most concede the value of other models. They all agree, however, that ethics is a rational and reflective process of deciding how we ought to treat each other.

Biographical Sketch: W. Wesley Eckenfelder

Wes Eckenfelder (born 1926) received his undergraduate education in civil engineering at Manhattan College and went on to become a giant in the field of environmental engineering. He is a prolific writer, the author of hundreds of technical papers and scores of books, many of which are texts that have helped to educate the next generation of environmental engineers. Ironically, he has never received an earned Ph.D. because the professors at the university where he enrolled did not approve his doctoral research proposal. Eckenfleder wanted to do mathematical modeling of biological systems, and the professors thought that this could not be done. He decided that he did not need the professors, and left the university, going on to be one of the first to apply Michalis–Menton kinetics to biological treatment and to use clever graphical techniques to explain and understand complex biological processes. He received an honorary doctorate from New York University for his significant contribution to environmental engineering.

Eckenfelder has one other attribute that was lacking in many academic types—he is down to earth and can understand and talk to treatment plant operators, who in the end have to make the plants work. This ability—to explain complicated materials in simple terms and often to commiserate with the operators—makes him one of the most beloved and respected environmental engineers of the age.

One example will illustrate this point. At a huge national meeting, Eckenfelder was presenting a paper about the removal of color in wastewater, and his data were reported using the *color number,* a technical term that describes the depth of color in a wastewater. During the question and answer period, he was asked what he meant by a color number of 2 and a color number of 5. Instead of putting down the questioner, he said: "If you buy a scotch and water at the bar, it will have a color number of about 2. If you mix your own scotch and water at home, it will have a color number of about 5." The house roared in laughter and appreciation.

Like other professions, engineers strive for excellence. This is articulated in codes of ethics, which extend the mandate beyond avoiding actions that are clearly wrong and move us to strive for engineering accomplishments that advance human endeavors. The engineering profession pushes us to seek ways to do what is right. Part of the formula for ethical behavior is to know who is affected by what we do. But for whom do we strive to do what is right? Certainly, the company, agency, or holders of contracts are our clients, but in the environmental realm, our clients are vast. They are all of the people who may be affected by our advice, those now living and those of future generations. Our advice and actions must be seen through the prism of sustainability. As such, our actions must not simply solve immediate problems. Rather, they must be viewed as to how they will play out in future generations. Our science is always shaded by uncertainty, which means that all environmental decisions will also be uncertain to some degree. Thus, we are not always certain about the extent of influence of our ethical decisions, so we may need make them conservatively, following a *precautionary principle.* Such a principle states that if the consequences of an action, such as the application of a new technology, are unknown but the possible scenario is sufficiently devastating, it is prudent to avoid the action altogether. For example, the precautionary principle applied to global warming ought to lead us to take action to prevent such a catastrophic event from occurring, even though there is still a margin of doubt that global warming is being caused by human activity. Of course, the precautionary approach is not foolproof. The greater the precaution, the more likely opportunities will be missed to improve things, such as trying out new technologies and medical procedures.

Biographical Sketch: Gordon Maskew Fair

Gordon Maskew Fair was born in 1894 South Africa, and after graduating from high school in Berlin, went to study at Harvard. He received a doctorate of engineering (Dr. Ing.) from the Techniche Hochschule in Berlin in 1951. Coming back to Harvard, he had a series of appointments on the faculty, finishing up as a distinguished professor and dean of the faculty of engineering.

Fair, along with John Geyer of Johns Hopkins University, published a textbook on sanitary engineering in 1954 that became the standard text for generations of students studying in this field. For the first time, Fair and Geyer approached water and wastewater treatment from a scientific perspective, using chemical and microbiological principles to approach treatment design. The book was reissued in 1966 with Daniel A. Okun as the third author and remains a text that many consider the standard for the field.

Fair was elected to the National Academy of Engineering and worked for many years with consulting engineers in various health-related projects. He considered his first priority to be that of educator, but his passion was public health and he traveled all over the world lecturing and working with emerging environmental programs to help others. Ironically, although he developed a strong program in environmental engineering at Harvard, it was dismantled after he retired. Fair died in 1970, leaving a legacy of having educated some of the best and finest in the profession. The National Academy of Engineers obituary, written by Abel Wolman, said in part: "Whether it be in the swamps of Sardinia, in the jungles of Brazil, in the lecture rooms of the Ecole Polytechnique in Paris, or the laboratories of the London School of Hygiene, the presence of Gordon Fair inspired all those with whom he came into contact. . . . He was a man of the age of reason, a classic rather than romantic, a man with whom one could discuss any subject with the assurance he would come away with a balanced view."

Cardinal virtues are virtues on which morality hinges (Latin: *cardo,* hinge): justice, prudence, temperance, and fortitude. Among them, justice is the key to sustainability. This is the empathic view and is basic to many faith traditions, notably half of the Christians' greatest commandment to love one's neighbor as thyself (the other half is to love God completely), which is broadly articulated as the Golden Rule and the Native Americans' and Eastern monks' axiom to "walk a mile in another's shoes." Actually, one of commonalities among the great faith traditions is that they share the empathetic precept; for example:[26]

- *Judaism,* Shabbat 31a, Rabbi Hillel: "Do not do to others what you would not want them to do to you."
- *Christianity,* Matthew 7:12: "Whatever you want people to do to you, do also to them."
- *Hinduism,* Mahabharata XII 114, 8: "One should not behave toward others in a way that is unpleasant for oneself; that is the essence of morality."
- *Buddhism,* Samyutta Nikaya V: "A state that is not pleasant or enjoyable for me will also not be so for him; and how can I impose on another a state that is not pleasant or enjoyable for me?"
- *Islam,* Forty Hadith of an-Nawawi, 13: "None of you is a believer as long as he does not wish his brother what he wishes himself."
- *Confucianism,* Sayings 15:23: "What you yourself do not want, do not do to another person."

Our competence can take us in the right direction. We must excel in what we know and how well we do our technical work. This is a necessary requirement of the engineering experience, but it is not the only part. Engineering schools have increasingly recognized that engineers need to be both competent and socially aware. The ancient Greeks referred to this as *ethike arêtai* ("skills of character"). The competence of the professional engineer is linked inherently to character.

Codes of Ethics

A good place to start to find one's *ethike arêtai* is adherence to a commonly accepted code of ethics. Much of the environmental engineering mandate[27] is encompassed under engineering professional codes in general and in the Code of Ethics of the American Society of Civil Engineers (ASCE) in particular.[28] As evidence, in its most recent amendment on November 10, 1996, the code incorporated the principle of sustainable development. The code lays out four principles that civil engineers abide by to uphold and to advance the "integrity, honor, and dignity of the engineering profession." Civil engineers are expected to engage in the profession by:

1. Using their knowledge and skill for the enhancement of human welfare and the environment
2. Being honest and impartial and serving with fidelity the public, their employers, and their clients

3. Striving to increase the competence and prestige of the engineering profession

4. Supporting the professional and technical societies of their disciplines

The code further provides seven fundamental canons:

1. Engineers shall hold paramount the safety, health, and welfare of the public and shall strive to comply with the principles of sustainable development in the performance of their professional duties.

2. Engineers shall perform services only in areas of their competence.

3. Engineers shall issue public statements only in an objective and truthful manner.

4. Engineers shall act in professional matters for each employer or client as faithful agents or trustees, and shall avoid conflicts of interest.

5. Engineers shall build their professional reputation on the merit of their services and shall not compete unfairly with others.

6. Engineers shall act in such a manner as to uphold and enhance the honor, integrity, and dignity of the engineering profession.

7. Engineers shall continue their professional development throughout their careers, and shall provide opportunities for the professional development of those engineers under their supervision.

The first canon is a direct charge to deal properly with possible environmental insults resulting from our projects and to ensure that our "client at large," the public, lives in a healthy environment. The sustainability clause requires that we make certain that we view our projects not only from a standpoint of their immediate completion, but also in terms of how they will affect others now and in the future. More on this later.

Professional preparation, including educational requirements, is designed to help engineers meet the second canon, regarding competence. This canon is also proscriptive in that it calls for specialization within the sub-disciplines of engineering. The remaining canons prescribe and proscribe activities to ensure trust. The code applies to all civil engineers, so even a structural engineer must "hold paramount" the public health and environmental aspects of any project and must seek ways to ensure that the structure is part of an environmentally sustainable approach. This is an important aspect of sustainability, in that environmental considerations are not deferred solely to the "environmental professions" but are truly an overarching mandate for all professions (including medical, legal, and business-related professionals). That is why environmental decisions must incorporate a wide array of perspectives while being based on sound science. The first step in an inclusive decision-making process, then, is to ensure that every stakeholder understands the data and information gathered sufficiently when assessing possible environmental contamination.

The codes of ethics for the various engineering disciplines share numerous similarities. One noteworthy exception is the emerging field of biomedical research. Arguably, biomedical engineering is at the forefront of numerous important breakthroughs; many, such as genomics and sensors, will quite likely change environmental science and engineering. With such groundbreaking research comes the potential for unforeseen ethical problems. Major public debates have been occurring over the risks of nanotechnologies, neurotechnologies, cellular research, and genetic engineering. One means of addressing these risks is to ensure that each engineer is properly educated and trained to deal with ethical issues as they arise during one's career. This is the *microethics model*. For example, the recently approved Biomedical Engineering Society code of ethics states:

> Biomedical engineering is a learned profession that combines expertise and responsibilities in engineering, science, technology, and medicine. Since public health and welfare are paramount considerations in each of these areas, biomedical engineers must uphold those principles of ethical conduct embodied in this Code in professional practice, research, patient care, and training. This Code reflects voluntary standards of professional and personal practice recommended for biomedical engineers.[29]

Note the emphasis on a voluntary approach and deference to other medical and engineering standards of practice. In fact, the first professional obligation of the biomedical codes falls short of the "paramount" canon of other engineering codes, asking only that their membership "use their knowledge, skills, and abilities to enhance the safety, health, and welfare of the public." Some argue, and we agree, that this is a necessary but not completely sufficient response to the growing biomedical engineering ethical challenges. We also need the profession to articulate guidance regarding societal issues (i.e., *macroethics*). Some technologies are growing so quickly that the ethics lags behind, so that if we do see some major problems before they emerge, it may be too late since the technology has been released unchecked. An argument for why biomedical engineering takes a comparably tentative approach to ethics is that it bridges two professions, medicine and engineering, and thus must adhere to codes of both the American Medical Association and the National Society of Professional Engineers. In fact, the code does include a mix of engineering, health care, research, and training obligations. So it may well end up that other engineering societies will need to be more direct in dealing with macroethical issues that arise from biomedical engineering.

Limitations of Codes of Ethics

All codes of the various engineering professions provide a first line of defense when ethical questions arise. With enough diligence and information, most ethical problems can be solved using the basic premises of the codes. But sometimes the codes are unclear about the details of circumstances, or at times the codes even give contradictory answers.

One problem with the code is that the "public" to whom engineers owe primary responsibility is not defined. For example, suppose that a certain pesticide is banned by the U.S. EPA, and it has been determined that it is a carcinogen. Chemical manufacturing companies often have huge financial investments in the production of new pesticides and they would be loathe to lose this investment just because the EPA does not give them permission to use a product in American agriculture. They are then tempted to sell the product overseas where no such bans exist and the sale would be perfectly legal. Decisions such as the sale of a banned pesticide require definition of just what "public" to which the engineer is responsible. A similar problem exists for engineers working in the armaments industry. Arguments for or against designing weapons center around whether their principle use will be to "destroy" or to "deter." Thus the "public" good associated with armaments depends on one's perceived "public." We often shift risks from one group of people to another.

Discussion: Risk Shifting—Organochlorine Pesticides

Why is it that, on the whole, older Americans have much more positive impressions of the molecule 1,1,1-trichloro-2,2-bis(4-chlorophenyl)ethane, best known as DDT, than do today's younger generations? DDT, introduced during World War II, has had an impressive record of reducing infectious diseases and destroying human parasites such as lice. Yet today, DDT is generally condemned as a threat to health and the environment. So is DDT "good" or "bad"?[30]

In our experience, younger respondents are prone to categorize DDT as bad. One of influences they mention is Rachel Carson's seminal work, *Silent Spring*,[31] which exemplified the negative trend in thinking about organochlorine pesticides in the 1960s, particularly that these synthetic molecules are threats to wildlife, especially birds (hence the "silent" spring), as well as to human health (particularly cancer).

DDT

Conversely, students are further removed in time from when Allied troops were devastated by tropical vector-borne diseases such as malaria, yellow fever, and typhus, and they are less likely to know that the chemist Paul H. Müller won the 1948 Nobel Prize for Physiology or Medicine for synthesizing DDT. To many, his discovery was one of the major triumphs of the twentieth century. But despite the fanfare, Müller was circumspect about his discovery, especially as it was a harbinger of greater reliance on chemical biocides. In his 1948 acceptance speech, he was prescient in articulating the seven criteria for an ideal pesticide:[32]

- Great insect toxicity
- Rapid onset of toxic action
- Little or no mammalian or plant toxicity
- No irritant effect and no or only a faint odor (in any case, not an unpleasant one)
- As wide a range of action as possible, covering as many Arthropoda as possible
- Long, persistent action (i.e., good chemical stability)
- Low price (= economic application)

Thus, the combination of efficaciousness and safety was, to Müller, the key. Disputes between the pros and cons of DDT are interesting in their own right. The environmental and public health risks *versus* the commercial benefits can be hotly debated. Our students rightfully are concerned that even though the use of a number of pesticides, including DDT, has been banned in Canada and the United States, we may still be exposed by importing food that has been grown where these pesticides are not banned. In fact, Western nations may still allow the pesticides to be formulated at home, but do not allow their application and use. However, the pesticide comes back in the products we import, known as the *circle of poisons.*

However, debates about risks *versus* risks are arguably even more important. The decision is not simply about taking an action (e.g., banning worldwide use of DDT), which leads to many benefits (e.g., less eggshell thinning of endangered birds and fewer cases of cancer). What it sometimes comes down to trading off one risk for another. Since there are yet to be reliable substitutes for DDT in treating disease-bearing insects, policymakers must decide between ecological and wildlife risks and human disease risk. Also, since DDT has been linked to a chronic effect such as cancer and endocrine disruption, how can these be balanced against expected increases in deaths from malaria and other diseases where DDT is part of the strategy for reducing outbreaks? Is it appropriate for economically developed nations in temperate climates to push for restrictions and bans on products that can cause major problems in the health of people living in developing tropical countries? Some have even accused Western nations of "eco-imperialism" when they attempt to foist Western-style solutions onto developing nations. That is, we are exporting fixes based on our values (anticancer, ecological) that are incongruent with the values of other cultures [primacy of acute diseases over chronic effects (e.g., thousands of cases of malaria are more important to some than a few cases of cancer) and certainly more important than threats to the bald eagle from a global reservoir of persistent pesticides].

Finding substitutes for chemicals that work well on target pests can be very difficult. This is the case for DDT. In fact, chemicals that have been formulated to replace possibly dangerous products have either been found to be more dangerous [e.g., aldrin and dieldrin (which have also subsequently been banned)] or much less effective in the developing world (e.g., pyrethroids). For example, in huts in tropical and subtropical environments that have been sprayed with DDT, fewer mosquitoes are found than in untreated huts. This probably has much to do with the staying power of DDT in mud structures compared to the higher chemical reactivity of pyrethroid pesticides.

In another example of risk shifting, the Allied Chemical Company had operated a pesticide formulation facility in Hopewell, Virginia, since 1928. The Hopewell plant had produced many different chemicals over its operational life. Reflecting the nascent growth of petrochemical revolution in the 1940s, the plant began to be used to manufacture organic[33] insecticides which had recently been invented, DDT being the first and most widely used. In 1949 it started to manufacture chlordecone (trade name Kepone), a particularly potent herbicide that was so highly toxic and carcinogenic (see Table 3.3) that Allied withdrew its application to the Department of Agriculture to sell this chemical to American farmers. It was, however, very effective and cheap to make, so Allied started to market it overseas (see the "circle of poison" discussion above).

Table 3.3   Properties of Chlordecone (Kepone)

Formula	Physicochemical properties	Environmental persistence and exposure	Toxicity
1,2,3,4,5,5,6,7,9,10,10-dodecachloroocta-hydro-1,3,4-metheno-2H-cyclobuta(*cd*)pentalen-2-one ($C_{10}Cl_{10}O$)	Solubility in water: 7.6 mg L^{-1} at 25°C; vapor pressure: less than 3×10^{-5} mmHg at 25°C; log K_{ow}: 4.50	Estimated half-life ($T_{1/2}$) in soils between 1 and 2 years, whereas in air it is much higher, up to 50 years. Not expected to hydrolyze or biodegrade in the environment. Also, direct photodegradation and vaporization from water and soil is not significant. General population exposure to chlordecone mainly through the consumption of contaminated fish and seafood.	Workers exposed to high levels of chlordecone over a long period (more than one year) have displayed harmful effects on the nervous system, skin, liver, and male reproductive system (probably through dermal exposure to chlordecone, although they may have inhaled or ingested some as well). Animal studies with chlordecone have shown effects similar to those seen in people, as well as harmful kidney effects, developmental effects, and effects on the ability of females to reproduce. No studies are available on whether chlordecone is carcinogenic in people. However, studies in mice and rats have shown that ingesting chlordecone can cause liver, adrenal gland, and kidney tumors. Very highly toxic for some species, such as Atlantic menhaden, sheepshead minnow, or Donaldson trout, with LC_{50}[a] values between 21.4 and 56.9 mg L^{-1}.

Source: United Nations Environmental Programme, *Chemicals: North American Regional Report,* Regionally Based Assessment of Persistent Toxic Substances, Global Environment Facility, UNEP, New York, 2002.

[a]LC_{50}, lethal concentration 50: 50% of a population exposed to a contaminant is killed.

chlordecone
(each line intersection is a carbon)

In the 1970s the U.S. Congress amended the Federal Water Pollution Control Act to establish the national pollutant discharge elimination system (NDPES).

One of the NPDES permit requirements was that Allied list all the chemicals that it was discharging into the James River. Recognizing the problem with Kepone, Allied decided not to list it as part of their discharge, and a few years later *tolled* the manufacture of Kepone to a small company called Life Science Products Co., set up by two former Allied employees, William Moore and Virgil Hundtofte. The practice of tolling, long-standing in chemical manufacture, involves giving all the technical information to another company as well as an exclusive right to manufacture a certain chemical—for the payment of certain fees, of course. Life Sciences Products set up a small plant in Hopewell and started to manufacture Kepone, discharging all of its wastes into the sewerage system.

The operator of the Hopewell wastewater treatment plant soon found that he had a dead anaerobic digester (that is, the anaerobic bacterial count dropped to zero). He had no idea what killed his digester and tried vainly to restart it by giving it antacids. In 1975, one of the workers at the Life Sciences Products plant visited his physician, complaining of tremors, shakes, and weight loss. The physician took a sample of blood and sent it to the Centers for Disease Control in Atlanta for analysis. What they discovered was that the worker had an alarmingly high 8 mg L^{-1} of Kepone in his blood. The state of Virginia immediately closed down the plant and took everyone into a health program. Over 75 people were found to have Kepone poisoning. It is not known how many of these people eventually developed cancer. The Kepone that killed the digester in the wastewater treatment plant flowed into the James River, and over 100 miles of the river was closed to fishing due to the Kepone contamination. The sewers through which the waste from Life Science flowed was so contaminated that it was abandoned and new sewers built. These sealed sewers are still under the streets of Hopewell, and serve as a reminder of corporate avarice.

A blatant example of risk shifting is that of environmental justice situations. This occurs when the overall population risk is lowered by moving contaminants to sparsely populated regions, but the risk to certain groups is in fact increased. This type of risk shifting can also occur internationally, such as when a nation decides that it does not want its population or ecosystems to be exposed to a hazardous substance, but still allows the manufacture and shipping of the substance outside its borders.[35]

Another problem with using the code as a primary source of ethical decision making is its applicability to nonprofessional situations. When engineers retire or take on other responsibilities, do they stop being engineers and are they no longer held to the requirements of the code of ethics? The vice president of Morton-Thiokol, manufacturers of the space shuttle *Challenger,* was asked by the company president to "take off his engineering hat and put on his management hat," and he subsequently agreed to the ill-fated launch. Was it ethically possible for him to take off the engineering hat?

The fact is that engineers owe a responsibility to society, much like physicians. The physician has a commitment to help if he or she possibly can. If a person becomes ill on an airplane flight, for example, the physician is ethically called to assist. It does not matter that he or she might be retired, or have a practice quite different from the im-

Biographical Sketch: Rachel Carson

After graduating from the Pennsylvania College for Women (now Chatham College) in 1929 Rachel Carson (1907–1964) studied at the Woods Hole Marine Biological Laboratory and received a master's degree in zoology from Johns Hopkins University in 1932. She was a writer, scientist, and in 1962 wrote perhaps the most influential book ever published in the environmental field, *Silent Spring*. The title comes from what she foresaw as the death and destruction of birds due to the extensive use at that time of chlorinated pesticides and she called for an end to their indiscriminate use. The reaction by the chemical industry to Carson's book was immediate and vitriolic. She was branded as anything from a flake to a Communist sympathizer. The words of Parke C. Brinkley, President of the National Agricultural Chemicals Association, exemplified the emotional tirades her book unleashed:

(Courtesy of the U.S. Fish and Wildlife Service.)

The great fight in the world today is between Godless Communism on the one hand and Christian Democracy on the other. Two of the biggest battles in this war are the battle against starvation and the battle against disease. No two things make people more ripe for Communism. The most effective tool in the hands of the farmer and in the hands of the public health official as they fight these battles is pesticides."[34]

But in the end, it was clear that her cause was just, and as an increasing amount of evidence piled up on how the upper food chain was being affected by these non-biodegradable pesticides, she become a hero in the environmental movement. She did not, unfortunately, live to see her life and work honored by the world.

mediate emergency needs, or even have not practiced medicine for a long time because of a career change. The physician is still ethically obligated to help.

This is also true for engineers. Being an engineer is a lifetime commitment to assist individuals and society whenever possible and proper. Being an engineer is like getting a tattoo—it does not wash off. One cannot stop being an engineer.

Another problem with codes of ethics is that they are often internally inconsistent. For example, the first canon admonishes engineers to hold paramount the health *and* safety *and* welfare of the public—all three at the same time and all three are paramount. But there are times when this cannot be done. For example, a traffic engineer may decide that allowing holiday traffic to use a bridge that is under construction enhances public welfare and presents a minor risk to public safety. This engineer is thus balancing these two requirements, which cannot be maximized concurrently.

Codes can also have contradictory statements. Take, for example, these two canons:

Canon 1: Engineers shall hold paramount the safety, health, and welfare of the public.

Canon 4: Engineers shall act for each employer or client as faithful agents or trustees.

The first cannon says that the welfare of the public is of greatest importance, and the second admonishes the engineer to be a faithful to his or her employer. In isolation, there is nothing wrong with either canon, but adhering to both can result in problems in certain situations.

Finally, the engineering codes of ethics have little to say about questions regarding responsibilities to nonhuman nature other than a need to consider actions on the basis of sustainability. The codes do not spell out what, if any, responsibility engineers have to nonhuman animals, plants, or places. The only concern would be that the actions of the engineers not diminish the welfare of the (human) public. If an engineering project causes the demise of a plant or animal species, the concern is not for that plant or animal but for future humans who may not be able to enjoy looking at this species or obtaining some benefit from its use.

The conclusion we come to is that the engineering code of ethics is a worthwhile first, and very rough, cut at making ethical decisions in engineering. Often, when engineers are confronted by ethical problems, a quick glance at an engineering code of ethics is enough to encourage a decision that the engineer can live with. But ethical problems are seldom straightforward, and the right actions are not obvious. There is a great deal of subtlety in ethics, and any set of guidelines such as a code of ethics cannot hope to cover all cases.

Making Ethical Decisions in Engineering

Engineering solutions are predictable. A mechanical engineer designing an automobile can use the heat transfer equations in calculations with little concern for their accuracy of applicability. If there is a difference in temperature, the heat will be transferred from the hot side to the cold side, and the rate of that transfer is governed by the properties of the material. This will always work. No exceptions. And if 100 engineers did the calculation, 100 of them would get the same answer.

But this is not how ethics works. If 100 engineers are confronted by a problem in ethics, one would expect at least 100 suggested solutions (and possibly a lot more if some people waffle). All that can be said is that some of these solutions are better than others. None is wrong, and none is right: some are just better than others. So the question is: How can we design a system for coming up with answers to engineering ethical problems that will more often than not identify alternatives that most of us will consider to be acceptable?

One suggestion is to use a stepwise procedure for ethical decision making:

1. *Find the relevant facts.* Some problems disappear when the facts are all in. Getting the facts can also avoid grave embarrassment.

2. *Determine the moral issues.* What exactly is bothering you? What wrong has been done or may be done? Is this a problem in engineering ethics, or is this a question of personal morality? If it is engineering ethics, is it a breach of the engineering code of ethics, or something more complex?

3. *Decide who is affected by the decision you have to make.* Include your own family, your friends, and others who will be affected by your final decision.

4. *Determine your alternatives.* Here is where you want to be creative and "think outside the box." Perhaps you can come up with some imaginative alternatives that will not harm anyone and will not compromise your own integrity.

5. *Calculate the expected outcomes of each possible action.* We cannot, of course, predict accurately what the future will hold and what people will do. What is important here is that you differentiate between those actions that will undoubtedly occur and those that may occur. Actions by people are not always predictable, yet in this phase of decision making it is good to try to make such predictions.

    Sometimes it is impossible to prove that something has happened, such as proving that the company had a chemical spill years ago. You would simply not be able to prove that the spill occurred, and hence going to the state with the story would result in pitting your word against that of others—a no-win situation.

6. *Determine the personal costs associated with each possible action.* We all have an obligation to do the right thing, but this obligation is limited by the costs that we might incur. For example, if a certain ethical action will more than likely result in your losing your job, this is a large cost, but it may be acceptable if the situation demands it. On the other hand, if the probable cost is the loss of your life, most rational people will agree that this cost is too high except in highly unusual circumstances.

7. *Get help.* Chances are good that your problem is not unique. Someone has had the same concern and had considered the ethics ramifications of the various actions. Getting help from more experienced people is the best way to calibrate your own notions as to what you ought to do. Another source of help is the code of ethics of your profession or discipline. If one of your alternatives is clearly and unequivocally listed as being unethical, you know that this is probably not a valid alternative.

8. *Considering the moral issues, practical constraints, possible costs, and expected outcomes, decide what action should be taken.* This is an approach familiar to most engineers, that is, *synthesis.* Factor and weigh each element of a moral decision and use your best judgement. Consider how your decision would appear to a "reasonable person" or to a professional review board after the fact. As recommended by Immuanuel Kant, determine the effect of your decision if it were to become a universal act. Or as the John Stuart Mill harm principle requires, consider the possible harm of your action (or inaction).

ENGINEERING AND LAW

Many federal and state statutes that address environmental quality have approached environmental protection from a "command and control" perspective rather than from a pollution prevention framework. For example, the 20 major federal statutes, hundreds of state and local ordinances, thousands of federal and state regulations, and even more federal and state court cases deal with pollution. The number of environmental laws has

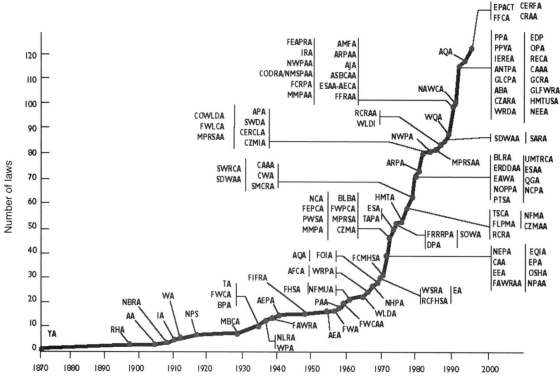

Figure 3.7 Growth of federal environmental legislation in the United States. (From U.S. Environmental Protection Agency, D. R. Shonnard, Chapter 3, http://www.epa.gov/oppt/greenengineering/ch3_summary.html#example, 2004; and D. T. Allen and D. R. Shonnard, *Green Engineering: Environmentally Conscious Design of Chemical Processes,* Prentice Hall, Upper Saddle River, NJ, 2002.)

grown exponentially in recent decades[36] (see Figure 3.7). In one way or another, this body of environmental law calls for action, such as the control of what can be discharged to the waters of the United States pursuant to the Clean Water Act's national pollutant discharge elimination system (NPDES) permit, or the amount of pesticides that can be found in food based on the Food Quality Protection Act.[37]

ENGINEERING AND EDUCATION

Inclusiveness is a requisite for environmental justice. A major challenge facing environmental engineering and the profession of engineering in general is how to improve cultural diversity. The engineering profession is not culturally representative of contemporary society. According to the U.S. Bureau of Labor Statistics, by the year 2008, minorities are expected to comprise upward of 40% of entrants to the workforce, but if current trends continue, their underrepresentation in technical professions will continue (see Figure 3.8).[38] African Americans presently make up less than 4% of the engineering, physical

Biographical Sketch: Edmund Muskie

Edmund Muskie (1914–1996) was a long-time senator from Maine. After getting his law degree from Cornell University, he served in the Navy during World War II and then entered politics, serving in the Maine legislature before being elected governor in 1955. He was the first Democrat to be elected to the U.S. Senate from Maine.

During the 1960s he championed legislation for cleaner water and air and was instrumental in the passage of both the Clean Air Act and the Clean Water Act. He became known as "Mr. Clean" both for his integrity and his stance on pollution issues. He was a front-runner for the Democratic presidential nomination but lost it when he broke down one snowy day defending his wife's integrity following a scurrilous editorial by the *Manchester Union Leader.* Much of the wisdom and effectiveness in the various environmental acts were the direct result of his work and energy. The main ideas in the control of pollution developed by Muskie and his staff have become the template for environmental legislation all over the world. Recently, more flexible and market-based approaches have been considered to augment this command and control approach. However, these only moderate the regulatory stance, and do not supplant it.

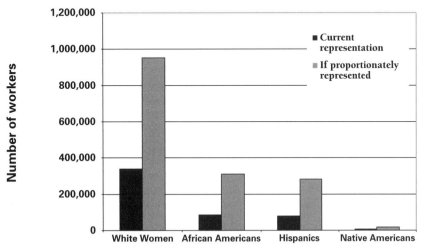

Figure 3.8 Representation of minority subpopulations in quantitative technical careers if participation continues at its present pace. (From P. B. Campbell, E. Jolly, L. Hoey, and L. K. Perlman, *Upping the Numbers: Using Research-Based Decision Making to Increase Diversity in the Quantitative Disciplines,* Education Development Center, Newton, MA, 2002.)

sciences, and economics workforce. Hispanic Americans make up less than 4% of the quantitative disciplines workforce, and Native Americans, less than half of 1%.[39] By increasing the representativeness of these groups in the engineering professions, we can make strides in addressing the problems of environmental injustices. But the problem goes back to middle and high school, or even elementary school preparation.

Two seemingly contradictory trends are occurring in the general public. The first is a divergence between the technologically literate and those not conversant in technical matters. Those who are trained in technical fields appear to be becoming more highly specialized and steeped in "techno-jargon," leaving the majority of people dependent on whatever the technologists say. There is another trend toward greater fluency of what were formerly specialized engineering terms in the larger public arena.

The first trend is not the same as the *digital divide;* the difference in access to and use of information technology by groups of different socioeconomic status, racial or ethnic group, and gender. The technical literacy gap is more fundamental than any single issue. Some fear that we are not properly educating our future citizenry to participate in an increasingly complex, technology-rich future. In the United States, there is much concern about students' lack of preparation in math and science as well as that of the present and future workforce. This is the mirror image of the problem of engineers' inadequate training in and appreciation for the humanities and social sciences that we discuss in this chapter. Engineers will definitely have to enhance their reach to include a greater number of perspectives in their projects, and simultaneously, we need to help the general public increase its appreciation for things technological. This confluence will not be easy. For example, the National Center for Education Statistics reports that in 1999, the United States lagged behind much of the developed world, and even a number of developing nations, in its middle school and high school students' achievement in mathematics and science (see Table 3.4). Even more alarming, the tests used in these comparisons stress environmental sciences and the math and science underpinning environmental engineering. The Trends in International Mathematics and Science Study (TIMSS) measures aptitude in arithmetic, algebra, geometry, data representation, analysis and probability, measurement, earth sciences, life sciences, physics, chemistry, environmental science, scientific inquiry, and the nature of science.

The other trend is that previously arcane and highly technical concepts and jargon are becoming increasingly mainstream. So, although many students do not seem to be motivated to participate fully in the increasing technological demands of society (e.g., taking rigorous math and science courses), they somehow are gaining a large repertoire of scientific expertise in their everyday lives. Children and teenagers are often "early adopters" of technical advances (e.g., technologies and terminologies). Viewed optimistically, we may be able to take advantage of this trend to increase minority representation in engineering.

So if trends continue, the engineering professions must reconcile any technical deficiencies indicated by the math and science gaps with the creeping technological savvy of the general public. We must provide a cadre of experts who can explain complicated engineering concepts in a straightforward manner, but at the same time, be prepared for a public that expects immediate, high-tech solutions to their problems.

Table 3.4 Mathematics and Science Achievement of Eighth-Graders in the 1999 Trends in International Mathematics and Science Study[a]

Mathematics		Science	
Nation	Average	Nation	Average
Singapore	604	Taiwan	569
Korea, Republic of	587	Singapore	568
Taiwan	585	Hungary	552
Hong Kong	582	Japan	550
Japan	579	Korea, Republic of	549
Belgium–Flemish	558	Netherlands	545
Netherlands	540	Australia	540
Slovak Republic	534	Czech Republic	539
Hungary	532	England	538
Canada	531	Finland	535
Slovenia	530	Slovak Republic	535
Russian Federation	526	Belgium–Flemish	535
Australia	525	Slovenia	533
Finland	520	Canada	533
Czech Republic	520	Hong Kong SAR	530
Malaysia	519	Russian Federation	529
Bulgaria	511	Bulgaria	518
Latvia	505	**United States**	**515**
United States	**502**	New Zealand	510
England	496	Latvia	503
New Zealand	491	Italy	493
Lithuania	482	Malaysia	492
Italy	479	Lithuania	488
Cyprus	476	Thailand	482
Romania	472	Romania	472
Moldova	469	Israel	468
Thailand	467	Cyprus	460
Israel	466	Moldova	459
Tunisia	448	Macedonia, Republic of	458
Macedonia, Republic of	447	Jordan	450
Turkey	429	Iran, Islamic Republic of	448
Jordan	428	Indonesia	435
Iran, Islamic Republic of	422	Turkey	433
Indonesia	403	Tunisia	430
Chile	392	Chile	420
Philippines	345	Philippines	345
Morocco	337	Morocco	323

Source: U.S. Department of Education, National Center for Education Statistics, J. D. Sherman, S. D. Honegger, and J. L. McGivern, *Comparative Indicators of Education in the United States and Other G8 Countries: 2002,* NCES 2003–026, NCES, Washington, DC, 2003.

[a] The Trends in the International Mathematics and Science Study (TIMSS), formerly known as the Third International Mathematics and Science Study, was developed by the International Association for the Evaluation of Educational Achievement to measure trends in students' mathematics and science achievement. The regular four-year cycle of TIMSS allows for comparisons of students' progress in mathematics and science achievement.

Biographical Sketch: Perry McCarty

Following the dismantling of the environmental engineering program at MIT in the late 1950s, Perry McCarty, with a 1959 MIT Ph.D. in hand, went west, and along with his colleague Rolf Eliassen, established the environmental engineering program at Stanford University. Over the past 45 years, Stanford has been perhaps the preeminent producer of outstanding faculty for environmental engineering departments throughout the United States and overseas. McCarty received his undergraduate degree from Wayne State University before going to MIT, where he worked with Ross McKinney, who himself left MIT to establish the program at the University of Kansas (Vallero is a product of the Kansas program).

McCarty specializes in biological processes and was one of the originators of the application of Michelis–Menton kinetics to biological wastewater treatment plant design. His later research efforts were in hazardous waste movement and attenuation through soils, and he was responsible for setting up the Western Region Hazardous Substance Research Center at Stanford. He is a member of the National Academy of Engineering and the recipient of many honors, including an honorary doctorate from the Colorado School of Mines.

Teaching Ethics in Engineering Schools

One of the problems inherent in trying to teach ethics to students is that usually the students have little societal or real-life context for the subject matter. Unless they are a nontraditional student with practical experience in the field, many of the cases and lessons presented are merely academic exercises or, at best, parables that we can only hope will be retained and applied once they begin practicing engineering. Similarly, the practicing engineer may be tempted to reject the moralizing of many ethics texts because they are not rooted in reality (at least the practical reality as perceived by the engineer). The practitioner is constantly balancing the bottom lines. The client, the employer, and the public expect results. The balance between profitability, design specifications, and career growth may not lend themselves to the simple case studies considered by ethicists. Professional life is more complicated than these *paradigms*.

This can be likened to driver's education training, where we teach the basics of driving a vehicle from a textbook (i.e., the "rules of the road"), augmented by hypothetical cases and scenarios to engage the student in "what ifs" (e.g., what factors led to a bad outcome, like a car wreck). Society realizes that new drivers are at risk and are placing other members of society at risk. Teenagers are seeking society's permission to handle an object with a lot of power (e.g., hundreds of horsepower) and a large mass (greater than a ton), with a potential to accelerate rapidly and travel at high speeds. To raise the consciousness (and we hope, their conscientiousness as well), we show films of what happens to drivers who do not take their driving responsibilities seriously. Similarly, in ethics class we also show films and discuss cases that scare the future engineer in hopes that this will remind him or her of what to do or not to do when an ethical situation arises. We do this in a safe environment (the classroom with a mentor who can share experiences), rather than relying on one's own experiences (i.e., simply immersing the new engineer in the "school of hard knocks").

But memory fades with time. Psychologists refer to this as *extinction,* which can be graphed much like a decay curve familiar to engineers (see Figure 3.9). Unless the event is very dramatic, it will eventually be forgotten. This may be why educators often use cases with extremely bad outcomes (e.g., the toxic gas release in Bhopal, India; the Hyatt walkway collapse, the Tacoma Narrows bridge failure) as opposed to less extreme, yet more likely cases, such as the engineer who must decide whether to avoid a conflict of interest in selecting bids for a project.

If a bad thing happens to someone else, like the scenarios in drivers' education films, they are not easily remembered, even if the results are gory.[40] Much more memorable are events that occur to ourselves. Anyone who has been in a car wreck will remember it for many years. The hope is that young people do not have to wreck a car to learn, but if they do wreck a car that the wreck that would be memorable but not harmful to the driver. This analogy holds for the newly minted engineer as well.

There are at least two ways of trying to make sure that future professionals remember the importance of ethical decisions: (1) using powerful cases, or (2) repeating the lessons. In the latter case the extinction curve is bumped up periodically (see Figure 3.10). But,

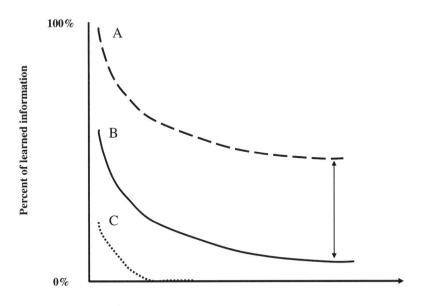

Figure 3.9 Hypothetical extinction curves. Curve A represents the most memorable case; curves B and C, less memorable. Curve C is completely forgotten with time. While the events in curves A and B are remembered, less information about the event is remembered in curve B because the event is less dramatic. The double-ended arrow represents the difference in the amount of information retained in long-term memory.

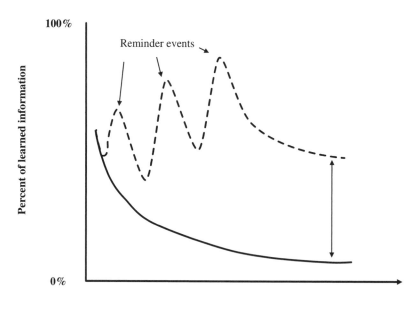

Figure 3.10 Hypothetical extinction curves. The solid line represents a single learning event with no reminders (reinforcements). The dashed line shows reminder events in addition to the initial learning event. The double arrow is the difference in retained information retained in long-term memory as a result of adding reminders.

like driving, most ethical education occurs "out there." Only when a professional is confronted with an actual ethical problem will what has been learned be put to the test. Learning engineering ethics, as is the case in learning to drive a car, results from a combination of formal study, interactive learning, and practice.

ENGINEERING AND TERRORISM

In our discussions of the engineering codes of ethics, we consider how we should "hold paramount" the public good. Another way of holding paramount the health, safety, and welfare of the public is that engineers minimize the risk of failure if failure can harm the health, safety, and welfare of the public. Civilian engineers (as opposed to military engineers), who build water and wastewater treatment plants, design our power grids, clean up hazardous waste sites, design highways, manufacture cars, protect our environment, and make possible the material advantages of civilization, are committed to minimizing failures, and these failures can be of four kinds: (1) mistakes and miscalculations, (2) extraordinary natural forces, (3) unpredictable occurrences, and (4) ignorance or carelessness. And, we suggest a fifth kind of failure that has recently emerged—intentional accidents.

Types of Engineering Failures

Mistakes and Miscalculations   Sometimes engineers make mistakes and their works fail due to their own miscalculations, as when a hyphen is omitted in a computer program and the satellite becomes inoperable. Some failures occur when engineers do not correctly estimate the load on a machine or structure. For example, many engineering educators have used the urban legend that the main library at a large midwestern university is sinking an inch per year because the engineers did not take into account the weight of the books when designing the foundation. (We now believe this is legend and not true.) Some stories are, however, tragically true. The Hyatt collapse ended badly with death and destruction due to the lack of appreciation of a change made during construction (a single cable replaced by a two-cable/nut system). The Citicorp Building miscalculation was caught before any structural failure occurred, but at great expense. We have also heard of costly mistakes where engineers did not properly consider material properties, such as not compensating for the amount of stretching of steel boilers from infrastructures in power plants. The boilers, which should have been suspended, actually stretched to the ground surface because insufficient clearance was included in the design. Such mistakes are completely avoidable if the first principles of the physical sciences and mathematics are properly applied. Avoiding such errors is reflective of engineering competence.

Extraordinary Natural Forces   Failure can also occur when factors of safety are exceeded due to extraordinary natural occurrences. Engineers can, with fair accuracy, predict the probability of failure due to natural forces such as wind loads and they design the structures for some maximum loading, but these natural forces can be exceeded. Engineers design for an acceptably low probability of failure—not for 100% safety and zero risk. However, tolerances and design specifications must be defined as explicitly as possible. Design is a process in optimization. The engineer must properly perceive and predict design outcomes within the tolerances demanded by *acceptable risk.* As we saw in the devastation from Hurricane Katrina, the U.S. Army Corps of Engineers perception of *acceptable risk* from a natural force differed from that of a large segment of people living in the U.S. Gulf Coast.

Unpredictable Occurrences   Although forgetting the weight of the books in the library is unforgivable, no engineer can predict all of the possible failure modes of every structure or other engineered device, and unforeseen situations can occur. A classical case is the Holy Cross College football team hepatitis outbreak in 1969.[41] Circumstances, which could not have been foreseen, occurred where water contaminated with hepatitis virus entered a drinking water system. A water pipe connected the college football field with the town, passing through a golf course. Children had opened a water spigot on the golf course, splashed around in the pool they created, and apparently discharged hepatitis virus into the water. A low pressure was created in the pipe when a house caught on fire and water was pumped out of the water pipes. This low pressure sucked the hepatitis-contaminated water into the water pipe. The next morning the Holy Cross football team drank water from the contaminated water line, and many came down with hepatitis. The case is memorable because it was so highly unlikely—a combination of circumstances that were nearly impossible to predict. Nevertheless, the job of engineers is to do just that, to try to predict the unpredictable and thereby to protect the health, safety, and welfare of the public.

Carelessness/Ignorance   Engineers also have to protect the public from its members' own carelessness. The case of the woman trying to open a 2-liter soda bottle by turning the aluminum cap the wrong way, with a pipe wrench no less, and having the cap fly off and into her eye is a famous example of unpredictable ignorance. She sued for damages and won, with the jury agreeing that the design engineers should have foreseen such an occurrence. This is also an example of learning and translating lessons from different disciplines. For example, laboratory safety dictates certain fail-safe devices that *assume* human error, such as reverse threading for certain explosive gas cylinders (e.g., hydrogen). If the soft drink container industry's engineers were made aware of the lab safety design specifications, perhaps even this unlikely event could have been foreseen and the fail-safe measures included in the design. (The new plastic caps have interrupted threads that cannot be stripped by turning in the wrong direction.)

In the design of water treatment plants, engineers are taught to design the plants so that it is easy to do the right thing and very difficult to do the wrong thing. Pipes are color-coded, valves that should not be opened or closed are locked, and walking distances to areas of high operator maintenance are minimized and protected. This is called making the treatment plant "operator-proof." Design engineers recognize that if something can be done incorrectly, sooner or later it will and that it is their job to minimize such possibilities.

Another such example occurred in the early 1970s, when jet-powered airliners were replacing propeller aircraft, the fueling system at airports was not altered, and the same trucks fueled both types of craft. The nozzle fittings for both types of fuels were therefore the same. A tragic accident occurred near Atlanta, where jet fuel was mistakenly loaded into a Martin 404 propeller craft. The engines failed on takeoff, resulting in fatalities. A similar accident occurred in 1974 in Botswana with a DC-4 and again near Fairbanks, Alaska, with a DC-6.[42] Engineers should have recognized that no amount of signs or training could prevent such tragedies. They had to modify the fuel delivery systems so that it was impossible to put jet fuel into a propeller-driven airplane, and vice versa. An example of how this can be done is the modification of the nozzles used in gasoline stations. The orifice in the gas tank is too small now to take the nozzles used for either leaded fuel or diesel fuel. (Drivers of diesel-engine cars can still mistakenly pump gasoline into their cars, however).

A Fifth Type of Failure   Every time something fails, whether a manufactured product (such as a tire blowout) or a constructed facility (such as a bridge collapse), it is viewed as an engineering failure. The job of engineers historically has been to predict the problems that can occur and to design so as to minimize these events, protecting people from design errors, natural forces, unforeseen events, and ignorance/carelessness.

This objective of engineers changed on September 11, 2001 with the attack on the United States by terrorists. Today, the engineer is also required to protect the health, safety, and welfare of the public from acts of terrorism. It had never occurred to most engineers that they have a responsibility to protect people from those who would want intentionally to harm other people or to destroy public facilities intentionally. This is a totally new failure mode in engineering. Sarah Pfatteicher calls such failures *intentional accidents,* failures resulting from intentional actions.[43]

Engineers now find themselves in the position of having to address these "intentional accidents." Military engineers of course have had to design against such destructive actions since the days of moats and castles but those were all structures built explicitly to withstand attack. Civilian engineers have never had to think in these terms, but are now asked to design structures for this contingency.

CASE STUDY: THE WORLD TRADE CENTER

The two World Trade Center towers were designed to withstand a hit by a Boeing 707, at that time the largest airplane in the skies, but the engineers did not consider the possibility of the airplane being loaded with jet fuel. The combined effect of the structural damage and intense fire was never considered in the design. The buildings were designed to withstand the effect of fire, but this calculation used the contents of the building for estimating the maximum heat that could be generated. The damage that could be caused by thousands of gallons of burning jet fuel never entered the design calculations. If this failure mode had been considered, the towers would probably not have been built, since it would not have been possible to design such structures.[44]

Was this an engineering failure? The reality of life is that we all fail. What we hope and prepare ourselves for is to design in enough factors of safety to compensate for uncertainty and even remotely possible events. We also hope to learn from failures and successes: those of others ("textbook" cases and those shared with us by our mentors) as well as our own.

Minoru Yamasaki, by most accounts, was a highly successful designer and a prominent figure in the modernist architectural movement of the mid-twentieth century. Tragically and ironically, Yamasaki may be best remembered for having designed the World Trade Center—a project that failed.

Yamasaki and Emery Roth and Sons designed the World Trade Center towers that were to become emblems of Western capitalism (see Figure 3.11). Could it be that Yamasaki's bold modern gothic architecture was an unforeseen invitation to those terrorists who despise all things modern?

There is a difference between failure and blame. *Failure* is simply a statement that the outcome has not, in some manner, met our expectations. We expect buildings to stand. Certainly, Yamasaki cannot be blamed, but the towers failed. In fact, the failure of architects for buildings is seldom structural and often aesthetic or operational (e.g., ugliness, an inefficient flow of people). Yamasaki strived to present an aesthetically pleasing structure. One may argue that his architectural success in creating a structure so representative of contemporary America was a factor in its failure, making it a prime target of terrorists. By extension, are we asking architects and engineers to make their works unappealing in appeasement to radicals? Engineers are blamed

Figure 3.11   World Trade Center dust and smoke moving eastward on September 11, 2001. (Photo courtesy of David Thom.)

when a building fails for structural, environmental, or operational reasons. The lead structural engineer for the towers, Leslie Robertson, has written a gut-wrenching account of what went into the design and how well the buildings met design criteria. He describes what was known and should have been known and whether this knowledge was incorporated into the design. What, if anything, could have been foreseen by the engineers to lessen the impact of 9/11? Les Robertson's anguish is illustrated by the last paragraph of his article:

> . . . the events of September 11 have profoundly affected the lives of countless millions of people. To the extent that the structural design of the World Trade Center contributed to the loss of life, the responsibility must surely rest with me. At the same time, the fact that the structures stood long enough for tens of thousands to escape is a tribute to the many talented men and women who spent endless hours toiling over the design and construction of the project . . . making us very proud of our profession. Surely, we have all learned the most important lesson—that the sanctity of human life rises far above all other values.[45]

The difference between now and September 10, 2001 is that we now know better what we are up against. Most assessments have agreed that the structural integrity of the towers was sufficient well beyond the expected contingencies. However, if engineers do not learn the lessons from this tragedy, they can rightfully be blamed; and the failure will be less a failure of applying physical sciences (withstanding unforeseen stresses and strains) than a failure of imagination. Engineers have been trained to use imagination to envision a better way. Unfortunately, now we must imagine things that were unthinkable before September 11, 2001. Success depends on engaging the social sciences in our planning, design, construction, and maintenance of our projects. This will help to inform us of contingencies not apparent when exclusively applying the physical and natural sciences.

All engineered facilities are vulnerable to attack by terrorists. Most of our large suspension bridges, for example, can be destroyed by placing charges at the points where the cables are anchored to the ground at the ends of bridges. Not only are these bridges vulnerable, but it is also not readily apparent that the security agencies fully understand their vulnerability. A recent visit to the Brooklyn Bridge revealed police stationed at the base of the towers, structures that would withstand almost any reasonable explosive charge, but no security personnel were on top of the bridge, where the real damage could be done.

Our public water supplies are especially vulnerable to terrorist attack, and this worries many engineers. The American Society of Civil Engineers has set up a study group through the Environmental and Water Research Institute to evaluate the vulnerability of water supplies. They have reason to worry.[46]

Sometime during the student unrest days in the 1970s, an engineering student decided to write a term paper on intentional contamination of water supplies. He went snooping around the local water treatment plant at night and was unchallenged, concluding that he could easily have driven a truck through the gate, discharged its contents into the clearwell, and escaped without being detected. He then evaluated the various toxins that might be used for such a purpose, and discovered one that was available in 55-gallon drums at the local supply house. This chemical is tasteless and odorless, yet could have killed the entire population of the city of 100,000 inhabitants. When we talked over his conclusions, the student was clearly upset and suggested that we destroy the paper and not show it to anyone.

Engineering and engineers have a new challenge—to prevent such "accidents" on civilian targets by terrorists bent on harm to the public. We recommend that the response to this threat requires a two-pronged approach: *technical* and *social*.

Technical Response of Engineering to Terrorism

Engineers have responded enthusiastically to the call for better technology to fight terrorism. This effort has included better sensing devices for explosives and weapons, better communication systems to warn and predict when terrorist activity might occur, better identification systems for suspect individuals, better ways of identifying used explosives, and many more such innovations. A recent issue of *Prism,* the magazine of the American Society of Engineering Education, had several articles summarizing technology that can be brought to bear in the fight against terrorism, and these developments no doubt will have a beneficial effect.[47] But the use of such technology also signals four potential problems.

First, the application of such technology will almost always diminish our liberty. Witness, for example, the facial identification program used on attendees at a recent Super Bowl. Protests by the manufacturer and the security people to the contrary that such equipment only identifies those faces on file, many people were upset about having their picture taken. We Americans are just not used to having people or machines knowing where we are and what we do, and we resent such an intrusion on our privacy. We believe, and rightly so, that equipment and procedures such as face identification have the potential for abuse and loss of our liberty. This conflict between security and liberty is an example of competing values in ethics. For example, security can be seen as a necessary end, but some of the means to achieving this end will detract from liberty. This points to one of the problems with utilitarian ethics. Similarly, the United States was founded on principles espoused by rights ethicists such as John Locke (1632–1704). In fact, three of these rights—life, liberty, and the pursuit of happiness—are foundations of the U.S. Constitution. Recently, however, the words of Benjamin Franklin have taken on added currency: "They that can give up essential liberty to obtain a little temporary safety deserve neither liberty nor safety."[48] Thus, engineers may see the balance between security and liberty as an "optimization problem" between the two variables.

One of the reasons Americans fly less in recent years arguably is the unreasonable security checks they have to endure. The effect of security at airports has been traumatic and expensive (estimates are at $1 billion per year) as well as bothersome. As a fellow passenger recently remarked while a deadly serious security agent was confiscating his plastic disposable razor: "Those bastards won."

Second, a malicious person with money and knowledge can get around any technology that we can implement. Witness the scourge of viruses in our computers. As soon as the antivirus programs are modified to take care of the latest virus, unknown persons with devious minds and malicious intent can develop new viruses that get around the defenses. So it will be with antiterrorist technology. Any technically sophisticated person can devise ways of getting around our present airline safety procedures or can purchase ground-based heat-seeking missiles that can blow up airplanes in flight.

The third problem is that terrorists can use our own technology to create havoc in our country; the prime example being the use of fully loaded airplanes on September 11. The terrorist organizations have used Internet banking to move money around, defense communication networks to keep in touch, and sophisticated explosives to kill innocent people—technologies we developed as part of our national defense. The more technology we produce, the more it has the potential to harm us.

The fourth problem with the development of antiterrorist technology is that we as a nation are not good at it. We have developed few resources that would protect us from low-technology assault. Traditionally, our research funds have been spent either to enhance our own health or to develop sophisticated weapons for countering threats from similarly technically sophisticated enemies. The National Institutes of Health and the

Department of Defense are our 800-pound research gorillas, swamping the efforts of the National Science Foundation. The Strategic Defense Initiative ("Star Wars"), the missile defense shield, is not for protecting us from suicide bombers, for example. We have done very little research on how to protect the public from terrorist threats. But even if we had, the terrorists would simply use that technology for their own purposes, thus negating its original intent.

We have to conclude, therefore, that better technology cannot be the only, or maybe not even a primary, engineering response to the threat of terrorism. Terrorists with money and skill can get around our technology, and they can use our own technology against us. Even if we were good at antiterror research, this would eventually fail to protect us.

We are not arguing against the use of antiterrorism technology but suggesting that such technology is only part of the answer. It is not possible for engineers to prevent all such deeds, just as it is not possible for engineers to make anything 100% safe from other types of failure. In a manner of speaking, this is analogous to our reliance on pharmaceuticals to treat diseases. One of the fears of medical researchers is that we may destroy most of the microbes, but the remaining strains rapidly become resistant to even the newest antibiotics. So physicians try not to prescribe these drugs indiscriminately, and to use a holistic approach to prevent certain contagious diseases. No matter what technology we employ, the possibility always remains that terrorists can cause great harm by using the very same technology against us. We cannot be completely "inoculated" against these diseases, nor can we expect "quick technological fixes" every time. Engineering technology is therefore only partially able to protect the health, safety, and welfare of the public from intentional destruction.

Social Response of Engineering to Terrorism

There are two approaches to solving engineering problems: attack the symptoms or attack the cause, and often the best solution to a problem is to change what is causing the problem. For example, in hazardous waste treatment, it is almost always cheaper and safer to travel "up the pipeline" to find out where the waste is coming from and then change the process to eliminate or minimize this hazardous material. The alternative option, which is almost always more expensive, is to treat the waste once it arrives at the treatment plant.

The first step in finding solutions to engineering challenges is to define the problem: to understand all the variables and constraints. Using this approach to combat terrorism, we first have to understand why people would cause intentional accidents. To gain such understanding, we have to develop a greater knowledge of diverse cultures and societies and ask why people would want to cause harm, what their motives are, and try to understand what drives their actions. There is always a driving force behind every action, especially actions that result in suicide. Understanding this driving force and taking multiple perspectives in the solution of engineering problems (including intentional accidents) can be extremely helpful in preventing such events.

When asked, many people can envision circumstances and causes for which they would be willing not only to kill but also to sacrifice themselves. This belief can and has been used against the disadvantaged and maligned people of the world especially those being manipulated mentally by "mentors" and their religious leaders. While certainly not justifying the malicious acts, we should try to understand them.

ENGINEERING AND WAR

The greatest job satisfaction for engineers is watching something they conceive, design, and construct actually perform as intended. Samuel Florman[49] beautifully describes this joy as an *existential pleasure*—existential in that the engineer is free of concerns for how the end result will be used. How a product of engineering is used does not matter, argues Florman, as long as it works as intended. This gives the engineer the existential freedom to do good engineering and to not be concerned about what the product or facility will eventually be used for or who uses it. The joy of engineering is to make something that works. Next are some examples of how engineers are clever problem solvers.

In Drammen, Norway, engineers faced a daunting problem. They wanted to build a road from the east shore of the Oslofjord to the west shore. The problem was that the west shore of the fjord was a sheer cliff. They would have to cut a road alongside the cliff, and not only would such a road be difficult to travel, but it would destroy the natural beauty of the shoreline. The solution was to drill a spiral roadway into the solid granite, and cars could drive up through the cliff much like a ramp at a huge parking deck. The engineers realized that the problem was not how to build the road, but how to get cars and people up the cliff. Form followed function. This was a case of clever engineering that solved a problem, and the solution had no negative effects.

But sometimes the very cleverness of an engineering solution can create other problems. In Bangladesh, the World Health Organization responded to the previously mentioned problem with arsenic in the drinking water by installing thousands of ion-exchange resin canisters (much like the activated-carbon canisters used in some homes to improve the taste of drinking water) that absorbed the arsenic ion. The system worked well until the villagers asked what they should do with the used canisters, which then contained a high concentration of arsenic and were clearly hazardous waste that needed careful handling and disposal. Unfortunately, the WHO engineers never thought this through, and now Bangladesh has tens of thousands of these canisters that will eventually cause acute and long-term human health problems if not handled properly.

Some engineering projects are so blatantly wrong that one wonders why they were ever undertaken. The classic case of the U.S. Army Corps of Engineers project for draining the Everglades is typical of this category of engineering. Wanting to increase arable land for development and farming, the Corps constructed miles of waterways that drained the groundwater from the swamp, at great cost to the American taxpayer. The lower groundwater, however, resulted in a dry condition that gave rise to huge forest fires in the Everglades. The most devastating effect was the loss in wildlife that depended on swampy conditions. To their credit, the Corps finally realized their mistake and spent many more millions of tax dollars to correct the mistakes they had made. Well-intentioned engineering such as draining the Everglades can have negative unintentional consequences that can lead to other problems.

Engineering is no different from other professions, in that some actions by professionals such as physicians and lawyers can also have unintended and negative consequences. For example, a physician might treat a patient for a symptom, but in so doing, create other, unintended health problems. Such physician-caused health problems are called *iatrogenic diseases.*

Court cases have been brought by patients and their attorneys over the unintended adverse effects from psychotropic drugs, weight loss aids, and other measures recommended by physicians. Likewise, companies have been sued when the products designed by their engineers led to unforeseen adverse effects. When the engineers lose such cases, the courts in a sense are contending that these products have been caused by preventable, "iatrogenic" designs. Drew Endy, an engineer at MIT, has proposed a new word that describes such unintentional problems created by engineering. He calls them *mechanikogenic problems,* the unintended detrimental outcomes of engineering. Just as the Corps of Engineers had good intentions in draining the Everglades, their actions unfortunately caused great harm that could be described as mechanikogenic damage to the Everglades ecology. One of the mechanikogenic effects of engineering technology is making possible the waging of modern warfare. Engineers who work only to find joy in solving problems can discover that their efforts have made possible the waging of war and that their technological inventions are used to cause harm to others. This is an unintended collateral effect of engineering, and warfare becomes a mechanikogenic problem.

War Engineering

Engineering is essential to the waging of war, and because warfare has been a part of human behavior since the earliest humans roamed the Earth, the first engineers were no doubt military engineers who worked on both defensive structures and offensive machines for warfare. One does not necessarily need lawyers, or accountants, or journalists to wage war, but engineers are indispensable. This idea was brilliantly exemplified by the unforgettable scene in Stanley Kubrick's movie *2001: A Space Odyssey.* Kubrick suggested that war and disagreement among our ancestors was nothing but bluster and snorting, but then some engineer (?) discovered that if a club was used to hit the enemy over the head, the enemy died. Technological war was born.

Engineers have always been valuable to the rulers of states and principalities. The famous engineers of antiquity were all concerned with devising and building engines of war or fortifications to deter the engines of war, and these engineers occupied a respected place in their societies. Frontius (first century B.C.), for example, was a dignitary in Rome, and Taccola became an eminent citizen of his native Siena.[50] Even in Imperial Rome there was a distinction made between the engineers and the artisans who had the skills needed by the engineers. Artisans who were armament workers were subject to strict control and were branded on the arm so that if they deserted the factories, they could be identified and returned, whereas engineers enjoyed higher social status.[51]

As warfare evolved and defense and attack became more sophisticated, so did the importance of the engineers to the war effort. The military developments during the Middle Ages included increasingly sophisticated fortifications used for defense as well as siege towers and devices (e.g., catapults) for throwing heavy objects used for offense.

Siege towers were huge wheeled structures, often as high as castle walls, constructed of oak timbers and protected by ox hides from fire and enemy missiles. The top floors often had engines for lobbing fire and stones across the walls, and a bridge from the top of the tower became the route of attack against the castle wall. The lower stories of the siege towers had a ram that swung back and forth on ropes and was sheathed in iron. The men operating the ram were protected from arrows and other missiles by the siege tower structure covered by hides.

Engineers who knew how to build such devices were indispensable to warlords and often worked on a contract basis. During the First Crusade, a local specialist at Nicaea was given the task of constructing the siege towers even though he was not a Crusader. He reportedly was paid well for his efforts. Frederick II thought so highly of the engineer Calamandrinus that he had him kept in chains to prevent his escaping and going to work for others. Nevertheless, Calamandrius did eventually escape and switch sides, being promised his freedom together with a house and a wife.[52]

Master Bertram, born in 1225, became a royal engineer for Henry III of England. In an early record, Bertram *le Engynnur* was one of six such men rewarded by the king for his services. By 1276 he was employed in making engines for the Tower of London and apparently took great pride in this work, taking personal charge of buying the oak, beech, and elm from local forests. Later in his career he was in charge of engines at sieges in Wales and in 1283 at Castell-Y-Bere, when he is referred to as *machinator* and *ingeniator*. Before his death in 1284, Bertram had also built some of the earliest Welsh castles. He was, in short, a talented mercenary engineer willing to sell his services to the highest bidders.[53]

The need to better understand ballistics was the driving force for the development of the mathematics of motion. The term *ballistics* comes from the Greek word *ballein,* meaning "to throw," and for centuries ballistics was the province of the military engineer.

Medieval artillery included the *euthyronon,* a bow with a mount, capable of firing large arrows or small spears; the *palintonon,* which fired stones along a curved trajectory; the *catapult*—with rigid arms, twisted skeins for torsion, capable of shooting spears or stones; and the most effective engine prior to cannons, the *trebuchet*—which had unequal arms with a weight on one end to create a fulcrum and teeter-totter effect. The trebuchet was amazingly accurate because the force used to shoot the missile was always the same. With a range of about 100 meters, the trebuchet catapult was able to destroy sections of castle walls by incessant pounding.[54]

During the eighteenth century the best engineers in the world were French. They had the theoretical skills to apply technology to warfare and understood the importance of technology in warfare. Because of France's geographical location, war for France was always possible and invasion was a constant reality. Most significant engineering advances were therefore directed to military purposes. Whereas in England inventions (in textiles, hardware, and machinery) arose spontaneously from entrepreneurial engineers, in France engineering was controlled from above by a government concerned with both defensive and offensive warfare. Because so many engineers in France worked for the government in military matters, the word *engineer* was presumed to mean "military engineer." Even the development of the most basic technology was controlled by the government. For example, the use of orthographic projection in the expression of engineering machinery and facilities was invented by Gaspar Monge in the 1770s while he was employed by the French government. However, because his work involved the drawing of fortifications, his book was censured by the government and he was not able to publish his discoveries until 1795.

The French engineer B. F. deBelidor actually invented the shell commonly attributed to the British Major Henry Shrapnel. Pontoon bridges, designed by F. J. Camus in 1710 and D'Herman in 1773, carried the French armies marching three abreast over rivers encountered on maneuvers. Gun carriages designed by C. F. Berthelot set a pattern for French artillery. Polygonal defenses, designed by the Marquis de Montalembert, supplanted the previous defensive ideas of Vauban. The semaphore telegraph, a series of towers in visual sight of each other, was invented by French engineer Ignace Tresaguet for the purpose of providing rapid communication for French armies. Awarded the title of *l'Ingénieur Télégraphe,* he connected Paris to Lille, Strasbourg, Brest, and Lyon; this communication system enabled Napoleon to sustain his conquest of Italy by linking Lyon with Turin, Milan, and Venice. Engineering education was in the government-run l'Ecoles Polytechnique, specially organized to produce military engineers.

Across the Atlantic Ocean, the struggle for independence created a critical need for military engineers, and on June 16, 1775, the Continental Congress resolved that there be "one chief engineer at the grand army, and that his pay be $60 per month; that two assistants be employed under him, and that the pay of each of them be $20 per month." [55] The Continental Congress also decided to establish a military academy for the purpose of educating American engineers for the war effort. Because the best engineers in the world at that time were French, the first engineers in both General George Washington's army and at West Point were French, and thus a knowledge of the French language was critical in engineering education.

While over the years many Academy graduates have had distinguished civilian careers, the primary purpose of the Academy's engineering programs continues to be the education of military engineers, and during their over-200-year history, they have done this with great distinction. The much-storied U.S. Army Corps of Engineers has a long and proud history of assisting the military in accomplishing its tasks. The heroic actions of the engineers during D-Day during World War II may have saved the troops on Utah Beach and assured the success of the invasion. The U.S. Navy Seabees were perhaps the most famous engineering unit in World War II, setting up entire harbors and other facilities in a matter of days, often under heavy bombardment.

American engineering at home, which was essential to the success of the Allied forces by developing such marvels as radar, sonar, and the atomic bomb, came to world prominence after World War II and has remained preeminent due in great part to funded research and development from the military establishment. In the last decade, the collapse of the Soviet Union created an absence of a threat and led many to question the need for large expenditures for military technological development, but since 2001 the concern for terrorism has renewed interest in the use of technology for both defensive and offensive purposes.

The defense establishment, which includes the government, universities, and private industry, continues to argue that it is necessary to have military forces and technology second to none. In the *1997 Quadrennial Defense Review,* the Department of Defense reaffirmed that "it is imperative that the United States maintain its military superiority. . . ." During that same year, the National Defense Panel reported that if the United States does "not lead the technological revolution we will be vulnerable to it," and in the Senate Armed Services Committee Report, the committee wrote that its priority is "to maintain a strong, stable investment in science and technology in order to develop superior technology that will permit the United States to maintain its current military advantages . . . and hedge against technological surprise."[56]

Biographical Sketch: Curtis LeMay

Curtis LeMay (1906–1990) received his civil engineering education at Ohio State University and in 1930 became a second lieutenant in the U.S. Army Air Corps. As World War II approached, he moved quickly up the chain, using his engineering education and skills to become a lieutenant colonel at the start of America's involvement in Europe. He was transferred to the bombardment division and in 1944 was sent to the Pacific theater as a major general in charge of the bomber command.

As the United States began to island hop toward Japan, air bases were established from which bombers could attack the Japanese mainland. LeMay received a new bomber, the B-29, designed for high-altitude bombing and he was pressed by his superiors to achieve "results." But the problem was that the bombers had to fly at over 30,000 feet to be clear of antiaircraft fire and thus their bombs were largely off target. LeMay discovered that the Japanese antiaircraft guns were designed for defending against airplanes at high altitudes and could not be moved quickly. If an airplane came in at a very low altitude, such as 3000 feet, the guns could not be moved quickly enough to fire effectively at the attackers. In addition, such low-level attacks were able to concentrate bombs on small areas.

Most of the Japanese cities were susceptible to fire, and LeMay decided that he would use the bombers to drop incendiary bombs and to create firestorms in the cities, a technique he learned from the British who unwittingly had created such a firestorm in Hamburg, and then intentionally firebombed Dresden, where the entire city and its inhabitants were effectively incinerated.

The new technique worked amazingly well in Japan, with city after city being destroyed by dropping incendiaries and creating violent firestorms. On May 9, 1945, LeMay's bombers firebombed Tokyo, killing at least 100,000 people in one night. LeMay justified these attacks by arguing that the Japanese had decentralized their war manufacturing, and thus the entire city was a legitimate military target. By the time the atomic bombs were dropped on Hiroshima and Nagasaki, there were no cities left to bomb on the Japanese islands since all significant population centers had been firebombed, with the loss of many lives.

At the end of the war, LeMay observed that it was a good thing that our side had won. If we had not, he believed that he would certainly have been tried as a war criminal.

For more on Curtis LeMay, see Ricard Rhodes' *The Making of the Atomic Bomb,* Simon and Shuster, New York, 1986.

Peace Engineering   Some engineers are now beginning to ask if there are alternative careers in engineering that do not require working for the military–industrial complex. They recognize that military engineering (in all its forms, including working for defense contractors and conducting research for the Department of Defense) is destined to be used for warfare, either defensive or offensive, and are unsure if they want to participate

Biographical Sketch: Wernher von Braun

 While Wernher von Braun (1912–1977) was a student at the Institute of Technology in Berlin, he became interested in rockets and worked in his spare time with some of Germany's most imaginative scientists. He graduated from the institute in 1932 with a degree in aeronautical engineering and looked forward to a career in aviation, including getting his pilot's license after graduation. He soon became convinced that if rocketry was to be used to explore space, more than just the application of current engineering technology was needed, and he enrolled at the University of Berlin to work toward a Ph.D. in physics. His dissertation was on the use of liquid fuels in rockets, which was a new concept. After graduation in 1934 he was hired by the German Army Ordnance Corps, where he began to conduct experiments using liquid fuel rockets. When war came, von Braun continued his work for the Army as a civilian employee and designed the dreaded V-2 rocket that terrorized London during the last months of the war. After Germany's capitulation, von Braun and 120 of his colleagues surrendered to U.S. forces to avoid capture by the Russians.

The German scientists and engineers were brought to the United States and began their work for the U.S. Army, designing and building the Redstone, Jupiter, Juno, and Saturn rockets. In 1957 one of von Braun's rockets put *Explorer 1,* the first American satellite, into orbit, after an embarrassing misfire and explosion by the Navy's Vanguard rocket. The von Braun rockets were used for the Mercury manned missions into space, and his Saturn rockets lifted the Apollo teams into orbit. When President Kennedy declared our intention to put a man on the moon, von Braun was given the job of building the rocket to do the job. His Saturn V was highly successful and is still used, decades later, for the space shuttles.

Another use of rockets was to serve as the vehicles for carrying nuclear warheads, and thousands were built and stored in silos all over the country. The rockets were thought to be a deterrent (such as arguments that they provided mutually assured destruction) for attack by the Soviet Union and remained active throughout the Cold War (about 1950 to about 1990). The United States continues to have an unspecified number of these rockets, armed with nuclear warheads, ready in case of attack. This indicates that technologies themselves are often neutral, and their applications determine whether they are "good" or "bad."

The ease with which Wernher von Braun became the leading rocket expert for the United States after serving in the same capacity for Germany during World War II illustrates how engineers are able to concentrate on their job without being concerned with what their efforts can produce. It also serves as a warning to consider the possible ramifications (good or evil) of the application of technical skills.

During the 1960s, a satirist, Tom Lehrer (also a math instructor at Harvard), recorded some wonderful songs that spoke to the craziness of the age. One of his songs was about Wernher von Braun, and began with these verses:

> Gather round while I sing you of Wernher von Braun
> A man whose allegiance
> is ruled by expedience
> Call him a Nazi and he won't even frown
> "Nazi, Shmazi" says Wernher von Braun
>
> Don't say that he's hypocritical
> Say rather that he's apolitical
> "Once the rockets are up, who cares where they come down?
> That's not my department" says Wernher von Braun.
>
> (Copyright Tom Lehrer, used with permission.)

in such work. Many believe that in the United States, the name of our war department—the Department of Defense—suggests that we would not intentionally and without provocation start wars and kill people. Until this century, few wars have been started by democracies.

These engineers are looking for alternatives that would allow them to use their skills in a positive and proactive way to promote peace. Perhaps they are beginning to define a third kind of engineering—*peace engineering*—the use of technical skills to promote peace. The decision to practice peace engineering can be based on any number of considerations, from religious to political, but probably will be based on a belief that as President Carter said in his Nobel Peace Prize acceptance speech, "War is evil." Carter acknowledged that war is sometimes necessary but that it is always evil. Peace engineering is born when engineers recognize that war is preventable and then decide to work proactively for peace.

Defining just what working for peace means is much more elusive than defining what it means to work for the military. Some engineers who work in the armaments industry or for the military argue that what they are doing is working for peace. They point to the winning of the Cold War with the Soviet Union as proof that a strong defense eventually results in a peaceful settlement of disputes. Others argue that world history disputes this claim, however, and instead shows that military buildups almost always result in conflicts. There is validity in both arguments. Consider the old adage "Where you stand depends on where you sit." Humans are adept at perceiving things in a way that fits their own values, so one must be careful to be truthful, especially to oneself.

Peace engineering is the proactive use of engineering skills to promote a peaceful and just existence for all people. Examples of how engineers have been and will continue to use their skills for this purpose include the Peace Corps, the World Bank, Pan-American Health Organization, and perhaps hundreds of nongovernmental organizations such as Engineers Without Borders. One way of promoting peace in engineering is to teach ethical skills alongside technical skills within our universities. Thousands of engineers have devoted their lives to this concept, but few have ever thought to call themselves by this title: *peace engineer.*

Some engineers come to peace engineering after having been, perhaps unknowingly, practicing military engineering. Because the Department of Defense funds large research projects at both private and public universities and contracts out the construction of weapons systems to corporations, many engineers work as military engineers without realizing the purpose of their work. The concept of *dual use* means that engineering and technological advances may serve both military and civilian goals. Examples of dual use technologies include global positioning systems, antilocking braking systems, and satellites. An acquaintance of the authors who worked for General Electric thought that what he was working on was a new technology for toasting bread, and found out only by accident that his engineering calculations actually were being used clandestinely in the Star Wars project promoted by President Reagan. His keen disappointment with the secrecy and deception led him to resign his job at General Electric.

Just as was the case with our friend at General Electric, engineers who are concerned about the end result of their work are beginning to ask some disquieting questions. For example, they wonder if by working either directly or indirectly for the military establishment, they are truly living up to their own code of ethics. Engineering, as a profession, states its purpose and objectives in a code of ethics, and at least in the United States, the code of ethics of almost every engineering discipline begins with the statement: "The engineer, in his professional practice, shall hold paramount the health, safety, and welfare of the public." The two key words are *shall* and *paramount*. There is no equivocating about this as the primary commitment of engineering, and the vast majority of engineers agree with this statement and practice their profession accordingly.

There is, however, a problem with this statement when it comes to engineers working in military positions, and it centers on the word *public*. What exactly is the "public"? Suppose that an engineer works for a company that designs and produces land mines. Is public the people who pay his or her salary? Has the public decided, through a democratic process, that the manufacture of land mines is necessary? Or is the public of record those people who will eventually have to walk over the ground in which these land mines have been planted and be killed and maimed by the explosives?

Because of such moral concerns, some engineers have changed careers to reflect their own interpretation of what *public* means in the engineering code of ethics and have devoted their professional lives to the use of technology in the pursuit of peace. Perhaps in time, peace engineering will evolve and mature and eventually take its proper place alongside the military engineering and the civilian engineering. Engineers, especially young engineers, have to understand that they can *choose* to work for peace. The engineer's "right of conscience" is a measure of just how to implement that choice. Frederic II is no longer here to chain them to their posts, and they can elect not to participate in furthering lethal technology but instead, choose to devote their careers to peace engineering.

An oft-seen bumper sticker declares: *If you want global peace, work for local justice*.[57] Certainly, one way of practicing peace engineering is to seek justice.

ENGINEERING AND JUSTICE

Credat emptor! Let the buyer trust! The client entrusts engineers to provide a quality product that meets defined objectives and adheres to credible design specifications.[58]

Biographical Sketch: Benjamin Ernest Linder

 Occasionally, a person has so much goodness and talent and good-heartedness and technical skill that it is difficult to capture the personality in a short paragraph. Such a person is Ben Linder (1959–1987).

Born in 1959, Linder studied mechanical engineering at the University of Washington and became an avid unicycle rider and designer. During his studies he managed to design novel new mechanisms for unicyles that are now widely used.

He also figured out how to install packs around the wheels of unicycles to carry tents, sleeping bags, and other camping equipment. During graduation at the University of Washington, he smuggled his unicycle into the ceremony and rode across the stage to get his diploma, to the enthusiastic applause of the audience.

Ben loved life and cared for others, and this is why he decided to go to Nicaragua on his own to try to help some of the poorer people develop their resources. He was instrumental in setting up water supplies and hydroelectric projects near the village of San José de Bocay. This part of Nicaragua was unfortunately in the middle of a civil war raging between the Sandinistas, who sought to form a socialist government in Nicaragua, and the Contras, groups of mercenaries supported by the Reagan administration. The fact that the Sandinista movement was based on Marxist ideals made for conflict between it and the United States. It also led to the support by many in the United States for the opposition group, the Contras. Both the Contras and the Sandinistas used violence and intimidation to try to take control of the country. Unfortunately, Linder found himself in the cross fire of this conflict.

In 1987 the Contras attacked the small hydroelectric project, killing Ben Linder and his two Nicaraguan friends. The autopsy showed that Linder was first wounded by a grenade and then shot at point blank range in the head.

These types of projects were favorite targets for the Contras. Before Linder was killed, other Americans working to help the people of Nicaragua were murdered. Marlin Fitzwater, the White House spokesman, was quoted in the *New York Times* as saying that Linder's death was his own fault, that anyone working in Nicaragua "put themselves in harm's way."

But most people understood what had been lost. Dan Rather on the CBS News perhaps said it best:

> Benjamin Linder was no revolutionary firebrand, spewing rhetoric and itching to carry a rifle through the jungles of Central America. He was a slight, soft-spoken, thoughtful young man. When, at 23, he left the comfort and security of the United States for Nicaragua, he wasn't exactly sure what he would find. But he wanted to see Nicaragua firsthand, and so he headed off, armed with a new degree in engineering, and the energy and ideals of youth. This wasn't just another death in a war that has claimed thousands of Nicaraguans. This was an American who was killed with weapons paid for with American tax dollars. The bitter irony of Benjamin Linder's death is that he went to Nicaragua to build up what his own country's dollars paid to destroy—and ended up a victim of the destruction. The loss

of Benjamin Linder is more than fodder in an angry political debate. It is the loss of something that seems rare these days: a man with the courage to put his back behind his beliefs. It would have been very easy for this bright, young man to follow the path to a good job and a comfortable salary. Instead, he chose to follow the lead of his conscience.

Linder kept a meticulous diary of his days in Nicaragua, and these notes as well as interviews with the men who killed him were the basis for a 1999 book. For more information, see *The Death of Ben Linder* by Joan Kruckewitt (www. sevenstories.com).

Beyond that, however, what we design, plan, and build must perform in a way that improves the natural and built environment. As discussed in Chapter 2, this mandate embodies empathy and justice. We must envision the ways that our work will affect present and future populations, communities, and neighborhoods. This includes not only the assurance that engineering specifications are met but includes how our work fits within the societal fabric. As exemplified by the Pruitt–Igoe debacle in St. Louis, even the most structurally sound building can be a failure if the designer does not fully consider the needs of the people who will inhabit it. This is not simply a theoretical problem. As demonstrated in the previous discussion, there are ample examples and cases of "well designed" projects that are societal failures.

One of the characteristics that the public expects of engineers is to be fair—that is, to show equal justice to all. An engineer's decision in siting undesirable land-use facilities offers one of the most compelling examples of such fairness.

When the cost of any undesirable land use such as municipal solid waste landfills refuse is calculated, the environmental and social costs are usually ignored. Little effort seems to factor into quantifying social costs of such facilities. Of all the concerns faced by neighbors near planned facilities that can be considered undesirable (LULUs, locally undesirable land uses), the one that most often is of foremost importance is the loss in property values.

Some studies have shown that there is little loss in property values due to landfills. These studies examined the changes in property values during the years of landfill operation and concluded that the rate of appreciation was approximately the same for houses near the landfill and in other parts of the community.[59] What was not measured, probably because it was so difficult to do, was the loss in property values once a landfill has been sited in a neighborhood. The difficulty comes because social scientists and engineers studying such property value loss cannot predict where the next landfill will be sited, and they can only look at the property values in a neighborhood after the fact.

One study, however, got around this problem by creating a virtual town, with various neighborhoods, and then pretending that a landfill would be sited at a specific location in the town.[60] The loss in property values was estimated by asking a group of professional property assessors to estimate the loss in value the day after the landfill site would have been made public. From this study, several conclusions can be drawn.

First, properties close to the landfill lost more value than properties farther away from the landfill, a conclusion that would have been expected. Second, the amount of property depreciation decreases with distance from the landfill (a *price gradient,* if you will). Third, the effect of the landfill can depress values of properties over significant distances, such as within two or three miles from the landfill site. Finally, the study showed that higher-valued properties lose a greater percentage of their worth than less valuable properties.

In the last observation lies the root of environmental injustice. If the objective of engineers and governmental planners is to reduce the cost of facilities and operations, it is only natural that they would choose to locate undesirable land-use projects in areas of depressed land value, and these areas are often the homes to the economically less advantaged.

The depreciation curves developed in this study describe the losses experience by homeowners who live near a site chosen for a landfill. Such depreciation could be alleviated by governmental compensation, and several strategies have been suggested.[61] The argument often used against this compensation is that it would be too costly. This claim can be refuted by analyzing the actual costs associated with the operation of landfills.

CASE STUDY: THE DURHAM LANDFILL

The city of Durham, North Carolina, intended to purchase 700 acres of land adjacent to the existing landfill in order to expand its landfill capacity. Only 200 acres of the tract would be used for a landfill, with the remaining to be preserved as buffer. The total capacity of the landfill would be about 7 million metric tons, and the landfill would be expected to serve the city for about 20 years. The design would follow federal guidelines, with leachate control and gas collection. The initial estimates of the tipping fee, based on the operation of the existing landfill, was about $35 per metric ton. This figure included expected cost of land, construction, operation, closure, and post-closure activities.

Approximately 30% of the cost of a landfill can be attributed to gas and leachate control and about 3% for monitoring.[62] Using the curves for property depreciation as developed in the study by Hirshfeld, Vesilind, and Pas,[63] and knowing the number and value of the homes around the proposed site, the loss in property values can be estimated. If this is added to the total cost of the landfill, the calculation shows that the landfill would actually cost about as follows:

Landfill feature	Cost per metric ton
Cost of land and operation	$35.00
Leachate and gas control	25.00
Monitoring	2.50
Property depreciation	1.50
Total cost of landfill	$64.00

| The important fact in this calculation is that the property depreciation cost is only about 2% of the total cost of the landfill.

This case shows that the argument that compensation would be too expensive is simply not persuasive. Any such relatively small cost can usually be passed on to landfill users with negligible economic impact to the community. The lack of will on the part of a community to consider this or other measures of fairness may be an indication that environmental fairness is not a sufficiently strong value compared to others, such as avoiding the outcry if a similar facility were to be sited in a neighborhood with locally high property values.

In the next chapter we address the ways that engineers can reduce the risks to the health, safety, and welfare of public in our professional lives.

REFERENCES AND NOTES

1. Plutarch, The Life of Marcellus, in *Plutarch's Lives,* A. K. Wardman, Elek, London, 1974.
2. W. Wisely, *The American Civil Engineer,* ASCE, New York, 1974.
3. Wisely, note 2.
4. G. Bowman, John Smeaton: Consulting Engineer, in *Engineering Heritage,* Vol. II, Dover Publications, New York, 1966.
5. H. Straub, *A History of Civil Engineering,* MIT Press, Cambridge, MA, 1952.
6. L. Graham, *The Ghost of the Executed Engineer,* Harvard University Press, Cambridge, MA, 1993.
7. P. Eddy, E. Potter, and B. Page, *Destination Disaster: From the Trimotor to the DC-10,* Times Books/Random House, New York, 1976.
8. J. H. Fielder, Floors, Doors, Latches, and Locks, in *The DC-10 Case: A Study in Applied Ethics, Technology, and Society,* J. H. Fielder and D. Birsch (Eds.) SUNY Press, Albany, NY, 1992.
9. Eddy et al., note 7.
10. French Government Report on the 1974 Paris Crash, in *The DC-10 Case: A Study in Applied Ethics, Technology, and Society,* J. H. Fielder and D. Birsch (Eds.), SUNY Press, Albany, NY, 1992.
11. See D. Birsch, Whistleblowing, Ethical Obligations, and the DC-10, in *The DC-10 Case: A Study in Applied Ethics, Technology, and Society,* J. H. Fielder and D. Birsch (Eds.), SUNY Press, Albany, NY, 1992; R. DeGeorge, Ethical Responsibilities of Engineers in Large Corporations: The Pinto Case, *Business Professional Ethics Journal,* 1(1):1–14, 1981; and S. Bok, Whistleblowing and Professional Responsibilities, in *Ethics Teaching in Higher Education,* D. Callahan and S. Bok (Eds.), Plenum Press, New York, 1980, pp. 277–295.
12. G. G. James, In Defense of Whistleblowing, in *Ethical Issues in Professional Life,* J. Callahan (Ed.), Oxford University Press, New York, 1988, pp. 315–321.
13. H. Cross, *Engineers and Ivory Towers,* Ayer Company Publishing, Manchester, NH, 1952.
14. A. Leopold, *A Sand County Almanac* (1949), Oxford University Press, New York, 1987.
15. E. Birmingham, Position Paper: Reframing the Ruins: Pruitt–Igoe, Structural Racism, and African American Rhetoric as a Space for Cultural Critique, Brandenburgische Technische Universität, Cottbus, Germany, 1998. See also C. Jencks, *The Language of Post-modern Architecture,* 5th ed., Rizzoli, New York, 1987.
16. A. von Hoffman, *Why They Built Pruitt–Igoe,* Taubman Centre Publications, A. Alfred Taubman Centre for State and Local Government, Harvard University, Cambridge, MA, 2002.

17. J. Bailey, A Case History of Failure, *Architectural Forum,* 122(9), 1965.

18. Bailey, note 17.

19. See, for example, D. A. Vallero, Teachable Moments and the Tyranny of the Syllabus: September 11 Case, *Journal of Professional Issues in Engineering Education and Practice,* 129(2): 100–105, 2002.

20. A. Fausto-Sterling, *Myths of Gender: Biological Theories About Women and Men,* Basic Books, New York, 1985, pp. 207–208.

21. U.S. Environmental Protection Agency, *Guidance on Data Verification and Validation: QA/G8,* EPA/240/R-02/004, U.S. EPA, Washington, DC, 2002.

22. Testimony of R. A. Goyer before the U.S. House of Representatives, Science Committee, October 4, 2001.

23. National Research Council, *Arsenic in Drinking Water,* National Academies Press, Washington, DC, 1999.

24. The interview occurred on the Cable News Network (CNN), but the names of the engineer and reporter are not known.

25. World Health Organization, Press Release WHO/55, Researchers Warn of Impending Disaster from Mass Arsenic Poisoning, WHO, Geneva, 2000.

26. B. Allenby, presentation of this collection as well as the later discussions regarding macro- and microethics, Emerging Technologies and Ethical Issues, National Academy of Engineering Workshop, Washington, DC, October 14–15, 2003.

27. Environmental engineering is a subdiscipline of civil engineering in the United States.

28. American Society of Civil Engineers, *Code of Ethics,* ASCE, Washington, DC, adopted in 1914, amended November 10, 1996.

29. Biomedical Engineering Society, *Code of Ethics,* BES, Landover, MD, approved February 2004.

30. We ask the following question in our classes: Is DDT bad or good? By and large, the initial response of liberal arts and engineering students alike is that it is bad. This makes for an energetic discussion, as some of the facts in this box are shared.

31. R. Carson, *Silent Spring,* Houghton Mifflin, Boston, 1962.

32. P. H. Müller, Dichlorodiphenyltrichloroethane and Newer Insecticides, Nobel Lecture, Stockholm, Sweden, December 11, 1948.

33. The term *organic* is sometimes unclear when it comes to pesticides. In this usage, the term means that these pesticide compounds contain at least one carbon-to-carbon or carbon-to-hydrogen covalent bond. In contemporary usage, the term *organic* can also mean the opposite of *synthetic* or even *natural,* such as pesticides that are derived from plant extracts such as pyrethrin from the chrysanthemum flower. This is another example of how even within the scientific community, we are not clear what we mean, making risk communications difficult.

34. F. Graham, The Mississippi River Kill, in *Environmental Problems,* W. Mason and G. Fokerts, Eds., William C. Brown, 1973.

35. J. D. Graham and J. B. Wiener, Confronting Risk Tradeoffs, in *Risk Versus Risk: Tradeoffs in Protecting Health and the Environment,* J. D. Graham and J. B. Wiener, Eds., Harvard University Press, Cambridge, MA, 1995.

36. Environmental Protection Agency Pollution Prevention Directive, May 13, 1990, quoted in H. Freeman et al., Industrial Pollution Prevention: A Critical Review, presented at the Air and Waste Management Association Meeting, Kansas City, MO, 1992.

37. S. Richardson, Pollution Prevention in Textile Wet Processing: An Approach and Case Studies, *Proceedings*: *Environmental Challenges of the 1990's,* EPA/66/9-90/039, U.S. EPA, Washington, DC, September 1990.

38. U.S. Bureau of Labor Statistics, *Civilian Labor Force, 1980–98,* U.S. Department of Labor, Washington, DC, 1999.

39. National Science Foundation, *Women, Minorities, and Persons with Disabilities in Science and Engineering: 2000,* NSF 00-327, NSF, Arlington, VA, 2000.

40. I (Vallero speaking here) almost began this discussion with the clause "it goes without saying." Although I sometimes use it, I am puzzled when I hear or read this caveat. I am tempted to ask: "Then why are you saying it?" However, for the sake of argument, let us treat this as an idiom. What it really means is: "It *should* not need to be stated, but principles of practice are so often violated that I *must* say this!" One thing that seasoned professionals learn is that textbooks, manuals, and handbooks are valuable, but only when experience and good listening skills are added to the mix can wise (and just) decisions be made. Not to get "preachy," but this is the sage advice offered by the great thinkers and philosophers for the past three millennia, akin to the advice of St. Peter (Acts 24:25 and II Peter 1:6), who linked maturity with greater "self-control" or "temperance" (Greek *kratos* for "strength"). Interestingly, St. Peter considered knowledge as a prerequisite for temperance. Thus, from a professional point of view, he seemed to be arguing that one can really only understand and apply scientific theory and principles appropriately after one practices them. This is actually the structure of most professions. For example, engineers who intend to practice must first submit to a rigorous curriculum (approved and accredited by the Accreditation Board for Engineering and Technology), then must sit for the Future Engineers (FE) examination. After some years in the profession (assuming tutelage by and intellectual osmosis with more seasoned professionals), the engineer has demonstrated the *kratos* (strength) to sit for the Professional Engineers (PE) exam. Only after passing the PE exam does the National Society for Professional Engineering certify that the engineer is a "professional engineer" and eligible to use the initials PE after one's name. The engineer is, supposedly, now schooled beyond textbook knowledge and knows more about why in many problems the correct answer is: "It depends."

41. L. J. Morse, J. A. Bryan, J. P. Hurley, J. F. Murphy, T. F. O'Brien, and T. F. Wacker, The Holy Cross Football Team Hepatitis Outbreak, *Journal of the American Medical Association,* 219:706–708, 1972.

42. Aviation Safety Network, http://aviation-safety.net/database/index.html, 2002.

43. S. Pfatteicher, Learning from Failure: Terrorism and Ethics in Engineering Education, *Technology and Society,* 21(2):8–12, 21, 2002.

44. Z. Bazant, Why Did the World Trade Center Collapse?—Simple Analysis, *Journal of Engineering Mechanics,* 128(1):2–6, 2002.

45. L. Robertson, Reflections on the World Trade Center, *The Bridge,* 32, 2002.

46. See B. C. Ezell, J. V. Farr, and I. Wiese, Infrastructure Risk Analysis of a Municipal Water Distribution System, *Journal of Infrastructure Systems,* 6(3):118–122, 2000; and Y. Y. Haimes, N. C. Matalas, J. H. Lambert, B. A. Jackson, and J. F. R. Fellows, Reducing Vulnerability of Water Supply Systems to Attack, *Journal of Infrastructure Systems,* 4(4):164–177, 1998.

47. American Society for Engineering Education, *Prism,* 11(6), 2002.

48. B. Franklin, *Historical Review of Pennsylvania,* 1759.

49. S. C. Florman, *The Existential Pleasures of Engineering,* St Martin's Press, New York, 1976.

50. W. H. G. Armytage, *A Social History of Engineering,* Westview Press, Boulder, CO, 1961.

51. E. Garrison, *A History of Engineering and Technology,* CRC Press, Boca Raton, FL, 1991.

52. D. Hill, *A History of Engineering in Classical and Medieval Times,* Open Court Publishing Company, LaSalle, IL, 1984.

53. Armytage, note 50.

54. J. Bradbury, *The Medieval Siege,* The Boydell Press, Woodbridge, UK, 1992.

55. J. E. Watkins, "The Beginnings of Engineering" *ASCE Transactions,* vol. 24, 1891, reproduced in *The Civil Engineer: His Origins* ASCE Committee on History and Heritage of American Civil Engineering, Historical Publication No. 1, 1970.

56. Martel, W. C. (2001) *The Technological Arsenal: Emerging Defense Capabilities* Smithsonian Institution Press, Washington.

57. This quote is an extension of Pope Paul VI's statement to the United Nations for the celebration of the "Day of Peace" on January 1, 1972: "If you want peace, work for justice."

58. Note the difference between professional ethics and individual ethics. The vendors' credo, *caveat emptor,* places the onus on the buyer (client). The professional credo is very different. Ours is *credat emptor,* roughly translated from Latin as "let the client trust." Arguably, the engineer's principal client is the public; and the client need know little about the practice of engineering, because owing to the expertise, authority has been delegated to the engineer. Just as society allows a patient to undergo brain surgery even if that person has no understanding of the fundamentals of brain surgery, so our society cedes authority to engineers for design decisions. With that authority comes a commensurate amount of responsibility, and when things go wrong, culpability. The first canon of most engineering ethical codes requires that we "hold paramount" the health and welfare of the public. The public is an aggregate, not an "average." So leaving out any segment violates this credo.

59. C. L. Pettit and C. Johnson, The Impact on Property Values of Solid Waste Facilities, *Waste Age,* 18(4):97–102, 1987; and J. R. Price, The Impact of Waste Facilities on Real Estate Values, *Waste Management and Research,* 6(4):393–400, 1988.

60. S. Hirshfeld, P. A. Vesilind, and E. I. Pas, Assessing the True Cost of Landfills, *Waste Management and Research,* 10:471–484, 1992.

61. See M. O'Hare, Not in My Block You Don't, *Public Policy,* 25(4):407–458, 1997; C. Zeiss and J. Atwater, Waste Facilities in Residential Communities: Impacts and Acceptance, *Journal of Urban Planning and Development,* 113(1):19–34, 1987; and R. Lang, Equity in Siting Solid Waste Management Facilities, *Plan Canada,* 30(2):5–13, 1990.

62. Solid Wastes Association of North America, *Training Course Manual: Managing Sanitary Landfill Operations,* SWANA, Silver Spring, MD, 1989.

63. Hirshfeld et al., note 60.

4

Direct Risks to Human Health and Welfare

The activist is not the man who says the river is dirty.
The activist is the man who cleans up the river.

Ross Perot (1930–   ),
U.S. entrepreneur and politician

Much of the original concern with environmental pollution was that it was aesthetically displeasing and a nuisance. Certainly, a few Ancients in Egypt, Rome, and elsewhere did notice the relationship between exposure to pollutants and poor health. However, it was not until the late 1970s that risk was to become the dominant means of determining environmental effects. Thus, this new paradigm called for the comparison of one person's or group's risks to the risk in other groups as a means of determining whether that person or group is being inordinately exposed or placed at risk. An inordinately high exposure or risk is a first indication of injustice. This puts engineers in a pivotal position in environmental justice. Engineers are key players, "activists" if you will, as they search for ways to reduce environmental risks to highly exposed and sensitive groups. This proactive role is best articulated in the first canon of our codes of ethics: We must hold paramount the health, safety, and welfare of the public. We spend our careers providing these public services and finding ways to ensure that what we design is safe and does not detract from the public good.

In this chapter we deal with two concepts important to all engineering: risk and reliability. The principal value added by environmental engineers and other environmental professionals is in the improvement in the quality of human health and ecosystems. Environmental professionals do not have a monopoly on risk reduction, and in fact, all engineers can play a role in the enhancement of environmental quality.

Engineers add value when we decrease risk, so risk is one of the best ways to measure the success of engineers whose projects address environmental injustices. By extension, reliability tells us and everyone else just how well we go about preventing pollution, lowering the amounts of pollutants to which people are exposed, protecting ecosystems, and reducing overall risk. What we design must continue to serve its purpose throughout its useful life, in a manner sensitive to public health and environmental quality.

As it is generally understood, risk is the chance that something will go wrong or that some undesirable event will occur. Every time we get on a lawn tractor, for example, we are taking a risk that we might be in an accident and damage the tractor, get hurt, injure others, or even die in a mishap. The understanding of the factors that lead to a risk is called *risk analysis* and the reduction of this risk (e.g., by following the safety procedures delineated in the owner's manual and staying off steep grades) is *risk management*. Risk management is often differentiated from *risk assessment,* which is comprised of the scientific considerations of a risk.[1] Risk management includes the policies, laws, and other societal aspects of risk.

Engineers engage constantly in risk analysis, assessment, and management. Engineers must consider the interrelationships among factors that put people at risk, suggesting that we are risk analysts. Engineers provide decision makers with thoughtful studies based on sound application of the physical sciences and, therefore, are risk assessors by nature. Engineers control things and, as such, are risk managers. We are held responsible for designing safe products and processes, and the public holds us accountable for its health, safety, and welfare. The public expects engineers to "give results, not excuses,"[2] and risk and reliability are accountability measures of engineers' success. Engineers design systems to reduce risk and look for ways to enhance the reliability of these systems. Consequently, every engineer deals directly or indirectly with risk and reliability.

Thus, environmental justice embodies the concept of risk and how it can be quantified and analyzed. It also considers ways of reducing risk by conscious and intended risk management and how to communicate both the assessment and management options to those affected.[3]

RISK AND RELIABILITY

> *Probable impossibilities are to be preferred to improbable possibilities.*
> Aristotle

Aristotle was not only a moral philosopher and natural philosopher (the forerunner to "scientist"); he was also a risk assessor. In the business of human health and environmental protection, we are presented with "probable impossibilities" and "improbable possibilities."

To understand these two outcomes, we must first understand the different connotations of *risk*. Aristotle's observation is an expression of probability. People, at least intuitively, assess risks and determine the reliability of their decisions every day. We want to live in a safe world; but *safety* is a relative term. The "safe" label requires a value judgment and is always accompanied by uncertainties, but engineers frequently characterize the safety of a product or process in objective and quantitative terms. Factors of safety are a part of every design. Environmental safety is usually expressed by its opposite term, risk.

Discussion: Probability—The Mathematics of Risk and Reliability

Probability is the likelihood of an outcome. The outcome can be bad or good, desired or undesired. The history of probability theory, like much modern mathematics and science, is rooted in the Renaissance. Italian mathematicians considered some of the contemporary aspects of probability as early as the fifteenth century, but did not need, or were unable, to devise a generalized theory. Blaise Pascal and Pierre de Fermat, famous French mathematicians, developed the theory after a series of letters in 1654 considering some questions posed by a nobleman, Antoine Gombaud, Chevalier de Méré, regarding betting and gaming. Other significant Renaissance and post-Renaissance mathematicians and scientists soon weighed in, with Christian Huygens publishing the first treatise on probability, *De Ratiociniis in Ludo Aleae,* which was specifically devoted to gambling odds. Jakob Bernoulli (1654–1705) and Abraham de Moivre (1667–1754) also added to the theory. However, it was not until 1812 with Pierre Laplace's publication of *Théorie analytique des probabilités,* that probability theory was extended beyond gaming to scientific applications.[4]

Probability is now accepted as the mathematical expression that relates a particular outcome of an event to the total number of possible outcomes. This is demonstrated when we flip a coin. Since the coin has only two sides, we would expect a 50–50 chance of either a head or a tail. However, scientists must also consider rare outcomes, so there is a very rare chance (i.e., highly unlikely, but still possible) that the coin could land on its edge (i.e., the outcome is neither a head nor a tail), A "perfect storm" of a confluence of unlikely events is something that engineers must always consider, such as the combination of factors that led to major disasters such as Hurricane Katrina and the toxic cloud in Bhopal, India, or the introduction of a seemingly innocuous opportunistic species (e.g., Iron Gates Dam in Europe) that devastates an entire ecosystem, or the interaction of one particular congener of a compound in the right cell in the right person that leads to cancer. As engineers, we also know that the act of flipping or the characteristics of the coin may tend to change the odds. For example, if for some reason the heads is heavier than the tails side or the aerodynamics is different, the probability could change.

The total probability of all outcomes must be unity (i.e., the sum of the probabilities must be 1). In the case of the coin standing on end rather than being a head or a tail, we can apply a quantifiable probability to that rare event. Let us say that laboratory research has shown that 1 in a million times ($1/1,000,000 = 0.000001 = 10^{-6}$), the coin lands on edge. By difference, since the total probabilities must equal 1, the other two possible outcomes (heads and tails) must be $1 - 0.000001 = 0.999999$. Again, we are assuming that the aerodynamics and other physical attributes of the coin give it an equal chance of being either a head or a tail, the probability of a head = 0.4999995 and the probability of a tail = 0.4999995.

Stated mathematically, an event (e) is one of the possible outcomes of a trial (drawn from a population). In our coin-toss case, all events, head, tail, and edge, together form a finite *sample space,* designated as $E = [e_1, e_2, \ldots, e_n]$. The lay public is not generally equipped to deal with such rare events, so by convention, they usually ignore them. For example, at the beginning of overtime in a football game, a tossed coin determines who will receive the ball and thus have the first opportunity to score and win. When the referee tosses the coin, there is little concern about anything other than heads or tails. However, the National Football League undoubtedly has a protocol for the rare event of the coin not being a discernable head or tail. In environmental studies, e could represent a case of cancer. Thus, if a population of 1 million people is exposed to a pesticide over a specific time period, and one additional cancer is diagnosed that can be attributed to that pesticide exposure, we would say that the probability of e (i.e., $p\{e\}$) = 10^{-6}. Note that this was the same probability that we assigned to the coin landing on its edge.

Returning to our football example, the probability of the third outcome (a coin on edge) is higher than "usual" since the coin lands in grass or artificial turf compared to landing on a hard flat surface. Thus, the physical conditions increase the relative probability of the third event. This is analogous to a person who may have the same exposure to a carcinogen as the general population, but who may be genetically predisposed to develop cancer. The exposure is the same, but the probability of the outcome is higher for this "susceptible" person. Thus, risk varies by both environmental and individual circumstances.

Events can be characterized a number of ways. Events may be discrete or continuous. If the event is forced to be one of a finite set of values (e.g., six sides of a die), the event is discrete. However, if the event can be any value [e.g., size of tumor (within reasonable limits)], the event is continuous. Events can also be independent or dependent. An event is *independent* if the results are not influenced by previous outcomes. Conversely, an event affected by any previous outcome is a *dependent* event.

Joint probabilities must be considered and calculated since in most environmental scenarios, events occur in combinations. So if we have n mutually exclusive events as possible outcomes from E that have probabilities equal to $p\{e_i\}$, the probability of these events in a trial equals the sum of the individual probabilities:

$$p\{e_i \text{ or } e_2 \cdots \text{ or } e_k\} = p\{e_1\} + p\{e_2\} + \cdots + p\{e_k\} \tag{4.1}$$

Further, this helps us to find the probabilities of events e_i and g_i for two independent sets of events, E and G, respectively:

$$p\{e_i \text{ or } g_i\} = p\{e_i\}p\{g_i\} \tag{4.2}$$

For example, a company record book indicates that a waste site has 10 unlabeled buried chemical drums: five drums that contain mercury (Hg), two drums that contain chromium (Cr), and three drums that contain tetrachloromethane (CCl_4). We can determine the probability of pulling up one of the drums that contains a metal waste (i.e., Hg or Cr). The two possible events (Hg drum or Cr drum), then, are mutually exclusive and come from the same sample space; so we can use equation (4.1):

$$p\{Hg \text{ or } Cr\} = p\{Hg\} + p\{Cr\} = \frac{5}{10} + \frac{2}{10} = \frac{7}{10}$$

Thus, we have a 70% probability of pulling up a metal-containing drum.

If we have another waste site that also has 10 unlabeled, buried drums—three drums that contain dichloromethane (CH_2Cl_2) and seven drums that contain trichloromethane ($CHCl_3$)—we calculate the probability of pulling up a chromium drum from our first site and a $CHCl_3$ drum from the second site. Since the two trials are independent, we can use equation (4.2):

$$p\{Cr \text{ and } CH_2Cl_2\} = p\{Hg\} + p\{Cr\} = \frac{2}{10} \times \frac{3}{10} = \frac{6}{100}$$

Thus we have 6% probability of extracting a chromium and a dichloromethane drum on our first excavation.

Another important concept for environmental data is that of conditional probability. If we have two dependent sets of events, E and G, the probability that event e_k will occur if the dependent event g has occurred previously can be shown as $p\{e_k|g\}$, which is found using Bayes' theorem:

$$p\{ek|g\} = \frac{p\{e_k \text{ and } g\}}{p\{g\}} = \frac{p\{g|e_k\}p\{e_k\}}{\Sigma_{i=1}^{n}p\{g|e_i\}p\{e_i\}} \tag{4.3}$$

A review of this equation shows that conditional probabilities are affected by a cascade of previous events. Thus, the probability of what happens next can be highly dependent on what occurred previously. For example, the cumulative risk of cancer depends on the serial (dependent) outcomes. Similarly, reliability can also be affected by dependencies and prior events. Thus, characterizing any risk or determining the reliability of our systems are expressions, at least in part, of probability.

Engineers are comfortable with equations, so another way to present probabilities to characterize risk and reliability is by showing a *probability density function* (PDF) for data. The PDF is created from a probability density; that is, when the data are plotted in the form of a histogram, as the amount of data increases, the graph increases its smoothness (i.e., the data appear to be continuous). The smooth curve can be expressed mathematically as a function, $f(x)$. This is the PDF. The probability distribution can take many shapes, so the $f(x)$ for each will differ accordingly. For example, in environmental matters, distributions commonly seen are normal, log-normal, and Poisson. The normal (Gaussian) distribution is symmetrical and is best known as the *bell curve,* given its shape (see Figure 4.1). The log-normal distribution is also symmetrical, but its *x*-axis is plotted as a logarithm of the values.

The Poisson distribution is a representation of events that happen with relative infrequency, but regularly.[5] Stated mathematically, the Poisson distribution function expresses the probability of observing various numbers of a particular event in a sample when the mean probability of that event on any one trial is very small. So the Poisson probability distribution characterizes discrete events that occur independent of one another during a specific period of time. This is useful for risk assessments, since exposure-related measurements can be expressed as a rate of discrete events: the number of times that an event happens during a defined time interval (e.g., the frequency (times per week) during

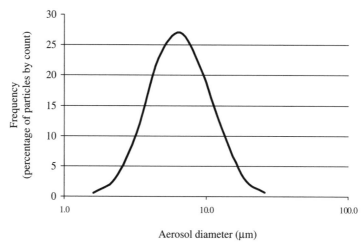

Figure 4.1 Distribution (log normal) of aerosols in the lower troposphere. (Data from L. Silverman, C. E. Billings, and M. W. First, *Particle Size Analysis in Industrial Hygiene,* Academic Press, New York, 1971.)

which a person eats shellfish that contain polychlorinated biphenyls (PCBs) in fish containing methyl mercury concentrations greater than 5.0 mg L^{-1}. The Poisson distribution describes events that take place during a fixed period of time (i.e., a rate), as long as the individual events are independent of each other. As the expected number of events or counts increases (i.e., the event rate increases), so does variability. Obviously, if we expect a count to equal 1, we should have little trouble picturing an observation of 2 or 0. If we expect a count equal to 50,000, counts of 49,700 and 50,300 are within reason. The range and variance of the latter, however, is much larger. The Poisson equation needed to compute the probability of a specific number of counts being observed over a defined time interval is

$$P_\lambda = \frac{e^{-\pi}\lambda^n}{n!} \tag{4.4}$$

where λ is the average or expected counts or events per unit time and n is the number of encounters.

Thus, the Poisson distribution is useful in a risk assessment to estimate exposures. It may be used to characterize the frequency with which a person (or animal or ecosystem) comes into contact with a substance (e.g., the number of times per day a person living near a wood treatment facility is exposed to pentachlorophenol. Assuming that based on existing data, the expected number of encounters is two per day, applying equation (4.4) with $\lambda = 2$, there is a 9% chance that a person will have 4 (i.e., $n = 4$) encounters with pentachlorophenol on a given day.

Risk itself is an expression of a probability (i.e., the chance of an adverse outcome). It is the probability of a consequence. So any calculation of environmental insult can be based on some use of probabilities.

Biographical Sketch: Abdul Q. Khan

Abdul Qadeer Khan was born in Bhopal, India, in 1936 to a middle-class Muslim family. At the time the population of Bhopal was made up of both Muslim and Hindus who lived an uneasy but peaceful coexistence. When India was partitioned in 1947, forming Pakistan, a huge migration occurred where Muslims moved north to Pakistan, and Hindus moved south to India. Abdul Khan, one of seven children, migrated with his family in 1952. They were harassed, beaten, and robbed during their trip and he ended up walking barefoot to Pakistan. The experience caused him to have a lifelong distrust and hatred of India.

Khan went to the University of Karachi in Pakistan and then to universities in West Germany and Belgium. After graduating with a PhD in metallurgical engineering Khan joined the staff of the Physical Dynamics Research Laboratory, or FDO, in Amsterdam, The Netherlands. FDO was a subcontractor for URENCO, a British-Dutch-German consortium specializing in the manufacture of equipment for the enrichment of uranium.

By 1974 India had developed and demonstrated their nuclear bombs and was threatening Pakistan. Taking advantage of an insurrection in Eastern Pakistan, the Indian army had already soundly defeated the Pakistani forces, resulting in the formation of a new country, Bangladesh.

Demoralized and stinging from their defeat in the Indo-Pakistani War, the political and military leaders of Pakistan concluded that they also had to have nuclear weapons to serve as a deterrent for the threat they perceived to be coming from India. In 1975 A.Q. Khan was reportedly asked by the then Prime Minister of Pakistan to develop a uranium-enrichment program for Pakistan. Using his position at FDO as an engineer working on uranium enrichment machinery, Khan began to steal classified documents and to send them to Pakistan. His actions eventually raised suspicions, but he escaped to Pakistan before he could be arrested. Once safely in Pakistan, he was asked by the government to establish the laboratory that would produce the nuclear device. Using his contacts and an unlimited budget, Khan's lab was able to enrich uranium by the late 1980s leading to the successful detonation of Pakistan's first nuclear device on 28 May 1998. For his efforts Khan became a national hero, living a life of privilege and amassing numerous honors and adulations.

In the late 1990s reports began to be made public accusing Khan and others at his lab of selling nuclear secrets, equipment and material to countries such as Libya, Iran, Malaysia, and North Korea, allegedly in return for tens of millions of dollars. In 2003 the Pakistani government began an investigation, at the prompting of the US government, into Khan and his lab. In January 2004 Khan was placed under house arrest and remains under detention to this day. President Musharraf of Pakistan subsequently pardoned Khan of any wrongdoing citing his service to Pakistan, but this was widely perceived as a face-saving exercise to deflect any blame from government officials, and to salvage the reputation of a popular national hero.

Abdul Khan's story demonstrates how engineers, using their technical skills and blinded by both wealth and nationalism, can forget their primary role of service to society. The sale of nuclear technology to countries that may use this knowledge to terrorize others was unconscionable.

Engineering success or failure is in large measure determined by comparing what we do to what our profession "expects" of us. Safety is a fundamental facet of our engineering duties. Thus, we need a set of criteria that tells us when our designs and projects are sufficiently safe. Four criteria are applied to test engineering safety:[6]

1. The design must comply with applicable laws.
2. The design must adhere to "acceptable engineering practice."
3. Alternative designs must be sought to see if there are safer practices.
4. Possible misuse of the product or process must be considered and estimated.

The first two criteria are usually more manageable than the latter two. The engineer can look up the physical, chemical, and biological factors to calculate tolerances and factors of safety for specific designs. Laws and regulations are promulgated to protect the public. Crossing these legal thresholds indicates the point at which the engineer has failed to provide adequate protection. Engineering practice standards go a step further. Much of the public and its lawyers would not be able to recognize this type of failure unless other engineers help to judge whether an ample margin of safety has been met. The margin is dictated by sound engineering principles and practice. However, finding alternatives and predicting misuse and mistakes require creativity and imagination. These criteria correspond closely with the five types of failure discussed in Chapter 3. In fact, a corollary to the fourth criteria could be added to contemporary engineering: intentional and willful misuse (e.g., intentional environmental injustices, terrorism), which is now a design criterion that must be considered by every engineer.

If one were to query a focus group as to whether risk can be quantified, the group is usually divided. At first thought, most respondents consider risk not to be quantifiable. The general consensus, at least in our unscientific queries, is that one person's risk is different from another's; risk is in the "eye of the beholder." Some of the rationale appears to be rooted in the controversial risks of tobacco use and daily decisions, such as choice of modes of transportation.

Discussion: Choose Your Route of Exposure

> Old man, look at my life, twenty four and there's so much more . . .
> Give me things that don't get lost.
> Like a coin that won't get tossed. . .
> *Old Man,* Neil Young (1945–   )

Young, the songwriter, seems be talking about risk, especially how it changes with age and how, when we are younger, our acceptance of risks may be rather high. Perhaps, since Young was about 25 years old when he wrote *Old Man,* he was displaying cognitive dissidence: on the one hand, wanting to take risks, but on the other, recognizing risk avoidance as a necessity in some matters (i.e., a coin that won't get tossed). An interesting phenomenon that supports this view seems to be taking place on today's college campuses. From some anecdotal observations, it would appear that students are more concerned about some exposure pathways and routes than others. It is not uncommon at Duke or Bucknell University, for example, to observe a student getting off her bicycle to smoke a cigarette and to take a few drinks of name-brand bottled water.

At first blush, one may conclude that the student has conducted some type of risk assessment, albeit intuitive, and has concluded that she needs to be concerned about physical fitness (biking) and the oral route of exposure to contaminants, as evidenced by the bottled water. For some reason, the student is not as concerned about the potential carcinogens in tobacco smoke as about the contaminants found in tap water. Or is it simply taste . . . or mass marketing?

With a bit more analysis, however, the apparent lack of concern for the inhalation route (tobacco smoke) may not be the case. The behavior may be demonstrating the concept of risk perception. The biking, smoking, and drinking activities seem to illustrate at least two principles regarding increased concern about risk: whether the student maintains some control over risk decisions, and whether the exposures and risks are voluntary or involuntary. The observation may also demonstrate the lack of homogeneity in risk perception. For example, risk perception appears to be age dependent. More frequently than the general population, teenagers and young adults perceive themselves to be invincible, invulnerable, and even immortal. Like most decisions, risk decisions consist of five components:

1. An inventory of relevant choices
2. An identification of potential consequences of each choice
3. An assessment of the likelihood of each consequence actually occurring
4. A determination of the importance of these consequences
5. Synthesis of this information to decide which choice is the best[7]

These perceptions change with age, as a result of experiences and of physiological changes in the brain. However, like the risks associated with the lack of experience in driving an automobile, a young person may do permanent damage while traversing these developmental phases. In fact, this mix of physiological, social, and environmental factors in decision making is an important variable in characterizing hazards. In addition, the hazard itself influences the risk perception. For example, whether the hazard is intense or diffuse, or whether it is natural or human-induced (see Figure 4.2) is a determination of public acceptance of the risks associated with the hazard. People tend to be more accepting of hazards that are natural in origin, voluntary, and concentrated in time and space.[8]

Other possible explanations are risk mitigation and sorting of competing values. The biker may well know that smoking is a risky endeavor and is attempting to mitigate that risk by other positive actions, such as exercise and clean water. Or she may simply be making a choice that the freedom to smoke

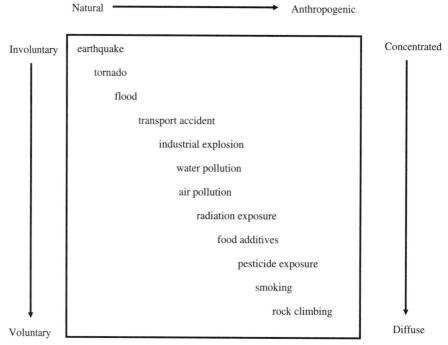

Figure 4.2   Spectrum of hazards. (From D. A. Vallero, *Paradigms Lost: Learning from Environmental Mistakes, Mishaps, and Misdeeds,* Butterworth-Heinemann, Burlington, MA, 2006; adapted from K. Smith, *Environmental Hazards: Assessing Risk and Reducing Disaster,* Routledge, London, 1992.)

outweighs other values, like a healthy lifestyle (students have reported that biking may well simply be a means of transportation and not a question of values at all).

It is likely that all of these factors affect different people in myriad ways, illustrating the complexities involved in risk management decisions.

Demographics is a determinant of risk perception, with certain groups more prone to "risk taking" and averse to authority. Teenagers, for example, are often shifting dependencies (e.g., from parents to peers). Later, the dependencies may be transitioning to greater independence, such as that found on college campuses. Eventually, these can lead to healthy, interdependent relationships. Engineers have to deal with these dynamics as a snapshot. Although individual persons may be changing, a population is often more static. There are exceptions, for example, if the mean age of a neighborhood is undergoing significant change (e.g., getting younger), there may a concomitant change in risk acceptance and acceptance of controls (e.g., changes to zoning and land use).

What people perceive as risks and how they prioritize those risks is only partly driven by the actual objective assessment of risk (i.e., the severity of the hazard combined with the magnitude, duration, and frequency of the exposure to the hazard). For example, the young student may be aware that cigarette smoke contains some nasty compounds but is not directly aware of what these are (e.g., polycyclic aromatic hydrocarbons and carcinogenic metal compounds). She has probably read the conspicuous warning labels many times as she held a pack in her hands, but they really have not "rung true" to her. She may never have met anyone with emphysema or lung cancer and may not yet be concerned with the effects on the unborn (i.e., *in utero* exposure). Psychologists also tell us that many in this age group have a feeling of invulnerability. Those who do think about it may also conclude that they will have plenty of time to end the habit before it does any long-term damage.

People seem to have their own "mathematics" when it comes to risk. If you visit a local hospital, you are likely to see a number of patients gathered near the hospital entrance in designated smoking areas. Here they are hooked up to IVs, pumps, and other miracles of medical technology and engaging simultaneously in one of the most potent health hazards: smoking. Of course, this is ironic. On the one hand we are assigning the most talented (and expensive) professionals to treat what is ailing them, yet they have made a personal decision to engage in very unhealthful habits. It seems analogous to training and equipping a person at great social costs to drive a very expensive vehicle, all the time knowing that the person has a nasty habit of tailgating and weaving in traffic. Even if the person is well trained and has the best car, this habit will increase the risk. However, there is another way to look at the smoking situation, that is, the person has decided (mathematically) that the sunken costs are dictating the decision. Intuitively, the smoker has differentiated long-term risks from short-term risks. The reason the person is in the hospital (e.g., heart disease, cancer, emphysema) is the result of risk decisions the person made years, maybe decades, ago. The exposure is long in duration and the effect is chronic. So the person may reason that the effects of today's smoking will only be manifested 20 years hence and has little incentive to stop engaging in the hazardous activity. Others see the same risk and use a different type of math. They reason that there is X odds that they will live a somewhat normal life after treatment, so they need to eliminate bad habits that will put them in the same situation 20 years from now. They both have the same data but reach very different risk decisions.

Another interesting aspect of risk perception is how it varies in scale and scope. Vallero recalls sitting in a meeting in the late 1970s among a group of highly trained engineers and scientists from the U.S. EPA and the Kansas Department of Health and Environment. Mind you, these are the two principal federal and state agencies, respectively, charged with protecting the environment. The meeting was to determine the appropriate ways to reduce the ambient concentrations of pollutants, especially particulate matter and carbon monoxide (CO). The meeting was held in a small room, and almost every person was smoking except Vallero. The room was literally smoke-filled. Here we were talking about the best approaches for bringing down ambient levels to the parts per million range (high by today's standards, but at detection limits in the 1970s) in a room that one could almost report as a percent (i.e., 10^{-6} *versus* 10^{-2}). The irony was lost on the participants, probably for good reason. They were not making personal decisions, they were making policy decisions. This is akin to saying "Do as I say, not as I do." Such compartmentalization is not foreign to engineers. We realize that some of our own homes would not meet many of the standards that we require for our clients' structures (e.g., accessibility, water and air quality, egress/exit, signage).

The compartmentalization concept is brought home some years back in a story shared by a former director of the Office of Management and Budget (OMB). He had been in a budget meeting earlier in the morning, where he was discussing a few multimillion-dollar projects and recommending some million-dollar increases in a number of them. Later, his wife called to update him on the new house they were building. One item was that the window contractor needed an additional $200 above the original estimate. The OMB director was outraged and starting ranting about how important it was to hold the line on such expenses. At this point, he was struck by the irony. Again, the policy decision has a different scale and scope than the individual decision (in the OMB director's case, 10^8 *versus* 10^2). But is this right? Is vigilance a function of size (dollars, risk, number of people), or must it be independent of scale? The neighbors do not care that you have a bigger project or one with much more societal import. This is their only neighborhood, these are their only children, and the risks are the ones that they will have to abide.

Another consideration is whether the scope flavors what we consider to be of value. For example, many communities are overrun with deer populations because suburban developments have infringed on the habitats of deer and their predators, changing the ecological dynamics. Certainly, the deer population increases present major safety problems, especially during rutting season when deer enter roadways and the likelihood of collisions increases. The deer also present nuisances, such as their invasions of gardens. They are even part of the life cycle of disease vectors when they are hosts to ticks that transmit Rocky Mountain spotted fever and Lyme disease. In this sense we may see deer as a "problem" that must be eradicated. However, when we come face to face with deer, such as when we see a doe and her fawns, we can appreciate the majesty and value of the individual deer.

Recently, I (Vallero speaking) have been taking numerous actions to make my garden unpleasant to these creatures, with little success (they particularly like eating my tomatoes before they ripen and enjoy nibbling young zucchini plants). I said some rather unpleasant things about the deer population. However, recently I was traveling on a country highway and noticed an emergency vehicle assisting a driver who had obviously crashed into something. The driver and responder were in the process of leaving the scene. Driving 20 meters further, I noticed a large doe trying to come to her feet to run into the woods adjacent to the highway, but she could not lift her back legs. I realized that the people leaving the scene must have concluded that the "emergency" was over, without regard to the deer that had been struck by the car. Seeing the wounded creature was a reminder that the individual, suffering animal had an intrinsic value. Such a value is lost when we see only the large-scale problem, without consideration of the individual. Incidentally, by the time I had turned around, another person had already called the animal control authorities. Coming into personal contact with that deer makes it *our* deer, just as knowing that there are too many dogs and cats in the nation does not diminish our devotion to *our* own dog or cat.[9] I still may take measures to prevent the deer herd from destroying my garden, but I would gladly open my garden to *that* deer I pitied on the side of the road.

Thus, we should be aware that what we are saying to people, no matter how technically sound and convincing to us as engineers and scientists, may be simply a din to our targeted audience, or even an affront to the values they cherish. Their value systems and the ways that they perceive risk are the result of their own unique blend of experiences.

The converse is also true. We may be completely persuaded based on data, facts, and models that something clearly does not cause significant harm, but those we are trying to convince of this finding may not buy it. They may think that we have some vested interest, or that they find us guilty by association with a group they do not trust, or that we are simply "hired guns." They may not understand us because we are using jargon and are not clear in how we communicate the risks. So do not be surprised if the perception of risk does not match the risk you have quantified.

Analyzing each part of Aristotle's seemingly counterintuitive and repetitious quote sheds some light on why risk is such a crucial part of any communication with the public, particularly the risk to those that have been or appear to have been exposed to disproportionately high concentrations of contaminants in their food, air, water, or homes. When the query continues with a bit more information, the number of advocates for the inability to quantify risks begins to dwindle. For example, when the numbers of tobacco users *versus* the incidence of cancer (or other health endpoints) are shown side by side for a population, the strength of association pushes the group to accept that risk as being quantitative. In the same vein, when automobile, rail, bike, and air travel mortality statistics are provided; people see that there is some numeric link between outcomes and behaviors, but they still perceive a greater risk from air travel. People can "do the math," but the math does not hold primacy over what they perceive to be risks. The weight of evidence includes some nonquantifiable factors. For example, they may simply wonder how a multiton metal object can stay airborne. In addition, of the various modes of transportation, air travel on a large plane gives the passenger the least control and fewest options if things go wrong (i.e., a perceived lack of control). So there is always the specter of "improbable possibilities" when managing risks.

Perception and Reality

Engineers generally would not disagree that the failure of the Tacoma Narrows Bridge or the devastation wrought by the 1906 earthquake in San Francisco were indeed *disasters*. But characterizing something as a disaster as opposed to the run-of-the-mill "failures" depends in large measure on how the public, or at least a substantial part of it, such as the media, perceive it. Failure occurs all the time. In fact, for all human endeavors failure is inevitable. Failure becomes a disaster when events in time and space lead to effects so severe that the events collectively are deemed to be a disaster. An event could also be classified as a disaster when engineers made such a miscalculation or left out some key information. Such mistakes may lead to the public perception that the failure was disastrous, compared to an even more severe outcome that was perceived as less preventable, or even inevitable.

For some events, we do not even recognize them as a disaster until well after the fact. Environmental justice disasters, in particular, may not be noticed for decades. This can be because the actual disease and negative effects are delayed (i.e., called the *latency period*. For example, the onset of cancer symptoms may not be diagnosed until decades

Biographical Sketch: Mary Amdur

Often the most important people in the lives of ordinary citizens are not recognized for their work. Such was the case with Mary Amdur (1922–1998).

She graduated from the University of Pittsburgh and went on to the Harvard Medical School, working with an eminent pulmonologist, Philip Drinker, the inventor of the iron lung that was used to help polio victims. While at Harvard Amdur began to do work in toxicology, especially chronic, sublethal doses of toxins. A principal motivation for her career was the premature death of her father at 40 from lung cancer linked to the steel works in Donora, Pennsylvania. She invented a pump that allowed her to expose guinea pigs to fine mists that carried the toxins into the lungs of the guinea pigs. She showed that young animals are affected sooner and more severely than older animals, and found that the sulfuric acid mist acts synergistically with fine particulates, helping to explain one of the potential ways that people might have died during the Donora air pollution episode. The two researchers also studied the animals that had died during the Donora air pollution episode, and these data confirmed their findings that age is an important factor. They went on to conduct studies with volunteer humans who were subjected to acid mist and who developed the same symptoms as the guinea pigs, clearly proving that acid mist is a dangerous air pollutant.

Unfortunately, the research conducted by Amdur and Drinker was supported financially by the American Smelting and Refining Company, which was not at all pleased with the findings. At one point company representatives allegedly attempted physical intimidation of Mary Amdur, using thugs who crowded her on an elevator and told her not to publish any more papers on the topic. Whatever pressure or intimidation was bought on Drinker seems to have worked. The actual facts are not known, but they resulted in Drinker's firing of Amdur and removal of his name from the papers that they had planned to publish jointly. One paper that at the time was ready for publication in the *Lancet,* the British medical journal, simply disappeared (this was before floppy disks and hard drive backups). Mary Amdur endured this intimidation and soon found a new job in the Harvard School of Public Health, where she stayed for 20 years, continuing to publish prolifically on the toxicology of air pollutants. Her work was well ahead of its time, and it took the world decades to catch up to her (e.g., it was not until the 1990s that the increase in mortality was linked statistically to episodes of elevated particulate levels, interestingly in large part to research by Douglas Dockery and other of Amdur's successors at Harvard[10]). Her courage and perseverance has been an inspiration to many environmental toxicologists.

after exposure to the carcinogen (e.g., asbestos workers may be exposed for decades before signs of mesothelioma or lung cancer). The lag in noticing problems may also reflect the current state of the science. For example, if a contaminant exists in soil at concentrations below those that can be detected by contemporary sampling and analytical methods, the unsafe conditions will not be reported. This was a factor in some infamous cases, like Times Beach and Love Canal. Even if the levels of dioxin and other contaminants had existed in the 1950s, the science would not have been sufficiently advanced to detect the problem. The lack of recognition of actual and pending disasters may also be the result of understudying and underreporting of the exposures and diseases, such as the relatively recent linkages between childhood exposures to the metal lead and neurological and developmental diseases.

The two problems (i.e., latency period and underreporting) can amplify one another in environmental justice situations. For example, certain workers may not want to jeopardize their livelihoods and are reluctant to report early symptoms of chronic diseases. Scientists historically have been more likely to study certain demographic groups (e.g., healthy workers) and have avoided others (children, women, and minorities). But when the results do flood in, such as the lead studies in the latter part of the twentieth century or the ongoing arsenic exposures in Bangladesh (see Chapter 3), they are perceived to be "public health disasters."

So risk perception is a crucial component of risk management. The engineer must be cognizant that sharing the same set of facts will be perceived differently by different groups. One group may see the facts as representing a problem that can easily be fixed, whereas another may perceive the same facts as representing an engineering or public health disaster. A notable example is the State University of New York, Stony Brook,[11] comparison of U.S. transportation fatalities in 1992. The study found that the modes of transportation had similar numbers of fatalities from accidents involving airplanes (775), trains (755), and bicycles (722). The public, however, considered air travel to have much higher risk associated with it than the risk from trains and certainly for bicycles. The researchers concluded that two driving factors may lead to these perceptions: (1) a single event in air crashes leads to a large loss of life, with much media attention; and (2) people aboard a large aircraft have virtually no control over their situation.

The increased anxiety resulting from highly visible failures and lack of control over outcomes leads to the greater perceived risk. These factors also seem to occur for environmental and public health risks. Certain terms are downright scary, like *cancer, central nervous system dysfunction, toxics,* and ominous-sounding chemical names like *dioxin, PCBs, vinyl chloride,* and *methyl mercury.* In fact, the chemicals listed *are* ominous! But some that are less harmful can also elicit anxieties and associated increased perceived risk, even from the well educated and erudite. For example, students at Duke have been asked for some years now as part of a pretest to an engineering ethics course to answer a number of questions. The first two questions on the exam are:

1. The compound dihydrogen monoxide has several manufacturing and industrial uses. However, it has been associated with acute health effects and death in humans, as a result of displacement of oxygen from vital organs. The compound has been found to form chemical solutions and suspensions with other substances, crossing cellular membranes, and leading to cancer and other chronic diseases in humans. In addition, the compound has been associated with fish kills when supersaturated with molecular oxygen, destruction of wetlands and other habitats, and billions of dollars of material damage each year. A prudent course of action dealing with dihydrogen monoxide is to:
   a. Ban the substance outright

    b. Conduct a thorough risk assessment, then take regulatory actions

    c. Work with industries using the compound to find suitable substitutes

    d. Restrict the uses of the substance to those of strategic importance to the United States

    e. Take no action except to warn the public about the risks

2. The class of compounds, polychlorinated biphenyls, had several manufacturing and industrial uses during the twentieth century. However, PCBs were associated with acute health effects and death in humans. The compound has been found to form chemical solutions and suspensions with other substances, crossing cellular membranes, and leading to cancer and other chronic diseases in humans. In addition, the compound has been associated with contaminated sediments, as well as wetlands and other habitats, and billions of dollars of material damage each year. A prudent course of action dealing with PCBs is to:

    a. Ban the substances outright

    b. Conduct a thorough risk assessment, then take regulatory actions

    c. Work with industries using compound to find suitable substitutes

    d. Restrict the uses of the substances to those of strategic importance to the United States

    e. Take no action except to warn the public about the risks

Everything in the question is factually correct. The two questions were intentionally worded similarly and the answers worded identically. The students are well-versed in math and science. On average, their scores on their Scholastic Achievement Tests are above 1400, and most have earned A's in high school or college chemistry, physics, and biology, and are on their way toward completing engineering and other technical degrees. Interestingly, the answers to the two questions differed very little. Most students appear to be influenced by the litany of negative effects to health and safety. The most frequent answer is b: conduct a risk assessment. Students seem to be heeding their teacher's relentless reminders that they get their facts straight before technical decisions (one of the themes of this book). Before we overcongratulate ourselves, as engineering educators, however, many of the students saw no difference between the two questions and several chose a: outright bans on both chemicals, the first of which is water!

Actually, the answers to the two questions should have been very different. We would recommend reply e for water and reply a for the polychlorinated biphenyls (simply because they have been banned since the 1970s following the passage of the Toxic Substances Control Act). We do note that water is not risk-free. In fact, it is a contributing factor in many deaths [drowning, electrocution, auto accidents, falls, especially water in its solid phase (i.e., ice) and workplace incidents such as steam-related accidents]. However, none of us could survive if we banned or placed major restrictions on its use!

Perception may be either higher or lower than actual risk. So, then, engineers must reconcile technical facts with pubic fears. What are the ethics of technical communication when it comes to risks? Like so many engineering concepts, timing and scenarios are crucially determinate. What may be the right manner of saying or writing something in one situation may be quite inappropriate in another. Our communication approaches will differ according to whether we need to motivate people to take action, alleviate undue fears, or simply share our findings clearly, whether they convey good news or bad.

Engineers may wish to avoid the business model in this case. Some have accused certain companies of using pubic relations and advertising tools to lower the perceived risks of their products. The companies may argue that they are simply presenting a counterbalance against unrealistic perceptions. Engineers must take care not to be manipulated by parties with vested, yet hidden interests. An emerging risk management technique is *outrage management,* described by Peter Sandman, a consultant to businesses and governmental agencies.[12] According to Sandman, the first step is to present a positive public image as a "romantic hero," pointing out all the good things the company or agency provides, such as jobs, modern conveniences, and medical breakthroughs. Although these facts may be accurate, they often have little to do with the decisions at hand, such as the type of pollution controls to be installed on a specific power plant near a particular neighborhood. Ethicists refer to such tactics in their extreme as *red herrings.*[13] Another way that a public image can be enhanced is to argue that the company itself is a "victim," suffering the brunt of unfair media coverage or targeted by politicians. If these do not work, some companies have confessed to being "reformed sinners," who are changing their ways. One of the more interesting strategies put forth by Sandman is that companies can portray themselves as "caged beasts." This approach is used to convince the public that even though in the past they have engaged in unethical pollution and unfair practices, the industry is so heavily regulated and litigated against that they are no longer able to engage in these acts. So the public is led to trust that this new project is different from the company's track record. There is obviously some truth underpinning this tactic, as regulations and court precedents have curtailed a lot of pollution. But the engineer must be careful to discern the difference between actual improvement and mere spin tactics to eliminate public outrage.

Holding paramount the health, safety, and welfare of the public gives the engineer no room for spin. On the other hand, the public does often exaggerate risks. Abating risks that are, in fact, quite low could mean unnecessarily complicated and costly measures. It may also mean choosing the less acceptable alternative (i.e., one that in the long run may be more costly and deleterious to the environment or public health). For example, in the preface to the American Council on Science and Health's recent report *America's War on "Carcinogens,"* George M. Gray, formerly the executive director of the Harvard Center for Risk Analysis and currently EPA's Assistant Administrator for Research and Development, warns:

> Public misperception of the magnitude of risks can have two important repercussions. First, people may make bad decisions for themselves and their families. If the costs of organic food, purchased to avoid the hypothetical cancer risks from pesticides, reduce total consumption of fruits and vegetables, a family will clearly be worse off if they ate the recommended amounts of conventionally grown produce. Second, people may exert pressure on government agencies to focus excessively on addressing negligible risks while placing too little effort on reducing larger risks.[14]

Gray's concerns raise the possibility that the members of the public may be wrong in their gauging of a project's risks, thus complicating the task of the engineer in presenting alternatives. In fact, the community may be choosing poorly in their assessment of *risk trade-offs*. The engineer's competency may run up against pleasing the client (both at least tacitly required in most engineering codes). The best alternative, such as siting a landfill in an unpopular location but in an ideal hydrological and environmental setting, is not simply going to be accepted by the neighbors. Nor can the engineer by *fiat* order the acceptance. This often calls for an arduous process of compromise wherein the engineer does not sacrifice what is dictated by expertise but reasonably and appropriately incorporates the needs of the community.

As shown in Table 4.1, risk assessment relies on problem identification, data analysis, and risk characterization, including cost/benefit ratios. Risk perception depends on thought processes, including intuition, personal experiences, and personal preferences. Engineers tend to be more comfortable operating in the middle column (using risk assessment processes), whereas the general public often uses the processes in the far right column. One can liken this to a "left-brained" engineer trying to communicate with a "right-brained" audience. It can be done, as long as preconceived and conventional approaches do not get in the way.

Our recent experience in a predominantly African American lower-socioeconomic-status community (i.e., an EJ community) in North Carolina is instructive. During one of the early scoping meetings regarding an environmental assessment, the plans for the early stages of the study were discussed. The engineers and scientists were explaining the need to be scientifically objective, to provide adequate quality assurance of the measurements, and to have a sound approach for testing hypotheses and handling data. We must admit that we thought going into the meeting that the subject matter was pretty "dry" and expected little concern or feedback. After the initial nod of approval to begin, we expected the neighborhood interest to pique only when the quality assured and validated data would be shared. However, during the scoping meetings, members of the community expressed concern about what we would do if we "found something." They wanted to know if we would begin interventions then and there. We were not prepared for these questions because we knew that the data were not truly acceptable until they had been validated and interpreted. So we recommended patience until the data met the scientists' requirements for rigor. The neighborhood representatives did not see it that way. At best, they thought we were naïve, and at worst, disingenuous. It seems that they had been "studied" before, with little action to follow these studies. They had been told previously some of the same things they were being told at our meeting. "Trust us!" We

Table 4.1 Differences between Risk Assessment and Risk Perception Processes

Analytical phase	Risk assessment processes	Risk perception processes
Identifying risk	Physical, chemical, and biological monitoring and measuring the event	Personal awareness
	Deductive reasoning	Intuition
	Statistical inference	
Estimating risk	Magnitude, frequency, and duration calculations	Personal experience
	Cost estimation and damage assessment	Intangible losses and nonmonetized valuation
	Economic costs	
Evaluating risk	Cost–benefit analysis	Personality factors
	Community policy analysis	Individual action

Source: Adapted from K. Smith, *Environmental Hazards: Assessing Risk and Reducing Disaster,* Routledge, London, 1992.

were applying rigorous scientific processes (middle column), which they had endured previously. Their concerns are explained by their experience and awareness (right-hand column). As a result, our flowcharts were changed to reflect the need to consider actions and interventions before project completion. This compromise was acceptable to all parties.

So both "lay" groups and our highly motivated and intelligent engineers and scientists can have difficulty in parsing perceived and real risks. The balance between risk assessment and risk perception will probably be a major challenge in many projects, especially in EJ communities. One more cautionary note: Sometimes, perception *is* reality.

To scientists and engineers at least, risk is a quantifiable concept: Risk equals the probability of some adverse outcome. Risks merely result from a straightforward function of probability and consequence.[15] The consequence can take many forms. In the medical and environmental sciences, it is called a *hazard*. Risk, then, is a function of the particular hazard and the chances of a person (or neighborhood or workplace or population) being exposed to the hazard. In the environmental business, this hazard often takes the form of toxicity, although other public health and environmental hazards abound.

Defining Risk[16]

The foregoing discussion is not to taken to mean that there is complete agreement within the scientific community on the meaning of risk. Most definitions *do* include a harmful outcome and the probability of that outcome occurring. That is, many definitions of risk include both the probability of an event *and* the consequences that could result from that event. This is a common definition in planning for catastrophic events, such as nuclear accidents or terrorist attacks. For example, Christine E. Wormuth, Senior Fellow of the International Security Program, has stated:

> In most formal discussions of risk assessment, risk is defined as the product of the probability that a certain event might occur . . . and the consequences that could result from such an event. The probability side of the equation is basically a combination of threats and vulnerabilities.[17]

There is some variation within engineering and technical circles regarding the definition of risk. Recently, Enrico Cameron and Gian Francesco Peloso[18] articulated a somewhat similar definition as that used by Wormuth, but place greater emphasis on the magnitude of the adverse consequence:

> Different events . . . can have adverse effects on human life, health, property, or the environment, and consequently constitute a risk. The concept of risk can be . . . considered as the product between the magnitude of such adverse effects, expressed numerically as the number of deaths, percentage increase in cancer cases, property value loss, and so on, and the likelihood that the event causing will occur or, alternatively, the subsequent likelihood that the consequences themselves will occur.

They later state:

> Risk r will simply be expressed as the product of magnitude m and likelihood l, so with the data coming from risk analysis n risks $r_i = m_i \cdot l_i$ can be calculated."

Biographical Sketch: Alice Hamilton

Alice Hamilton (1869–1970) graduated from both the Fort Wayne College of Medicine in Indiana and the University of Michigan School of Medicine. Following her medical degree in 1893, she did internships in Munich and Leipzig (being allowed to sit in a class of all men as long as she did not make herself conspicuous).

Some of her early experiences were working in a settlement house in Chicago, where she started educational programs and clinics for the destitute. In 1902 she recognized the connection between waste disposal and typhoid fever and was able to initiate changes in the city health department. In her work with the poor, she noted the connection between unsafe conditions at work and the health of the workers, and in 1910, became the director of the new Occupational Disease Commission.

From there she moved in 1919 to the faculty of the Harvard Medical School, where she founded the program in occupational medicine (and was the only female member of the faculty, and was appointed only on the condition that she not join the faculty club.). She was a leading participant in two occupational controversies, the leaded gasoline debate and the health of the radium dial painters (known as the "radium girls"). In the leaded gasoline debate, she showed how lead can accumulate in the bones and fought against industry claims that there is a natural threshold of lead in the human body. She fought unsuccessfully the introduction of lead to gasoline in the 1920s, and her work was not vindicated until the 1970s, when lead in gasoline was finally banned (ironically, more for air pollution control reasons than human health). In the radium dial painters controversy, Hamilton's epidemiological studies showed how radiation exposure to women painting glow-in-the-dark watch dials was causing a high incidence of cancer.

Hamilton is acknowledged to be the founder of occupational medicine, and during her long lifetime received many honors and awards. In addition to her work in lead and radium, she paved the way for understanding numerous other environmental contaminants, including mercury (mad hatter's disease), organic solvents, and microbes (e.g., connecting typhoid fever to sewage). In 1944 she was listed in *Men of Science,* which must have caused her to chuckle.

Within this connotation of risk, there is a choice of whether to include the probability of the event or the consequence of the event. In many environmental risk assessments, it is the consequence, such as the added cancer cases or number of deaths (mortality rates) that are included in the risk equation. Also, the shorthand in this text and numerous other environmental risk documents is to present risk as a unitless value, e.g., one in a million or 10^{-6}. However, the consequence is understood. Although the risk is stated as a unitless fraction, such as a cancer risk of 10^{-6}, that numerical expression implies that the units, in fact, are number of cancer cases (i.e., one added case per million people exposed).

The National Research Council's Committee on Risk Perception and Communication[19] has defined risk as:

> . . . the product of a measure of the size of the hazard and its probability of occurrence. Regardless of how numerical estimates are made the essence of the distinction between hazard and risk is that "risk" takes probability explicitly into account.

This definition "adds the hazard and its magnitude the probability that the potential harm or undesirable consequence will be realized."

The National Research Council's Committee on Risk Characterization[20] has defined risk as:

> A concept used to give meaning to things, forces, or circumstances that pose danger to people or to what they value. Descriptions of risk are typically stated in terms of the likelihood of harm or loss from a *hazard* and usually include: an identification of what is "at risk" and may be harmed or lost (e.g., health of human beings or an ecosystem, personal property, quality of life, ability to carry on an economic activity); the hazard that may occasion this loss; and a judgment about the likelihood that harm will occur.

The importance of both the type of adverse outcome and its probability of occurrence is succinctly captured by Rasmussen:[21]

> The term risk usually expresses not only the potential for an undesired consequence but also how probable it is that such a consequence will occur. . . . . A mathematical definition of risk commonly found in the literature is

> Risk (Consequence/unit time) = Frequency (event/unit time)

> $\times$ Magnitude (consequence/event)

Thus, there are numerous ways to express risk quantitatively. All have an expression of probability and either explicitly or implicitly an expression of consequence.

The difference between hazard and risk can be demonstrated by two students in an engineering ethics class. Jan has made A's in all of her engineering and elective courses, including prerequisites for the ethics course. She has taken copious notes, has completed all of her homework assignments, and participates in study groups every Thursday evening. Dean, on the other hand, has taken only one of the three prerequisite courses, receiving a D. He has completed only half of his homework assignments and does not participate in study groups. Jan and Dean share the same hazard (i.e., flunking the course). However, based on the data, we would consider their risks of flunking to be very different, Dean's being much greater. Of course, this does not mean that Dean *will* flunk, or even that Jan *will* pass. It merely indicates that the probability is more likely that Dean will fail the course than will Jan. Even an A student has the slim chance of failing the course (e.g., may experience testing anxiety, may have personal problems the week of the final), just as a failing student has a slim chance of passing the course (e.g., becomes motivated, catches up on homework, reaches a state of illumination, correctly recognizes a pattern on the answer sheet). This is why there is seldom a "sure thing" (i.e., 100% probability) in risk assessment. However, the risk difference between Jan and Dean can be very large: say, 0.0001 for Jan and 0.85 for Dean.

The example also illustrates the concept of risk mitigation. For example, if Dean does begin to take actions, he can decrease the probability (i.e., risk). Perhaps by participating in a study group he decreases the risk of flunking to 50%, and by also catching up on his homework, the risk drops to 20%. These two risk abatement actions lowered his risk by 65%.

To illustrate further the difference between hazard and risk, let us consider an environmental example: a "highly exposed person" *versus* a person with very low exposure. Leinad works in a lead foundry, is removing lead-containing paint from his home walls, drinks from a private well with average lead concentrations of 10 mg L^{-1}, and in his spare time breaks down automobile batteries to remove the lead cores. Enraa is of the same gender and age as Leinad, but Enraa's only exposure to lead is from the public drinking water supply, which on average is 0.001 mg L^{-1}. Lead is well known to be neurotoxic: It causes damage to the central and peripheral nervous systems of mammals, including humans. The hazard in this instance is neurotoxicity. The hazard is identical for Leinad and Enraa, nervous system disorders. However, the neurotoxic risk to Leinad is orders of magnitude higher than the neurotoxic risk to Enraa.

The chemical concentration is part of the risk equation. However, the actual exposure (beyond mere ambient concentration or even dose) is influenced by activities (e.g., working, touching, drinking, and breathing in different situations). Several of Leinad's activities would be greater than the 99th exposure percentile. A good source of information about such activities is the U.S. Environmental Protection Agency's *Exposure Factors Handbook,*[22] which summarizes statistical data on the different activities and other factors related to how people are exposed to contaminants, including:

- Drinking water consumption
- Soil ingestion
- Inhalation rates
- Dermal factors, such as skin area and soil adherence factors
- Consumption of fruits and vegetables, fish, meats, dairy products, and homegrown foods
- Breast milk intake
- Human activity factors
- Consumer product use
- Residential characteristics

The handbook provides the recommended exposure values for the general population as well as for highly exposed and environmentally susceptible subpopulations. Such differences are especially crucial for environmental justice projects. Often, the default is to calculate average exposures and risks, but actual conditions may be at levels one or two standard deviations higher than measures of central tendency (mean, median, or mode), out in the tail of the distribution. After all, environmental justice communities are, by definition, exposed to contaminants disproportionately compared to the general population. Certain minority subpopulations have higher body burdens of persistent toxicants than the burdens found in the general population. For example, subsistent fishing and hunting is more common in Inuit populations in the Arctic regions of North America. Tissue concentrations of PCBs and toxic compounds in fish and top predators (e.g., polar bears) have increased dramatically in the past five decades.[23] Thus, the PCB body burden of the Inuit has also increased. Merely advising a change in activities, such as no longer hunting or fishing, may not only be infeasible (e.g., these may substantially represent the food source), but such recommendations may militate against traditional, even religious or spiritual, mores of the people. Thus, decisions about acceptable levels of exposures to PCBs for Inuit people must take into account the already elevated levels, and risk management must incorporate numerous social factors.

Risks appear in a complex social arrangement; they do not occur in a vacuum. Taking care of one risk can, if we are lucky, ameliorate another risk, such as when pollution control equipment removes particles and in the process also removes heavy metals that are sorbed to the particles. This means that not only are risks to heart and lung diseases reduced, but neurological risks are also reduced because of the decrease in exposures to lead, mercury, and other neurotoxic metals. Conversely, reducing one risk can, if we are unlucky, increase other risks, such as when solid waste is incinerated, eliminating the possibility of long-term risks from contaminated groundwater while escalating the concentrations of products of incomplete combustion in the air as well as creating bottom ash with very high concentrations of toxic metals.

Another problem occurs when one exposed group is exchanged for another. For example, to address the concern of possible exposures of building inhabitants to asbestos in building materials, we are likely to create occupational asbestos exposures to workers called in to remove the materials. This is always a consideration in engineering management: that is, the acceptable amount to which workers in a remediation or emergency response project should be exposed. Obviously, risk abatement measures such as respirators, protective clothing, and other measures must be part of any hazardous situation. Another example of risk shifting is that of environmental justice situations, when the overall population risk is lowered by moving contaminants to sparsely populated regions, but the risk to certain groups is in fact increased. This type of risk shifting can also occur internationally, such as when a nation decides that it does not want its population or ecosystems to be exposed to a hazardous substance but still allows the manufacture and shipping of the substance outside its borders[24] (see the discussion of risk shifting in Chapter 3).

Another example of risk shifting is a decision made in one part of world's impact on another remote region. Consider *persistent organic pollutants* (POPs), most of which are organochlorine compounds. In addition to the global problem of long-range transport, these compounds present abundant lessons on how to address local problems with risk trade-offs. As mentioned, subsistence anglers and hunters receive heavy doses of these substances in their food. Pregnant and lactating women in these regions often have elevated concentrations of PCBs, dioxins, and other POPs in their fats and breast milk. What can we learn from this? First, the engineer must ensure that recommendations are based on sound science. Although seemingly obvious, this lesson is seldom easy to put into practice. Sound science can be trumped by perceived risk, such as when a chemical with an ominous-sounding name is uncovered in a community, leading the neighbors to call for its removal. However, the toxicity may belie the name. The chemical may have very low acute toxicity, has never been associated with cancer in any animal or human studies, and is not regulated by any agency. This hardly allays the neighbors' fears. The engineer's job is not done by declaring that removal of the chemical is not necessary, even though the declaration is absolutely right. The community deserves clear and understandable information before we can expect any capitulation.

Second, removal and remediation efforts are never entirely risk-free. To some extent they always represent risk shifting in time and space. A spike in exposures is possible during the early stages of removal and treatment, as the chemical may have been in a place and form that made this less available until actions were taken. Due in part to this initial exposure, the concept of *natural attenuation* has recently gained greater acceptance within the environmental community. However, the engineer should expect some resistance from the local community when they are informed that the best solution is to do little or nothing but to allow nature (i.e., indigenous microbes) to take its course (doing nothing could be interpreted as intellectual laziness!).

Third, the mathematics of benefits and costs is inexact. Finding the best engineering solution is seldom captured with a benefit/cost ratio. Opportunity costs and risks are associated with taking no action (e.g., the recent Hurricane Katrina disaster presents an opportunity to save valuable wetlands and to enhance a shoreline by *not developing and not rebuilding* major portions of the Gulf region). The costs in time and money are not the only reasons for avoiding an environmental action. Constructing a new wetland or adding sand to the shoreline could inadvertently attract tourists and other users who could end up presenting new and greater threats to the community's environment.

Health costs are also not simply a matter of benefits *versus* cost. In addition, they often require that one risk be traded for another. Stakeholders must be fully aware of the pros and cons to make informed decisions.

Discussion: Informed Consent

If you have had a medical procedure recently, you were probably asked to sign a form that says that you understand the risks associated with the procedure. The probability of each adverse outcome (harm) is delineated with a percentage or some other expression of odds. For example, your operation may have a 1 to 5% chance of fever and extended hospital stay, a 0.1% chance of some hearing loss, and a 0.0001% chance of death. In other words, epidemiologists have found that complications in the type of surgery you are about to receive results in the death of 1 in a million cases. This is not the same as *your* risk, which is a function of your own vulnerabilities and strengths. It is probably a general reflection of *all* cases. If you are 25 years of age and in good health, your individual risk of death is much lower than that of an 89-year-old cancer patient. So your stratum of the population may have a 1 in a 100 million chance of death, and the elderly, ill stratum a 1 in 500 chance. However, there is still a chance that the older person will live and you won't. In statistics, *you* don't matter.

So if such information is so easily misunderstood, why is it given to everyone (old, young, educated, illiterate, citizen status, etc.)?[25] The answer is something known as *informed consent.* Much of this goes back to aftermath of Nazi Germany's unethical treatment of prisoners of war, especially the "medical" procedures to which they were subjected. Other unethical medical treatments took place in the United States under the banner of "research," such as the withholding of treatment and dishonesty of researchers in "treating"[26] syphilis in African Americans in Tuskegee, Mississippi (which began in 1930 and lasted 42 years), mistreatment of mentally handicapped patients, and sterilization of poor women of childbearing age. Such inhumane and inhuman practices cried for increased scrutiny and accountability of medical and other scientific research involving human subjects. One of the mandates that has become well established is that patient, client, or subject be thoroughly informed about any risks. As we see in this and the next chapter, however, risk assessment and management are quite complicated, even for those of us in the business of risk.

Certainly, people need to be informed about the risks of important decisions, and medical decisions are right near the top of most decisions we make. However, what is the sufficient amount of information needed to make such a decision? If we want to ask a third-grader to be included in a clinical trial, prudence and practical experience tells us that she will not be sufficiently prepared intellectually and morally for such a decision. We delegate such decisions to her guardian, as defined by regulation and the courts. However, who says that the guardian is sufficiently informed?

There is another problem that does not seem to get much attention. How many people are needlessly frightened away from certain necessary procedures because they see these percentages? When is professional care compromised in the name of full disclosure? Full disclosure is valuable only if the information is interpreted properly and precisely by the person making the decision.

We hope only a small number of people are so easily impressionable and susceptible. Is it possible that those seeing these odds of harm may simply give up or be so overcome with fear that the medical procedure is compromised? Of course, this is not an argument for the "ignorance is bliss" approach, but it is a warning that physicians (and engineers) are in a position of trust and have an enormous impact on their patients' (and clients') attitudes and outlooks.

One of the most difficult risk numbers for engineers to explain is that of the 100-year flood. Informed consent dictates that we tell people living in a certain part of town that the likelihood of a flood reaching their property line in one such flood is expected every 100 years. However, the people living there know good and well that they have had five floods in the past 20 years! We have listened to officials from U.S. Army Corps of Engineers, the Federal Emergency Management Administration, and various state agencies trying to explain this conundrum to local people. They may start by invoking the "geologic time" argument (i.e., yes, your five floods may have occurred over the past 20 years, but in a few million years there will be many more 20-year spans with no floods). This may be followed by a short course in statistics. The 100-year flood is a mathematical construction. It is a measure of central statistical tendencies, exactly like an arithmetic mean. Although we can sympathize with the engineer trying to explain this concept, we are not the least bit surprised when the local people are not impressed.

Informed consent spills over into all of our engineering specifications and accountability. Liners that leak, water treatment that fails, odorless landfills that stink, and pesticides that lead to previously unexpected health effects are all fuel for the public's skepticism and discontent. We informed, hopefully with the best available data and knowledge, but subsequent events show that we missed a few things. Sometimes, these are important. Sometimes, they are the difference between success and failure. But the public may believe that their consent was breached. They may feel that it was really "misinformed consent."

A tool needed to address contravening risk is optimization, with which engineers are quite familiar. Unfortunately, the greater the number of contravening risks that are possible, the more complicated our optimization routine becomes.

The concept of risk trade-off is a very common phenomenon in everyone's life. For example, local governments enforce building codes to protect health and safety. Often, these added protections are associated with indirect, countervailing risks. For example, the costs of construction may increase safety risks via income and stock effects. The *income effect* results from pulling money away from family income to pay the higher mortgages, making it more difficult for the family to buy other items or services that would have protected them. The *stock effect* results when the cost of the home is increased and families have to wait to purchase a new residence, so they are left in substandard housing longer.[27] Such countervailing risks are common in environmental decisions, such as arguments for greater amounts of open space and green areas in communities, with the overall effect of increasing median housing costs and making housing less affordable. Arguing for major environmental standards is tantamount to arguing for increased risks from income and stock effects by imposing increased environmental controls. In fact, some people opposing higher-density housing are in effect calling for less standing housing stock. Many of these same people would sign petitions calling for more affordable housing, but in the interests of protecting the environment in their own neighborhoods, they are in fact making housing less affordable. This is but one of example of how the planner and engineer are frequently asked to optimize two or more conflicting variables in environmental justice situations.

Reliability: A Metric of Socially Responsible Engineering

Like risk, reliability is an expression of probability, but instead of conveying something bad, it expresses the likelihood of a good, or at least a *desired,* outcome. *Reliability* is the extent to which something can be trusted. A system, process, or item is reliable as long as it performs the designed function under the conditions specified during a certain time period. In most engineering applications, reliability means that what we design will not fail prematurely. Or, stated more positively, reliability is the mathematical expression of success; that is, reliability is the probability that something that is in operation at time zero (t_0) will still be operating until the designed life (time $t = t_l$). As such, it is also a measure of an engineer's social accountability. People in neighborhoods near the proposed location of a proposed facility want to know if it will work and will not fail. This is especially true for those facilities that may affect the environment, such as incinerators, treatment facilities, landfills and power plants. Similarly, when environmental cleanup is being proposed, people want to know how certain the engineers are that the cleanup will be successful.

Biographical Sketch: Herbert Needleman

The heavy metal lead (Pb) is an extremely potent neurotoxin (i.e., a substance that damages nerve cells). Like several other heavy metals, lead interferes with physiological processes because, when ionized, divalent lead (Pb^{2+}) acts in many ways like divalent calcium (Ca^{2+}). Due to its larger size and other chemical differences, however, Pb^{2+} induces biological effects that differ from those of Ca^{2+}. For example, during gestation and in early childhood, the developing brain is harmed when Pb^{2+}, competing with Ca^{2+}, induces the release of a neurotransmitter in elevated amounts and at the wrong time (e.g., during basal intervals, when a person is at rest). Thus, at high lead exposures, a person may have abnormally high amounts of brain activity (when it should be lower), and conversely, when a neural response is expected, little or no increase in brain activity is observed. This may induce chronic effects when synaptic connections in the brain are truncated during early brain development.

Lead also adversely affects the release of the transmitter glutamate, which is involved in brain activities associated with learning. The N-methyl-D-aspartate (NMDA) receptor seems to be selectively blocked when lead is present. Other lead effects include the activation of protein kinase C (PKC) because PKC apparently has a greater affinity for lead than for the normal physiological activator, divalent calcium. This, complicates and exacerbates the other neurotransmitter effects and harms the cell's chemical messaging (i.e., second-messenger systems), synthesis of proteins, and genetic expression.

All these neurological effects, especially in the developing brain, began to be documented in earnest by the medical community only within the last half century. Enter Herbert Needleman, for whom the pervasive effects of lead and its compounds on the health of children has been a lifelong concern. Needleman, a pediatrician at the University of Pittsburgh Medical Center, discovered that a correlation existed between the amount of lead in the teeth of infants and their intelligence at age 16, as measured by their IQ scores. His research has shown a dose–response correlation between lead dose and IQ: that is, the higher the lead content, the lower the IQ in these teenagers. In a series of follow-up studies, Needleman determined that lead poisoning had long-term implications for a child's attentiveness, behavior, and school success.

But Needleman was not the type of scientist who simply published his papers and waited for others to implement his findings. He recognized that immediate action was required while scientific assessments continued (advice we have often heard from members of the environmental justice communities[28]). Needleman initiated a campaign to remove tetraethyllead from gasoline, to phase out lead-based paints, and to reduce exposure in houses where kids can chew on the paint chips.[29] The results have been dramatic, with average blood lead levels in this country dropping an estimated 78% from 1976 to 1991.

Not surprisingly, the lead industry has been highly critical of Needleman and his research, and even has alleged scientific fraud and misconduct charges, against which Needleman has defended himself successfully. Throughout his professional life, Needleman has remained a consistent advocate for the cause of eradicating pediatric lead poisoning.

The probability of a failure per unit time is the *hazard rate,* a term familiar to environmental risk assessment, but many engineers may recognize it as a *failure density,* or $f(t)$. This is a function of the likelihood that an adverse outcome will occur, but note that it is not a function of the severity of the outcome. $f(t)$ is not affected by whether the outcome is very severe (such as pancreatic cancer and loss of an entire species) or relatively benign (muscle soreness or minor leaf damage). The likelihood that something will fail at a given time interval can be found by integrating the hazard rate over a defined time interval:

$$P\{t_1 \leq T_f \leq t_2\} = \int_{t_1}^{t_2} f(t)\, dt \tag{4.5}$$

where T_f is the time of failure. Thus, the reliability function $R(t)$ of a system at time t is the cumulative probability that the system has not failed in the time interval t_0 to t_t:

$$R(t) = P\{T_f \geq t\} = 1 - \int_0^t f(x)\, dx \tag{4.6}$$

Engineers must be humble, since everything we design *will* fail. We can improve reliability by extending the time (increasing t_t), thereby making the system more resistant to failure. For example, proper engineering design of a landfill barrier can decrease the flow of contaminated water between the contents of the landfill and the surrounding aquifer (e.g., a velocity of a few micronmeters per decade). However, the barrier does not eliminate failure completely [i.e., $R(t) = 0$]; it simply protracts the time before the failure occurs (increases T_f).[30]

Equation (4.2) illustrates built-in vulnerabilities such as unfair facility siting practices or the inclusion of inappropriate design criteria; like cultural bias, the time of failure is shortened. Like pollution, environmental injustice is a type of inefficiency. If we do not recognize these inefficiencies upfront, we will pay by premature failures (e.g., lawsuits; unhappy clients; a public that has not been well served in terms of our holding paramount their health, safety, and welfare).

Reliability engineering, a discipline within engineering, considers the expected or actual reliability of a process, system, or piece of equipment to identify the actions needed to reduce failures, and once a failure occurs, how to manage the effects expected from that failure. Thus, reliability is the mirror image of failure. Since risk is really the probability of failure (i.e., the probability that our system, process, or equipment will fail), risk and reliability are two sides of the same coin. Recall from our discussion of the five types of failure in Chapter 3 that it may come in many forms and from many sources. Injustice is a social failure. A tank leaking chemicals into groundwater is an engineering failure, as is exposure to carcinogens in the air, water, and food. A system that protects one group of people at the expense of another is a type of failure. So if we are to have reliable engineering, we need to make sure that whatever we design, build, and operate is done with fairness. Otherwise, these systems are, by definition, unreliable.

The most common graphical representation of engineering reliability is the *bathtub curve* (Figure 4.3). The U shape indicates that failure is more likely to occur at the beginning (infant mortality) and near the end of the life of a system, process, or equipment. Actually, the curve indicates engineers' common proclivity to compartmentalize. We are tempted to believe that the process begins only after we are called on to design a solution. Indeed, failure can occur even before infancy. In fact, many problems in environmental justice occur during the planning and idea stage. A great idea may be shot down before it is born.

Injustices can gestate even before an engineer becomes involved in a project. This "miscarriage of justice" follows the physiological metaphor closely. Historically, certain groups of people have been excluded from preliminary discussions, so that if and when they do become involved, they are well beyond the "power curve" and have to play catch-up. The momentum of a project, often being pushed by project engineers, makes participation very difficult from some groups, so we can modify the bathtub distribution accordingly. Figure 4.4 shows that the rate of failure is highest during gestation. This may or may not be the case, since identifying the number of premature failures is extremely difficult to document with any degree of certainty.

Another good way to visualize reliability as it pertains to socially responsible engineering is to link potential causes to effects. *Cause-and-effect diagrams* (also known as *Ishikawa diagrams*) identify and characterize the totality of causes or events that contribute to a specified outcome event. A *fishbone diagram* (see Figure 4.5) arranges the categories of all causative factors according to their importance (i.e., their share of

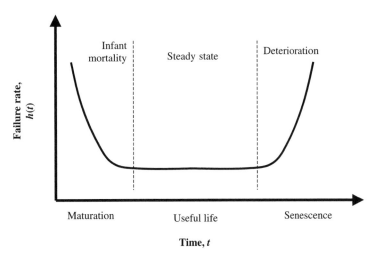

Figure 4.3 Prototypical reliability curve: the bathtub distribution. The highest rates of failure, $h(t)$, occur during the early stages of adoption (infant mortality) and when the systems, processes, or equipment become obsolete or begin to deteriorate. For well-designed systems, the steady-state period can be protracted (e.g., decades). (From D. A. Vallero, *Paradigms Lost: Learning from Environmental Mistakes, Mishaps, and Misdeeds,* Butterworth-Heinemann, Burlington, MA, 2006.)

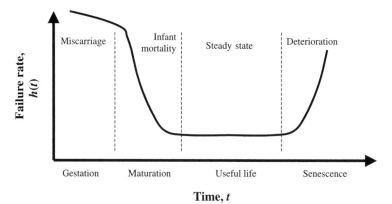

Figure 4.4 Prototypical reliability curve with a gestation (e.g., idea) stage. The highest rate of failure, $h(t)$, can occur even before the system, process, or equipment has been made a reality. Exclusion of people from decision making or failure to get input about key scientific or social variables can create a high level of hazard. (From D. A. Vallero, *Paradigms Lost: Learning from Environmental Mistakes, Mishaps, and Misdeeds,* Butterworth-Heinemann, Burlington, MA, 2006.)

the cause). The construction of this diagram begins with the failure event to the far right (i.e., the "head" of the fish), followed by the "spine" (flow of events leading to the failure). The "bones" are each of the contributing categories. This can be a very effective tool in explaining failures to communities, especially if the engineer constructs the diagrams with input from neighbors. Even better, the engineer may construct the diagrams in real time in a community meeting. This will help prevent recurring accidents, releases, and other failures in which contributing causes have been ignored (an all too common occurrence in environmental justice communities, whose members often lament that "nobody listened").

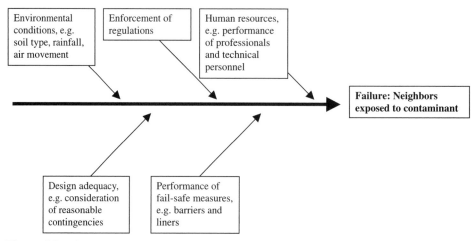

Figure 4.5 Fishbone reliability diagram, showing contributing causes to an adverse outcome (exposure to an environmental contaminant).

The premise behind cause-and-effect diagrams such as the fishbones and fault trees is that all the causes have to connect through a logic gate. This is not always the case, so another more qualitative tool may need to be used, such as the Bayesian belief network (BBN). Like the fishbone, the BBN starts with a failure (see Figure 4.6). Next, the most immediate contributing causes are linked to the failure event. The next group of factors that led to the immediate causes is then identified, followed by the remaining contributing groups. This diagram helps to catalog the contributing factors and also compares how one group of factors affects the others. Again, this can be an effective tool for gathering information about causes from neighbors as well as from government agencies, industries, and other stakeholders.

The engineering and scientific communities often use the same terms for different concepts. This is the case for reliability. Environmental engineering and other empirical sciences commonly use the term *reliability* to indicate quality, especially for data derived from measurements, including environmental and health data. In this use, reliability is defined as the degree to which measured results are dependable and consistent with respect to the study objectives (e.g., stream water quality). This specific connotation is sometimes called *test reliability,* in that it indicates the consistency of measured values over time as well as, how these values compare to other measured values, and how they differ when other tests are applied. Like engineering reliability, test reliability is a matter of trust. As such, it is often paired with test validity, that is, just how near the true value (as indicated by some type of known standard) the measured value is. The less reliable and valid the results, the less confidence scientists and engineers have in interpreting and using them. This is very important in engineering communications generally, and risk communications specifically.

The engineer must know just how reliable and valid the data are and must communicate this properly to clients and the public. This means that however discomfiting,

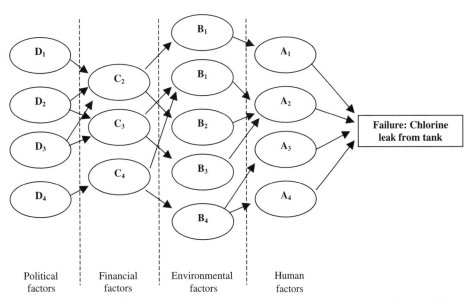

Figure 4.6   Bayesian belief network, with three groups of contributing causes leading to a failure.

the engineer must "come clean" about all uncertainties. Uncertainties are ubiquitous in risk assessment. The engineer should take care not to be overly optimistic or overly pessimistic about what is known and what needs to be done. Full disclosure is simply an honest rendering of what is known and what is lacking for those listening to make informed decisions. Part of the uncertainty involves conveying the meaning; we must communicate the potential risks clearly. A word or phrase can be interpreted in many ways. Engineers should liken themselves to physicians writing prescriptions. Be completely clear; otherwise, confusion may result and lead to unintended, negative consequences.

As evidenced from our discussions of disasters, risk trade-offs, and reliability, the concept of safety is laden with value judgments. Thus, ethical and just environmental decisions must rely on both sound science and quantifiable risk assessment, balanced with social fairness. In the following section we introduce a number of the elements of sound science, together with case studies to demonstrate what happens when credible risk assessment is not matched with justice.

THE ELEMENTS OF ENVIRONMENTAL RISK

Risk is a quantifiable engineering concept, and in its simplest form, risk (R) is the product of the hazard (H) and the exposure (E) to that hazard:

$$R = H \times E \tag{4.7}$$

Environmental risk assessment consists of a number of steps, described below.

Hazard Identification

A *hazard* is anything with the potential for causing harm. Ice is a slipping hazard. Sharps (e.g., syringe needles) are infection hazards. Pesticides are health hazards. A hazard is an intrinsic property of a substance, product, or process (i.e., a concept of potential harm). For example, a chemical hazard is an absolute expression of a substance's properties, since all substances have unique physical and chemical properties. These properties can render the substance to be hazardous. Conversely, equation (4.7) shows that risk can only occur with exposure. So if you are walking on a street in the summer, the likelihood of your slipping on ice is near zero. Your total slipping risk is not necessarily zero (e.g., you could step on an oily surface or someone may have thrown out ice from a freezer). If you are not in a medical facility, your infection risk from sharps may be near zero, but your total infection risk is not zero (e.g., you may be exposed to the same infection from a person sneezing in your office). If you do not use pesticides, your pesticide health risk is also lower. However, since certain pesticides are persistent and can remain in the food chain, your exposure is not zero. Also, even if your pesticide exposure is near zero, your cancer risk is not zero, since you may be exposed to other cancer hazards.

Engineers and scientists working in environmental areas consider a number of hazards; the most common is toxicity. Other important environmental hazards are shown in Table 4.2, such as landfills, storage facilities, and hazardous waste sites. Hazards can be expressed according to the physical and chemical characteristics, as in Table 4.2, as well as in the ways they may affect living things. For example, Table 4.3 summarizes some of the expressions of biologically based criteria of hazards. Other hazards, such as flammability, are also important to environmental engineering. However, the chief hazard in most environmental justice cases has been toxicity.

The first means of determining exposure is to identify *dose,* the amount (e.g., mass) of a contaminant that comes into contact with an organism. Dose can be the amount administered to an organism (called the *applied dose*), the amount of the contaminant that enters the organism (the *internal dose*), the amount of the contaminant that is absorbed by an organism over a certain time interval (the *absorbed dose*), or the amount of the contaminants or its metabolites that reach a particular "target" organ (the *biologically effective dose* or simply *bioeffective dose*), such as the amount of a hepatotoxin (a chemical that harms the liver) that finds its way to liver cells or a neurotoxin (a chemical that harms the nervous system) that reaches the brain or other nervous system cells. Theoretically, the higher the concentration of a hazardous substance that comes into contact with an organism, the greater the adverse outcome expected. The pharmacological and toxicological gradient is called the *dose–response curve* (see Figure 4.7). Generally, increasing the amount of the dose means a greater incidence of the adverse outcome.

Dose–response assessment generally follows a sequence of five steps:[31]

1. Fitting the experimental dose–response data from animal and human studies with a mathematical model that fits the data reasonably well

2. Expressing the upper confidence limit (e.g., 95%) line equation for the mathematical model selected

3. Extrapolating the confidence limit line to a response point just below the lowest measured response in the experimental point (known as the *point of departure*): the beginning of the extrapolation to lower doses from actual measurements

4. Assuming that the response is a linear function of dose from the point of departure to zero response at zero dose

5. Calculating the dose on the line that is estimated to produce the response

The curves in Figure 4.7 represent those generally found for toxic chemicals.[32] Once a substance is suspected of being toxic, the extent and quantification of that hazard is assessed.[33] This step is frequently referred to as a *dose–response evaluation* because this is when researchers study the relationship between the mass or concentration (i.e., dose) and the damage caused (i.e., response). Many dose–response studies are ascertained from animal studies (*in vivo* toxicological studies), but they may also be inferred from studies of human populations (epidemiology). To some degree, "Petri dish" (i.e., *in vitro*) studies, such as mutagenicity studies like the Ames test[34] of bacteria, complement dose–response assessments, but they are used primarily for screening and qualitative or, at best, semiquantitative analysis of responses to substances. The actual name of the Ames test is the Ames *Salmonella*/microsome mutagenicity assay, and it shows the short-term reverse mutation in histidine-dependent *Salmonella* strains of bacteria. Its main use is to screen for a broad range of chemicals that induce genetic aberrations leading to genetic mutations. The process works by using a culture that allows only those bacteria whose genes

Table 4.2   Hazards Defined by the Resource Conservation and Recovery Act

Hazard type	Criteria	Physical/chemical classes in definition
Corrosivity	A substance with an ability to destroy tissue by chemical reactions.	Included are acids, bases, and salts of strong acids and strong bases. The waste dissolves metals and other materials, or burns the skin. Examples include rust removers, waste acid, alkaline cleaning fluids, and waste battery fluids. Corrosive wastes have a pH < 2.0 or > 12.5. The U.S. EPA waste code for corrosive wastes is D002.
Ignitability	A substance that oxidizes readily by burning.	This group includes any substance that combusts spontaneously at 54.3°C in air or at any temperature in water, or any strong oxidizer. Examples are paint and coating wastes, some degreasers, and other solvents. The U.S. EPA waste code for ignitable wastes is D001.
Reactivity	A substance that can react, detonate, or decompose explosively at environmental temperatures and pressures.	A reaction usually requires a strong initiator [e.g., an explosive like TNT (trinitrotoluene)], confined heat (e.g., saltpeter in gunpowder), or explosive reactions with water (e.g., Na). A reactive waste is unstable and can react rapidly or violently with water or other substances. Examples include wastes from cyanide-based plating operations, bleaches, waste oxidizers, and waste explosives. The U.S. EPA waste code for reactive wastes is D003.
Toxicity	A substance that causes harm to organisms. Acutely toxic substances elicit harm soon after exposure (e.g., highly toxic pesticides causing neurological damage within hours after exposure). Chronically toxic substances elicit harm after a long period of time of exposure (e.g., carcinogens, immunosuppressants, endocrine disruptors, and chronic neurotoxins).	Toxic chemicals include pesticides, heavy metals, and mobile or volatile compounds that migrate readily, as determined by the toxicity characteristic leaching procedure (TCLP): a TC waste. TC wastes are designated with the waste codes D004 through D043.

Table 4.3 Biologically Based Classification Criteria for Chemical Substances

Criterion	Description
Bioconcentration	The process by which living organisms concentrate a chemical contaminant to levels exceeding the surrounding environmental media (e.g., water, air, soil, or sediment).
Lethal dose (LD)	A dose of a contaminant calculated to expect a certain percentage of a population of an organism (e.g., minnow) exposed through a route other than respiration (dose units are milligrams of contaminant per kilogram of body weight). The most common metric from a bioassay is the lethal dose 50 (LD_{50}), wherein 50% of a population exposed to a contaminant is killed.
Lethal concentration (LC)	A calculated concentration of a contaminant in the air that when respired for 4 hours (i.e., exposure duration = 4 h) by a population of an organism (e.g., rat) will kill a certain percentage of that population. The most common metric from a bioassay is the lethal concentration 50 (LC_{50}), wherein 50% of a population exposed to a contaminant is killed. (Air concentration units are milligrams of contaminant per liter of air.)

Source: P. A. Vesilind, J. Peirce, and R. F. Weiner, *Environmental Engineering,* 3rd ed., Butterworth-Heinemann, Boston, 1993.

revert to histidine interdependence to form colonies. As a mutagenic chemical is added to the culture, a biological gradient can usually be determined. That is, the more chemical that is added, the greater the number and size of colonies on the plate. The test is widely used to screen for the mutagenicity of new or modified chemicals and mixtures. It is also a "red flag" for carcinogenicity, since cancer is a genetic disease and a manifestation of mutations.

The toxicity criteria include both acute and chronic effects, and include both human and ecosystem effects. These criteria can be quantitative. For example, a manufacturer of a new chemical may have to show that there are no toxic effects in fish exposed to concentrations below 10 mg L^{-1}. If fish show effects at 9 mg L^{-1}, the new chemical would be considered to be toxic.

A contaminant is acutely toxic if it can cause damage with only a few doses. *Chronic toxicity* occurs when a person or ecosystem is exposed to a contaminant over a protracted period of time, with repeated exposures. The essential indication of toxicity is the dose–response curve. The curves in Figure 4.7 are sigmoidal because toxicity is often concentration dependent. As the doses increase, the response cannot stay mathematically linear (e.g., the toxic effect cannot double with each doubling of the dose). So the toxic effect continues to increase but at a decreasing rate (i.e., decreasing slope). Curve A is the classic cancer dose–response; that is, any amount of exposure to a cancer-causing agent may result in an expression of cancer at the cellular level (i.e., no safe level of exposure). Thus, the curve intercepts the *x*-axis at 0.

Curve B is a classic noncancer dose–response curve. The steepness of the three curves represents the potency or severity of the toxicity. For example, curve B is steeper

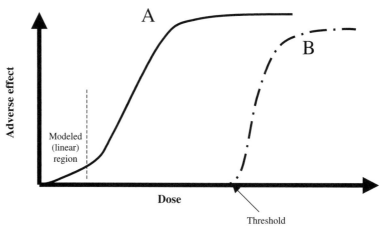

Figure 4.7 Prototypical dose–response curves. Curve A represents a no-threshold curve, which predicts a response (e.g., cancer) even if exposed to a single molecule (one-hit model). As shown, the low end of the curve, below which experimental data are available, is linear. Thus, curve A represents a linearized multistage model. Curve B represents toxicity above a certain threshold [no observable adverse effect level (NOAEL) is the level below which no response is expected]. Another threshold is the no observable effect concentration (NOEC), which is the highest concentration where no effect on survival is observed (NOEC$_{survival}$) or where no effect on growth or reproduction is observed (NOEC$_{growth}$). Note that both curves are sigmoidal in shape because of the saturation effect at high doses (i.e., less response with increasing dose). (Adapted from D. A. Vallero, *Environmental Contaminants: Assessment and Control,* Elsevier Academic Press, Burlington, MA, 2004.)

than curve A, so the adverse outcome (disease) caused by chemical in curve B is more potent than that of the chemical in curve A. Obviously, potency is only one factor in the risk. For example, a chemical may be very potent in its ability to elicit a rather innocuous effect, like a headache, and another chemical may have a rather gentle slope (lower potency) for a dreaded disease such as cancer.

With increasing potency, the range of response decreases. In other words, as shown in Figure 4.8, a severe response represented by a steep curve will be manifested in greater mortality or morbidity over a smaller range of dose. For example, an acutely toxic contaminant's dose that kills 50% of test animals (i.e., the LD$_{50}$) is closer to the dose that kills only 5% (LD$_5$) and the dose that kills 95% (LD$_{95}$) of the animals. The dose difference of a less acutely toxic contaminant will cover a broader range, with the differences between the LD$_{50}$ and LD$_5$ and LD$_{95}$ being more extended than that of the more acutely toxic substance.

The major differentiation of toxicity is between carcinogenic and noncancer outcomes. The term *noncancer* is commonly used to distinguish cancer outcomes (e.g., bladder cancer, leukemia, or adenocarcinoma of the lung) from other maladies, such as neurotoxicity, immune system disorders, and endocrine disruption. The policies of many regulatory agencies and international organizations treat cancer differently than noncancer effects, particularly in how the dose–response curves are drawn. As we saw in the introduction to the dose–response curves, there is no safe dose for carcinogens. Cancer dose–

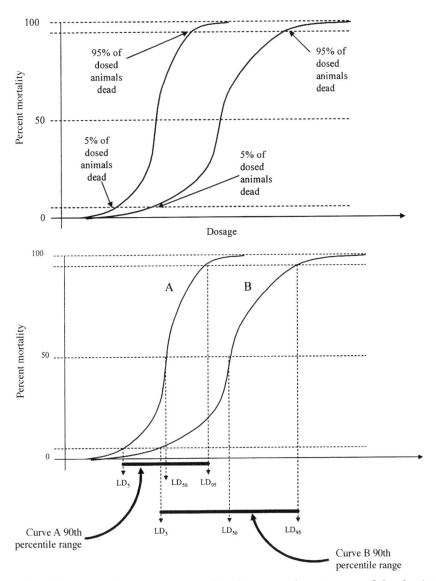

Figure 4.8 The greater the potency or severity of response (i.e., steepness of the slope) of a dose–response curve, the smaller the range of toxic response (90th percentile range shown in bottom graph). Also, note that both curves have thresholds and that curve B is less acutely toxic based on all three reported lethal doses (LD_5, LD_{50}, and LD_{95}). In fact, the LD_5 for curve A is nearly the same as the LD_{50} for curve B, meaning that about the same dose, contaminant A, kills nearly half the test animals, but contaminant B has killed only 5%. Thus, contaminant A is much more acutely toxic. (From D. A. Vallero, *Environmental Contaminants: Assessment and Control,* Elsevier Academic Press, Burlington, MA, 2004.)

response is almost always a nonthreshold curve (i.e., no safe dose is expected, whereas theoretically at least, noncancer outcomes can have a dose below which the adverse outcomes do not present themselves). So for all other diseases, safe doses of compounds can be established. These are known as *reference doses* (RfDs), usually based on the oral exposure route. If the substance is an air pollutant, the safe dose is known as the *reference concentration* (RfC), which is calculated in the same manner as the RfD, using units that apply to air (e.g., $\mu g\ m^{-3}$). These references are calculated from thresholds below which no adverse effect is observed in animal and human studies. If the models and data were perfect, the safe level would be the threshold, known as the *no observed adverse effect level* (NOAEL).

The term *noncancer* is very different from *anticancer* or *anticarcinogens*. Anticancer *procedures* include radiation and drugs that are used to attack tumor cells. *Anticarcinogens* are chemical substances that work against the processes that lead to cancer, such as antioxidants and essential substances that help the body's immune, hormonal, and other systems to prevent carcinogenesis.

In reality, the hazard identification and dose–response research is inexact and often has much uncertainty. Chief reasons for this uncertainty include variability among the animals and people being tested, as well as differences in response to the compound by different species (e.g., one species may have decreased adrenal gland activity, while another may show thyroid effects). Sometimes studies indicate only the lowest concentration of a contaminant that causes the effect—the *lowest observed adverse effect level* (LOAEL)—but the NOAEL is unknown. If the LOAEL is used, one is less certain how close this is to a safe level where no effect is expected. Often, there is temporal incongruence, such as most of the studies taking place in a shorter time frame than in real world exposures. Thus, acute or subchronic effects have to be used to estimate chronic diseases. Similarly, studies may have used different ways to administer the doses. For example, if the dose is oral but the pollutant is more likely to be inhaled by humans, this route-to-route extrapolation adds uncertainty. Finally, the data themselves may be weak because the study may lack sufficient quality, or the precision, accuracy, completeness, and representativeness of the data are unknown. These are quantified as *uncertainty factors* (UFs). *Modifying factors* (MFs) address uncertainties that are less explicit than UFs. Thus, any safe level must consider these uncertainties, so the RfD moves closer to zero; that is, the threshold is divided by these factors (usually, multiples of 10):

$$RfD = \frac{NOAEL}{UF \times MF} \tag{4.8}$$

Uncertainty can also come from error. Two errors can occur when information is interpreted in the absence of sound science. The first is the *false negative,* reporting that there is no problem when one in fact exists. The need to address this problem is often at the core of the positions taken by environmental and public health agencies and advocacy groups. They ask such questions as the following:

- What if the leak detector registers zero, but in fact, toxic substances are being released from the tank?
- What if this substance really does cause cancer but insufficient testing is the basis for this conclusion?

- What if people are being exposed to a contaminant, but via a pathway other than the ones being studied?
- What if there is a relationship that differs from the laboratory when this substance is released into the real world, such as the difference between how a chemical behaves in the human body by itself as opposed to when other chemicals are present (i.e., the problem of "complex mixtures")? This can either make for more toxicity (synergism) or less toxicity (antagonism).

The other concern is, conversely, the *false positive*. This can be a major challenge for public health agencies with the mandate to protect people from exposures to environmental contaminants. For example, what if previous evidence shows that an agency had listed a compound as a potential endocrine disruptor, only to find that a wealth of new information is now showing that it has no such effect? This can happen if the conclusions were based on faulty models or on models that work well only for lower organisms but subsequently developed models have taken into consideration the physical, chemical, and biological complexities of higher-level organisms, including humans. False positives may force public health officials to devote inordinate amounts of time and resources to deal with so-called *non-problems*. False positives also erroneously scare people about potentially useful products. False positives, especially when they occur frequently, create credibility gaps between engineers and scientists and the decision makers. In turn, the public, those whom we have been charged to protect, lose confidence in us as professionals.

To reduce the occurance of both false negatives and false positives, environmental risk assessment is in need of high-quality scientifically based information. Put in engineering language, the risk assessment process is a *critical path* in which any unacceptable error or uncertainty along the way will decrease the quality of the risk assessment and, quite likely, will lead to a bad environmental decision.

Reliable risk assessment begins with an understanding of the intrinsic properties of compounds, which render them more or less toxic. For example, polycyclic aromatic hydrocarbons (PAHs) are a family of large, flat compounds with repeating benzene structures. This structure makes them highly hydrophobic (i.e., fat soluble) and difficult for an organism to eliminate (since most blood and cellular fluids are mainly water). This property also enhances the PAHs' ability to insert themselves into the deoxyribonucleic acid (DNA) molecule, interfering with transcription and replication. This is why some large organic molecules can be mutagenic and carcinogenic. One of the most toxic PAHs is benzo[*a*]pyrene, which is found in cigarette smoke, combustion of coal, coke oven emissions, and numerous other processes that use combustion.

After a compound is released into the environment, its chemical structure can change substantially. Further, compounds change when taken up and metabolized by organisms. For example, methyl parathion, an insecticide used since 1954, has been associated with numerous farmworker poisonings. It has also been associated with health problems in environmental justice communities. Methyl parathion can cause rapid, fatal poisoning through skin contact, inhalation, and eating or drinking. Due to its nature, it can linger in homes for years after its application. People living in low-income housing projects are exposed disproportionately to methyl parathion. Although methyl parathion is heavily restricted, residents and landlords have been able to obtain it, since it is one of the most effective ways to deal with cockroaches. Exposures have led to illnesses and even reports of death. In addition, the parent compound breaks down after the pesticide is applied. It may become less toxic, but it can also be transformed to more toxic metabolites, a process known as *bioactivation*. Figure 4.9 shows how methyl parathion can change in the environment, and Figure 4.10 illustrates the metabolism of methyl parathion in rodents.

Like many environmental toxicants, methyl parathion degradation involves catalysis. Organic catalysts, such as hydrolases, are known as enzymes. Note that these reactions can generate by-products that are either less toxic (i.e., detoxification) or more toxic (i.e., bioactivation) than the parent compound. For methyl parathion, the metabolic detoxification pathways are shown as 2 and 3 in Figure 4.10 and the bioactivation pathway as 1. Methyl paroxon is more toxic than methyl parathion. Note that these reactions occur within and outside an organism, so a person may be exposed to the more toxic by-product some time after the pesticide has been applied. In other words, it is possible that

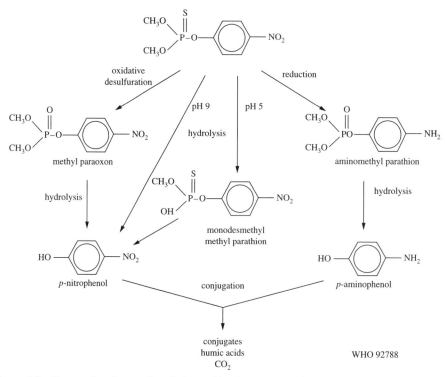

Figure 4.9   Proposed pathway of methyl parathion in water. Environmental factors, including pH, available oxygen, and water, determine the pathway. [From World Health Organization, International Programme on Chemical Safety, *Environmental Health Criteria 145: Methyl Parathion,* WHO, Geneva, 1993; A. W. Bourquin, R. L. Garnas, P. H. Pritchard, F. G. Wilkes, C. R. Cripe, and N. I. Rubinstein, Interdependent Microcosms for the Assessment of Pollutants in the Marine Environment, *International Journal of Environmental Studies,* 13(2):131–140, 1979; and R. Wilmes, Parathion-methyl: Hydrolysis Studies, Bayer AG, Institute of Metabolism Research, Leverkusen, Germany, 1987, 34 pp. (unpublished report PF 2883, submitted to WHO by Bayer AG).]

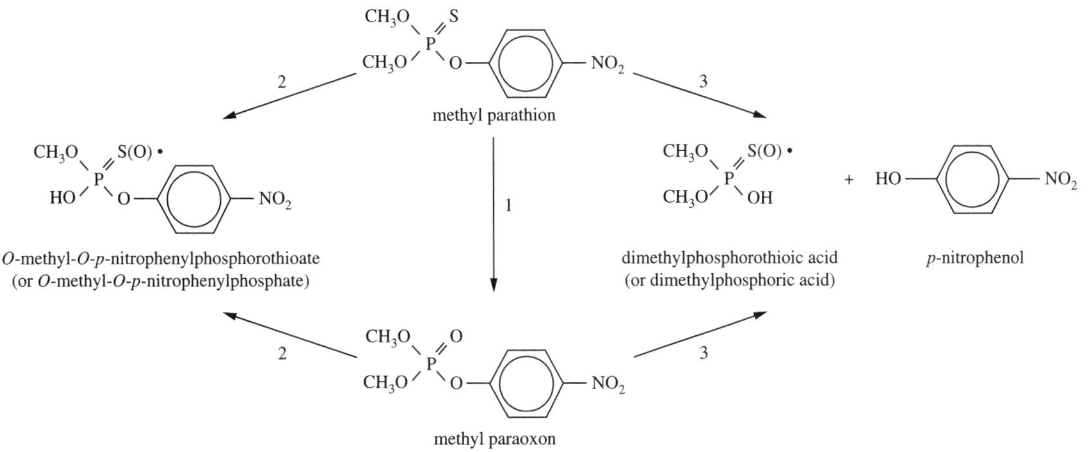

Figure 4.10   Sometimes, chemicals become more toxic as a result of an organism's (in this instance, rodents) metabolism. For example, methyl parathion's toxicity changes according to the degradation pathway. During metabolism, the biological catalysts (enzymes) make the molecule more polar by hydrolysis, oxidation, and other reactions. Bioactivation (pathway 1) renders the metabolites more toxic than the parent compound, and detoxification (pathways 2 and 3) produces less toxic metabolites. The degradation product, methyl paraoxon, may be metabolized in the same pathways as those for methyl parathion. This results in the oxygen analog, designated as (0)*. [From International Agency for Research on Cancer, Methyl Parathion, in *Miscellaneous Pesticides,* IARC, Lyon, France, 1983, pp. 131–152 (IARC Monographs on the Evaluation of the Carcinogenic Risk of Chemicals to Humans, Vol. 30).]

the risk of health effects is increased with time until the less toxic by-products (e.g., *p*-nitrophenol) replace the more toxic substances (e.g., methyl paroxon).

Figure 4.11 indicates that the exposure and toxicity properties are affected by extrinsic conditions, such as whether the substances are found in the air, water, sediment, or soil, along with the conditions of these media (e.g., oxidation-reduction, pH, and grain size). For example, the metal mercury is usually more toxic in reduced and anaerobic conditions because it is more likely to form alkylated organometallic compounds, such as monomethyl mercury and the extremely toxic dimethyl mercury. These chemically reduced mercury species are likely to form when buried under layers of sediment where dissolved oxygen levels approach zero. Ironically, engineers have unwittingly participated in increasing potential exposures to these toxic compounds. With the good intention of attempting to clean up contaminated lakes in 1970s, engineers recommended and implemented dredging programs. In the process of removing the sediment, however, the metals and other toxic chemicals that had been relatively inert and encapsulated in buried sediment were released to the lake waters. In turn, the compounds were also more likely to find their way to the atmosphere (see Figure 4.11). This is a lesson for engineers to take care in considering as many physical, chemical, and biological characteristics of the compound as possible, as well as the environment where it exists.

Some of these important physical and chemical characteristics that are part of hazard identification are listed in the Appendix. For the most toxic substances, the principal components of a hazard are its persistence, its ability to accumulate in organisms, and its ability to elicit a biological response. Let us consider each of these components briefly.[35]

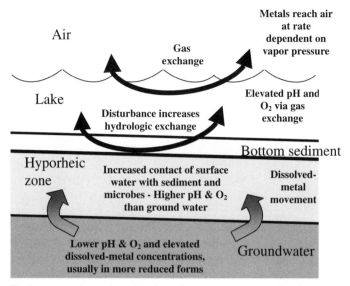

Figure 4.11   Exchanges and reactions involve all environmental media, including groundwater, sediment, and surface water. Human activities such as dredging can increase the exchanges between media and enhance the mobility of dissolved metallic compounds. (Adapted from U.S. Geological Survey and D. A. Vallero, *Environmental Contaminants: Assessment and Control,* Elsevier, Academic Press, Burlington, MA, 2004.)

Persistence

The likelihood that a substance will cause a problem is influenced by how long it lasts, so persistence is important in both parts of the risk equation (i.e., hazard and exposure). Substances that once released remain in the environment are more likely to continue to cause problems or to be a threat. Persistence is commonly expressed as the chemical half-life ($T_{1/2}$) of a substance (i.e., the time it takes to degrade one-half of the mass). The U.S. Environmental Protection Agency considers a compound to be persistent if its $T_{1/2}$ value in water, soil, or sediment is greater than 60 days, and very persistent if the $T_{1/2}$ value is greater than 180 days. In air, the compound is considered persistent if its $T_{1/2}$ value is greater than two days. Some of the most notoriously toxic chemicals are also very persistent.

CASE STUDY: TIMES BEACH

Pick up an organic chemistry textbook published before 1975 and you might find that it estimates the half-life of dioxin to be a few months. In other words, if 20 milligrams (mg) of dioxin is released, in a few months you will have only 10 mg. Unfortunately, these estimates were based on best-case conditions in a laboratory, such as a thin film exposed to light over a large surface area. When organic compounds such as dioxin reach the environment, such rosy scenarios are the exception, not the rule. For example, PCBs, dioxins, and other aromatic compounds have a strong affinity for soil, especially the clay-sand organic matter. This means that instead of a few months, the half-lives can be increased to hundreds of years. When dioxins were found at Love Canal in New York and shortly after at numerous other sites, chemists and environmental scientists were thrust into hyperdrive in an effort to understand these substances. As is often the case, outrage and frustration were the norm for neighbors where the compound was discovered in soil. The experiences at Love Canal, Times Beach, Missouri and the Valley of the Drums in Kentucky were key events that led to the passage of the Comprehensive Environmental Response, Compensation and Liability Act, better known as the Superfund, in 1980.

Times Beach, about 17 miles west of St. Louis, was an unlikely spot for a hazardous waste controversy. Up to the 1970s, it was a popular resort community along the Meramec River. With few municipal resources, the roads in the town were not paved and dust on the roads was controlled by spraying oil. For two years, 1972 and 1973, the contract for the road spraying went to a waste oil hauler named Russell Bliss. The roads were paved in 1973 and the spraying ceased.

Bliss obtained his waste oil from the Northeastern Pharmaceutical and Chemical Company in Verona, Missouri, which manufactured hexachlorophene, a bactericidal chemical. In the production of hexachlorophene, the company had to remove and dispose of considerable quantities of dioxin-laden waste. A significant amount of the dioxin was contained in the "still bottoms" of chemical reactors, and the company found that having it burned in a chemical incinerator was expensive. The company was taken over by Syntex Agribusiness in 1972, and the new company decided to contract with Russell Bliss to haul away the still bottom waste without telling Bliss what was in the oily substance. Bliss mixed it with other waste oils, and this is what he used to oil the roads in Times Beach, unaware that the oil contained high concentration of dioxin (greater than 2000 ppm), including the most toxic congener, 2,3,7,8-dibenzo-p-dioxin (TCDD).

dioxin structure

2,3,7,8-dibenzo-*p*-dioxin (TCDD)

Bliss also sprayed oil to control dust, especially in horse arenas. He used the dioxin-laden oil to spray the roads and horse runs in nearby farms. In fact, it was the death of horses at these farms that first alerted the Centers for Disease Control to sample the soil at the farms. They found the dioxin but did not make the connection with Bliss. Finally, in 1979, the U.S. EPA became aware of the problem when a former employee of the company told them about the sloppy practices in handling the dioxin-laden waste. The EPA converged on Times Beach in "moon suits" and panic set in among the populace. The situation was not helped by the message from the EPA to the residents of the town. "If you are in town it is advisable for you to leave, and if you are out of town do not go back." In February 1983, on the basis of an advisory from the Centers for Disease Control, the EPA permanently relocated all of the residents and businesses at a cost of $33 million. Times Beach was by no means the only problem stemming from the contaminated waste oil. Twenty-seven other sites in Missouri were also identified by the EPA as being contaminated with dioxins.

The concern with dioxin, however, may have been overstated. As a previous accident in Seveso, Italy, had shown, dioxin is not nearly as acutely toxic to humans as originally feared, causing some to conclude that is unlikely that the damage to human health in Times Beach was anywhere near the catastrophe originally anticipated. Even some EPA officials later admitted that the evacuation and bulldozing of the community was probably unnecessary. But given the knowledge of dioxin toxicity in 1979, the decision to detoxify the site was not unreasonable. For one thing, the carcinogenicity of TCDD was later better established and found to be very high (slope factors $> 10^5$ for inhalation, ingestion, and dermal routes). The psychological toll of such decisions is more difficult to measure. Some years after the decision, a former resident was said to have committed suicide because, according to his wife, he was unable to cope with losing his home. Later, a local official who was party to issuing the advisory admitted the uncertainties of the social costs and that the health danger was possible, but not certain.

After everyone had moved out of Times Beach, the houses were razed and Syntex Corporation was required to build an incinerator for burning the contaminated soil. The Superfund site was eventually decontaminated at a cost of over $200 million, and the site now is a beautiful riverside park.

The concept of persistence elucidates the notion of trade-offs that are frequently needed as part of many responses to environmental insults. It also underlines the importance that good science is necessary but never sufficient to provide an acceptable response to environmental justice issues. Let us consider the pesticide DDT [1,1,1-trichloro-2,2-bis(4-chlorophenyl)ethane ($C_{14}H_9Cl_5$)]. DDT is relatively insoluble in water (1.2 to 5.5 mg L^{-1} at 25°C) and is not very volatile (vapor pressure: 2.0×10^{-7} mmHg at 25°C).[36] Looking at the water solubility and vapor pressures alone may lead one to believe that people and wildlife are not likely to be exposed in the air or water. However, the compound is highly persistent in soils, with a $T_{1/2}$ value of about 1.1 to 3.4 years, so it may still end up in drinking water in the form of suspended particles or in the air sorbed to fine particles. DDT also exhibits high bioconcentration factors (on the order of 50,000 for fish and 500,000 for bivalves), so once organisms are exposed, they tend to increase body burdens of DDT over their lifetimes. In the environment, the parent DDT is metabolized mainly to DDD and DDE.[37]

The physicochemical properties of a substance determine how readily it will move among the environmental compartments (i.e., to and from sediment, surface water, soil, groundwater, air, and in the food web, including humans). So if a substance is likely to leave the water, it is not persistent in water. However, if the compound moves from the water to the sediment, where it persists for long periods of time, it must be considered environmentally persistent. This is an example of how terminology can differ between chemists and engineers. Chemists often define persistence as an intrinsic chemical property of a compound, while engineers see it as both intrinsic and extrinsic (i.e., a function of the substrate, energy and mass balances, and equilibria). So engineers usually want to know not only about the molecular weight, functional groups, and ionic form of the compound, but also whether it is found in the air or water, and the condition of the substrate in the media (e.g., pH, soil moisture, sorption potential, organic matter content and microbial populations). The movement among phases and environmental compartments is known as *partitioning*. Many toxic compounds are semivolatile (i.e., at 20°C and 101 kPa atmospheric pressure, vapor pressures $= 10^{-5}$ to 10^{-2} kPa) under typical environmental conditions. The low vapor pressures and low aqueous solubilities mean that they will have will low fugacities; that is, they lack a strong propensity to flee a compartment, (e.g., to move from the water to the air). The most common water-to-air fugacity measure is the *Henry's law constant*. Henry's law states that the concentration of a dissolved gas is directly proportional to the partial pressure of that gas above the solution:

$$p_a = K_H[c] \tag{4.9}$$

where K_H is the Henry's law constant, p_a is the partial pressure of the gas, and $[c]$ is the molar concentration of the gas; or

$$p_a = K_H C_W \tag{4.10}$$

where C_W is the concentration of gas in water.

Henry's law expresses the proportionality between the concentration of a dissolved contaminant and its partial pressure in the open atmosphere at equilibrium. That is, the Henry's law constant is an example of an *equilibrium constant,* which is the ratio of concentrations when chemical equilibrium is reached in a reversible reaction, the time when the rate of the forward reaction is the same as the rate of the reverse reaction. Most of the time, when a partitioning coefficient is given, it is assumed to be an equilibrium constant. For environmental partitioning, the amount of chemical needed to reach equilibrium is usually very small (i.e., very dilute solutions and other mixtures).

A direct partitioning between the air and water phases is the air–water partitioning coefficient (K_{AW}):

$$K_{AW} = \frac{C_A}{C_W} \qquad (4.11)$$

where C_A is the concentration of a gas in the air. The relationship between the air–water partition coefficient and Henry's law constant for a substance is

$$K_{AW} = \frac{K_H}{RT} \qquad (4.12)$$

where R is the gas constant (8.21×10^{-2} L · atm mol^{-1} K^{-1}) and T is the temperature (K).

Under environmental conditions, most toxic substances have very low K_H values since K_H is proportional to the concentration of a dissolved contaminant and its partial pressure in the atmosphere at equilibrium. There are many exceptions, however; such as the relatively water soluble compounds with high vapor pressures (e.g., alcohols, benzene, toluene, and many organic solvents). Since Henry's law is a function of aqueous solubility and vapor pressure, estimating the tendency for a substance's release in vapor form, K_H is a good indicator of the fugacity from the water to the atmosphere.

Another common expression of partitioning is the octanol–water coefficient (K_{ow}). The K_{ow} value indicates a compound's likelihood to exist in the organic *versus* aqueous phase. A rule to keep in mind is that "like dissolves like." The configuration of a molecule determines whether it is polar or nonpolar. *Polar compounds* are electrically positive at one end and negative at the other. The water molecule, for example, is highly electropositive at the hydrogen atoms and electronegative at the oxygen atom. Other molecules, like fats, are not polar (i.e., do not have strong differences between the positive and negative ends). If a relatively *nonpolar compound* is dissolved in water and the water comes into contact with another substance (e.g., octanol), the nonpolar compound will move from the water to the octanol. Its K_{ow} reflects just how much of the substance will move from the aqueous and organic solvents (phases) until it reaches equilibrium. For example, if at a given temperature and pressure a chemical is at equilibrium when its concentration in octanol is 100 mg L^{-1} and in water is 1000 mg L^{-1}, its K_{ow} is 100 divided by 1000, or 0.1. Since the range is so large among various environmental contaminants, it is common practice to express log K_{ow} values. So, for example, in a spill of equal amounts of two insecticides, DDT (log $K_{ow} \approx 7$) and methyl parathion (log $K_{ow} \approx 3$), the DDT has much greater affinity for the organic phases than does the methyl parathion (four orders of magnitude). This does not mean than a greater amount of either of the compounds is likely to stay in the water column, since they are both hydrophobic, but it does mean that they will vary in the time and mass of each contaminant moving between phases. The time it takes to reach equilibrium (i.e., the kinetics) is different (see Figure 4.12).

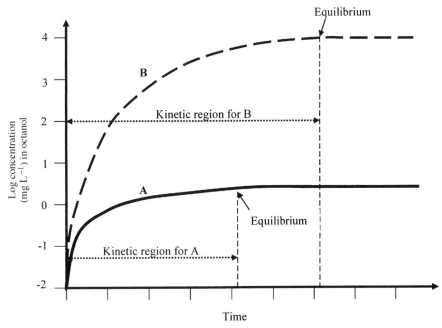

Figure 4.12 Hypothetical diagram of the relative concentrations of two compounds in octanol with time. In this instance, compound A is more soluble than compound B in octanol. Note the steeper slope for the compound A kinetics compared to the kinetics for compound B. This is an indication that the compound A's affinity for the organic phase is greater than that of compound B. The time units are not shown since these can vary considerably depending on the environmental conditions.

Even low K_H and K_{AW} compounds, however, can be transported long distances in the atmosphere when sorbed to particles. Fine particles can behave as colloids and stay suspended for extended periods of time, explaining in part why low-K_H compounds can be found in the most remote locations relative to their sources, such as in the Arctic regions. This is important, for example, when explaining to indigenous populations why they may be exposed to contaminants that are not produced near them.

Sorption is the partitioning of a substance from the liquid to the solid phase and is an important predictor of a chemical's persistence. If the substrate has sufficient sorption sites, such as many clays and organic matter, the substance may become tightly bound and persistent. The properties of the compound and those of the water, soil, and sediment determine the rate of sorption. The soil partition coefficient (K_d) is the experimentally derived ratio of a contaminant's concentration in the solid matrix to the contaminant concentration in the liquid phase at chemical equilibrium. Another frequently reported liquid-to-solid phase partitioning coefficient is the organic carbon partitioning coefficient (K_{oc}), which is the ratio of the contaminant concentration sorbed to organic matter in the matrix (soil or sediment) to the contaminant concentration in the aqueous phase. Thus, K_{oc} is derived from the quotient of a contaminant's K_d value and the fraction of organic matter (OM) in the matrix:

$$K_{oc} = \frac{K_d}{OM} \qquad\qquad (4.13)$$

Many toxic substances are expected to be strongly sorbed, but K_{oc} varies from substrate to substrate.

It is important to keep in mind the difference between chemical persistence and environmental persistence. For example, one can look at Henry's law, solubility, vapor pressure, and sorption coefficients for a compound and determine that the compound is not persistent. However, in real-life scenarios, this may not be the case. For example, there may be a repository of a source of a nonpersistent compound that leads to a continuous, persistent exposure of a neighborhood population. Or a compound that is ordinarily not very persistent may become persistent under the right circumstances (e.g., a reactive pesticide that is tracked into a home and becomes entrapped in carpet fibers). The lower rate of photolysis (degradation by light energy) indoors and the sorptive characteristics of the carpet twill can lead to dramatically increased environmental half-lives of certain substances.

CASE STUDY: FROM CANCER ALLEY TO TOXIC GUMBO[38]

Even before the devastation that followed Hurricane Katrina, the lower Mississippi River industrial corridor was a disaster waiting to happen. It is home to a predominantly low-income minority (African American and Hispanic American) community who are being exposed to many pollutants.[39] This 80-mile-long region, known as Cancer Alley, between Baton Rouge and New Orleans has experienced releases of carcinogens, mutagens, teratogens (birth-defect agents), and endocrine disruptors in the atmosphere, soil, groundwater, and surface water. More than 100 oil refineries and petrochemical facilities are located in this region. It has been reported that per capita release of toxic air pollutants is about 27 kilograms (kg), nine times greater than the U.S. average of only 3 kg.[40] The U.S. average 260 kg of toxic air pollutants per square mile is dwarfed by the more than 7700 kg per square mile in the industrial corridor.

One particular carcinogen of concern is vinyl chloride. In the 1970s, cases of liver cancer (hepatic angiosarcoma) began to be reported in workers at polymer production facilities and other industries where vinyl chloride was present. Since then, the compound has been designated as a potent human carcinogen (inhalation slope factor = 0.3 kg · day mg^{-1}).

vinyl chloride

Vinyl chloride may at first glance appear to be broken down readily by numerous natural processes, including abiotic chemical and microbial degradation (see Figure 4.13), but numerous studies have shown that vinyl chlo-

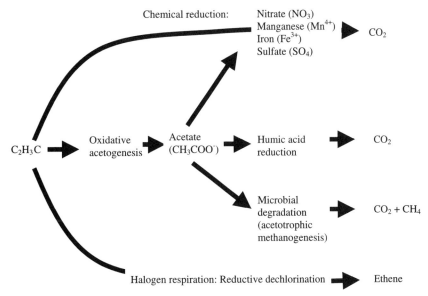

Figure 4.13 Biodegradation pathways for vinyl chloride. (From U.S. Geological Survey, Microbial Degradation of Chloroethenes in Ground Water Systems, Toxic Substances Hydrology Program: Investigations, http://toxics.usgs.gov/sites/solvents/chloroethene.html, accessed November 29, 2004.)

ride concentrations can remain elevated over long periods of time. In fact, under environmental conditions, vinyl chloride can be extremely persistent, with an anaerobic $T_{1/2}$ value in soil greater than two years. It can also be difficult to treat with conventional engineering methods. For example, aerobic degradation in sewage treatment plants and surface water in an isolated bacteria culture with vinyl chloride concentrations of 20 to 120 mg L^{-1} needs a minimum of 35 days to degrade the compound completely. Nontraditional treatment methods, such as attack by hydroxyl radicals, can significantly reduce the half-life.[41] In heavily polluted areas such as Cancer Alley, vinyl chloride repositories can remain intact for decades, serving as a continuous potential source. These repositories can actually be compounds other than vinyl chloride, but which break down to form the compound (e.g., chloroethylene solvents degrade to vinyl chloride). With its high vapor pressure (2300 mmHg at 20°C) and high aqueous solubility (1100 mg L^{-1}), the chances of people being exposed via the air or drinking water once vinyl chloride is formed can be considerable.

Local environmental and neighborhood groups have begun arming themselves with environmental data, such as the emissions and other release information in the Toxic Release Inventory (TRI) that show the inordinately high toxic chemical release rates near their communities. Local communities have challenged nearby industries with possible health effects linked to chemical exposures. For example, residents in Mossville, Louisiana, argue that several health problems in their community could be linked to chemical releases by 17 industrial facilities located within 1 kilometer of the community. These confrontations led to a number of advocates writing the 2000 report, *Breathing Poison: The Toxic Costs of Industries in Calcasieu Parish, Louisiana,* which called for "pollution reduction, environmental health services, and a fair and just relocation for consenting residents."[42] These efforts have gained the attention of national media and regulatory agencies and have been emblematic of the environmental justice movement.

Hurricane Katrina: The Perfect Storm and the Toxic Gumbo

In many ways, the category 4 hurricane that made landfall along the Gulf coast of Louisiana, Mississippi, and Alabama on August 29, 2005 was literally and metaphorically the *perfect storm.* That is, all of the conditions for a disaster occurred simultaneously. Even worse, engineering failures of every type preceded, co-occurred, and followed the storm. The vulnerability of New Orleans was no surprise to many scientists and engineers. For example, in *Geological Hazards* (Greenwood Press, Westwood, CT), earth scientist Timothy Kusky warned that "if a category 4 hurricane ever hits New Orleans directly, the dikes will be breached and destroyed, and thousands will perish." In fact, the cities on the delta of the Mississippi River (or any river system for that matter) are engaged in a constant struggle against nature.

The perfect storm required the confluence of meteorological events (high-intensity hurricane, 100-year flood), sociological conditions (traditionally poor and mistrusting populace), and political mistakes (botched evacuation planning, delayed response due to red tape and bureaucracy, corruption, and lack of coordination among responding agencies). In addition to the loss of life and property, the immediate- and long-term environmental effects have slowly begun to be fully understood.

Possibly the best characterization was by a nonscientist, New Orleans native singer Aaron Neville. In an appearance on the *Tonight Show* a week after the storm, Neville described the scenario as a *toxic gumbo.* The floodwaters contain the typical pathogens and vectors (e.g., rats and insects) following flooding. In addition, due to the industries and commercial enterprises in the region, numerous petroleum refineries and pipelines, pesticide manufacturers, chemical plants, and other sources of toxic pollutants were added to the exposures. Again, this is no surprise, since the Gulf has some of the highest exposures to toxic contaminants in the nation, (as we have just discussed with regard to vinyl chloride). Further exacerbating the contamination is the amount of time it took to pump out the water. The longer the water remained in the New Orleans basin, the greater the number of pipe breaks, line failures, and chemical releases. Fires started and gases were released in the days following the hurricane, so that thousands of first responders from around the nation had to be called in to suppress fires and repair lines; all this while trying to evacuate people and begin recovery.

We are reminded of the old adage, "When you are up to your neck in alligators, it is too late to think about draining the swamp." Conversely in this case, we would have to restate the adage as: "When you are up to your neck in toxic gumbo, it is too late to think about saving the wetlands." Wetlands are a natural part of hydrologic systems, especially in deltas and backwater areas. Humans had to make a cognizant (using the word advisedly) decision to destroy these natural systems and replace them with structures such as buildings and roads. This is doubly bad for the hydrology since the systems now have vulnerable land uses (and thus, people using the land), and the ability to infiltrate and remove water is exponentially decreased (due to the great increase in impervious surfaces, the loss of plant life, and the elimination of the integration of surface water and groundwater systems). Before the development, the streams were connected to each other and to the aquifers, allowing for efficient water removal. So the engineers and planners who allowed and even designed these new land uses were key players in the disaster.

Another aspect of the perfect storm was poor city and regional planning. In fact, the New Orleans disaster shares some common elements with one of the worst industrial disasters in Bhopal, India. First, engineers and planners seem to adopt a "one size fits all" mentality. Like Bhopal, where a U.S. company transplanted a Western type of pesticide plant in a completely different culture, with little regard for the social differences, the design and siting of industrial facilities in the Mississippi Delta were little different from that in an upland. Second, land-use planning failed in not considering the possible impact on people living next to a facility. The discontinuity should have been obvious. In Bhopal, squatters lived right at the company property line. In New Orleans, entire neighborhoods were in the shadow of heavy industry. Third, the adjacent residents had little voice in decisions that directly affect their health. This would have been the case even without Katrina, although in a less acute manner, since the residents were being exposed to contaminants every day. Fourth, the agencies that were supposed to be protecting public health and the environment were actually collaborating with those who were presenting the threat. In Bhopal, the local and national Indian governmental agencies were more concerned about encouraging industry than about protecting the most vulnerable. Unfortunately, it appears that this may also have been the case in New Orleans.

The major similarity between the aftermaths of Bhopal and Katrina is the disproportionate effect on those who were already at a cultural and social disadvantage. One interesting observation is the disproportionate impact on women and children in New Orleans (some preliminary estimates were that four times more women than men were among the "refugees" and those who lost their lives). This may well be the result of socioeconomic conditions, such as the unstable conditions of lower-socioeconomic-status families, including higher percentages of single heads of households. Characterizing such vulnerabilities *in any community* must be part of emergency and contingency planning.

So then, what should engineers do now? We might be tempted to respond by what we do best, that is, design and build. That is only the right thing to do if we think first. Designing and building must be the outgrowth of good planning and a thoughtful consideration of the events that led to the disaster. For example, "hardening" the levees and dams is not the solution, at least it is not the entire solution. Hindsight does approach 20/20. It seems obvious now that those engineers and planners who called for stronger levees deserve credit, but this is but a small part of the solution. Such engineering projects are successful only when they are integrated into an overall plan. There are positive signs in this regard. Numerous engineering experts have called for the construction of wetlands; and many are calling for aggressive and well-enforced land-use controls.

The United States has a checkered past when it comes to planning. Some have feared that land-use planning is too much like a Soviet-style centralized planning program. We have often foregone strong land-use controls, even in vulnerable settings like wetlands, coast lines, and sensitive habitat, in the interest of unfettered uses by landowners. This is understandable, and some would argue that it is guaranteed by the Constitution. Land-use controls are *de facto* takings as eminent domain, which must only be for the public good and for which the landowner is justly compensated. The New Orleans and other Gulf disasters demonstrate that wise land-use planning would certainly have provided for the public good. But the sticky issue is just how the landowners should be compensated. Actually, zoning and other ordinances have stood the tests of legal challenges for several decades, so it seems that a major reason for the lack of strong planning is a mix of politics and economics. Unfortunately, to get elected often requires immediacy. It takes much courage to run on long-term issues such as wetland protection, especially when others are offering short-term benefits (e.g., attracting industry, which in turn reduces taxes and offers jobs). The good news for engineers and other environmental professionals is that we are called to make the right decisions irrespective of politics. In reality, however, engineers all too often are beholden in some way to politics and economics (e.g., as line and staff employees in governmental agencies, as contractors to the city, and as beneficiaries of the short-term decisions). However, a glance at our first professional canon trumps these influences. We must hold paramount the health, safety, and welfare of the *public,* not the politicians and our bosses.

Actually, the concept of prevention is built into many engineering systems. For example, public impoundments usually proscribe any residential and most commercial buildings in the 100-year floodplain. Recall that the 100-year flood is a purely statistical concept. The Federal Emergency Management Agency characterizes a *100-year flood* as one whose magnitude is expected to be equaled or exceeded once on the average during any 100-year period. Thus, it is not a flood that will occur once every 100 years, but it is the flood elevation that has a 1% chance of being equaled or exceeded each year. It is less a temporal concept than a hydrological phenomenon.

Thus, the 100-year flood could occur more than once in a relatively short period of time, even several times in a single year. The 100-year flood is the standard applied by most federal and state agencies; for example, it is used by the National Flood Insurance Program (NFIP) as the benchmark for flood-plain management and to determine the need or the eligibility for flood insurance. It is used by the U.S. Army Corps of Engineers and the Bureau of Reclamation to prevent encroachment near impoundments and lakes. It is also a common standard for local, regional, and state land-use planning. The challenge is that people become used to artificial "100-year" floods. Had the dikes and levees not existed in New Orleans, for example, much of the area would be in a "zero-year" floodplain (i.e., under water).

So, is it wise engineering practice to build another, better dike and levee system? As usual the answer depends on one's perspective. We recommend that a larger view in time and space be used instead of what is politically expedient. Or as Benjamin Franklin would suggest, should we apply an ounce of prevention (land-use proscriptions) instead of a future pound of cure (repairing and rebuilding again in the vulnerable delta)? Prevention comes in various forms and scales. The largest scale is planetary. The intensity of the storm begs the question of what is the role of global climate change, if any, in breeding large hurricanes and other extreme meteorological events. From a purely thermodynamic perspective, one would suspect the answer to be "yes." If the buildup of greenhouse gases has led to greater amounts of stored energy (i.e., infrared—heat—converted from incoming solar radiation), there is more energy that needs to be released. Hurricanes are simply the result of two energy systems on Earth, heat and motion. The heat is derived almost completely from the sun and is converted and transferred through complex systems in the atmosphere. Motion is the result of the Earth's rotation (i.e., the Coriolis effect), where the air is deflected, so that in cyclones (low-pressure systems) and anticyclones (high-pressure systems) are formed. When greater amounts of heat are formed, the mechanical and thermodynamic systems must become more intense, and if this logic is correct, the storms that result will probably become increasingly violent. Of course, there is much debate but a general consensus in the scientific community about the grounding assumption; that is, does the buildup of carbon dioxide, methane, and other gases in the Earth's atmosphere *really* increase global temperatures? Like many uncertainties in science, this is *important if true.*

It is not that engineers will not have plenty to design and build, such as constructed wetlands and hardened facilities within the floodplain. The key is finding the proper balance of knowing what to build and what to avoid building. Perhaps the biggest lesson for engineers is the need to approach everything we do from a perspective of sustainability. When we design, plan, and build, how does this fit with and affect other parts of the systems within which our projects will exist? We must do complete life-cycle analyses and design with the ends in mind. Otherwise, our work will merely be a patchwork or worse yet, end up being a toxic gumbo.

Bioaccumulation

Some substances are eliminated easily, while others build up in an organism's tissues. The likelihood that a substance will find its way into the food web is another important aspect of its hazard. Toxicokinetic models predict the dynamics of uptake, distribution, depuration, and elimination of contaminants within organisms. Persistence and bioaccumulation are interdependent. If the substance is likely to be sorbed to organic matter (i.e., high K_{oc} value), it will have an affinity for tissues. A substance that partitions from the aqueous to the organic phase (i.e., high K_{ow} value) is likely to be stored in fats of higher-trophic-level organisms (e.g., carnivores and omnivores). The *bioconcentration factor* (BCF) is the ratio of the concentration of the substance in a specific genus to the exposure concentration, at equilibrium. The exposure concentration is the concentration in the environmental compartment (almost always surface water). The BCF is similar to the *bioaccumulation factor* (BAF), but the BAF is based on the uptake of the organism from both water and food. The BCF is based on direct uptake from the water only. A BCF of 500 means that an organism takes up and sequesters a contaminant to concentrations 500 times greater than the exposure concentration. Generally, any substance that has a BAF or BCF > 5000 is considered to be highly bioaccumulative, although the cutoff point can differ depending on the chemicals of concern, the regulatory requirements, and the type of ecosystem in need of protection.

It is important to note that genera will vary considerably in reported BCF values and that the same species will bioaccumulate different compounds at various rates. The amount of bioaccumulated contaminant increases generally with the size, age, and fat content of the organism and decreases with increasing growth rate and efficiency. Bioaccumulation also is often higher for males than for females and in organisms that are proficient in storing water. Top predators often have elevated concentrations of persistent, bioaccumulating toxic substances (known as PBTs).

The propensity of a substance to bioaccumulate is usually inversely proportional to its aqueous solubility, since hydrophilic compounds are usually more easily eliminated by metabolic processes. In fact, the first stages of metabolism often involve adding or removing functional groups to make it more water soluble. Generally, compounds with log $K_{ow} > 4$ can be expected to bioaccumulate. However, this is not always the case. For example, very large molecules [e.g., cross-sectional dimensions > 9.5 angstroms (Å) and molecular weights > 600] are often too large to pass through organic membranes (i.e., known as *steric hindrance*). Since, in general, the larger the molecule, the more lipophilic it becomes, some very lipophilic compounds (i.e. log $K_{ow} > 7$) will actually have surprisingly low rates of bioaccumulation, due to steric hindrance.

Bioaccumulation not only makes it difficult to find and measure toxic compounds but complicates how people and ecosystems can become exposed. For example, a release of a persistent, bioaccumulating substance can interfere with treatment plant efficiencies and greatly increase human exposures to pollutants that would otherwise have to be removed and treated. (This is another example of risk shifting, discussed at length in Chapter 3.)

Biological Response

Various organisms respond to environmental insults in different ways, so even if a substance persists and is taken up by an organism, its hazards are still dependent on the response of the organism after it comes into contact with the substance. This is the essence of the hazard; that is, does the chemical, physical, or biological agent elicit an adverse response?

This response is measurable. When a contaminant interacts with an organism, substances such as enzymes are generated as a response. Thus, measuring such substances in fluids and tissues can provide an indication or *marker* of contaminant exposure and biological effects resulting from the exposure. The term *biomarker* includes any such measurement that indicates an interaction between an environmental hazard and a biological system.[43] In fact, biomarkers may indicate any type of hazard: chemical, physical, and biological. An exposure biomarker is often an actual measurement of the contaminant itself or any chemical substance resulting from the metabolism and detoxification processes that take place in an organism. For example, measuring total lead (Pb) concentration in the blood, urine, or hair may be an acceptable exposure biomarker for people's exposures to Pb. However, other contaminants are better reflected by measuring chemical by-products, such as compounds that are rapidly metabolized upon entering an organism. Nicotine, for example, is not a very good indicator of smoking, but the metabolite, cotinine, can be a reliable indicator of nicotine exposure. Similarly, when breath is analyzed to see if someone has been drinking alcohol, the alcohol itself (i.e., ethanol) is

Biographical Sketch: John Snow

John Snow, the eldest son of a farmer, was educated at a private school and then apprenticed to a surgeon at the age of 14 in preparation for his career in medicine. After attending the Hunterian School of Medicine in London he became a public health physician, and in 1838 he became a member of the Royal College of Surgeons and, in 1850, the Royal College of Physicians.

In those days, London was plagued almost annually by cholera epidemics, and the medical field was trying desperately to understand what caused this dreaded disease. Some public health officials such as Edwin Chadwick thought that it was the bad odors that caused disease, whereas others thought it was "miasma," vapor that rises out of rotting stuff and seeps into rooms and infects people while they are asleep (hence, nightcaps). The possibility that it was contaminated water was not widely believed since the microorganisms that cause cholera and other infectious disease had not been identified. But John Snow thought he knew and set out to prove that cholera was waterborne.

In 1853 when a particularly vicious epidemic struck London, he decided to do a special study by noting on a map the locations of the people who had died. This became the very first "spot map" in the history of epidemiology and allowed Snow to pinpoint the water pump on Broad Street as the most likely source of the infection. He convinced the city fathers to remove the handle on the pump and the epidemic subsided. It was more than 30 years later that Koch identified the pathogen causing cholera, but by that time the germ theory had become widely believed and embraced as the etiology of contagious diseases.

not usually a good indicator, but various metabolites, such as acetaldehyde, that have been formed as the body metabolizes the ethanol are excellent markers.

Exposure to ethanol by the oral pathway (i.e., drinking alcoholic beverages) illustrates the continuum of steps between exposure and response (see Figure 4.14). Table 4.4 gives examples of the types of biomarkers for a specific type of exposure (i.e., maternal alcohol consumption). Interestingly, the response and biomarkers for alcohol consumption are similar to those for some environmental contaminants, such as Pb, mercury (Hg), and PCBs.

Exposure biomarkers are also useful as an indication of the contamination of fish and wildlife in ecosystems. For example, measuring the activity of certain enzymes, such as ethoxyresorufin-*O*-deethylase (EROD), in aquatic fauna *in vivo* indicates that the organism has been exposed to planar halogenated hydrocarbons (e.g., certain dioxins and PCBs), PAHs, or similar contaminants. The mechanism for EROD activity in the aquatic fauna is the receptor-mediated induction of cytochrome P450–dependent monooxygenases when exposed to these contaminants.[44] The biological response does not necessarily have to respond to chemical stress. Stresses to environmental quality can also come about from ecosystem stress (e.g., loss of important habitats and decreases in the size of the population of sensitive species).

A substance may also be a *public welfare hazard* that damages property values or physical materials, expressed for example as its corrosiveness or acidity. The hazard may be inherent to the substance, but like toxicity, a welfare hazard usually depends on the situation and conditions where the exposure may occur.

Situations are most hazardous when a number of conditions exist simultaneously; witness the hazard to firefighters using water in the presence of oxidizers. The challenge to the engineer is how to remove or modify the characteristics of a substance that renders it hazardous, or to relocate the substance to a situation where it has value.

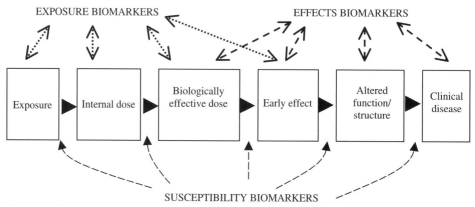

Figure 4.14   Continuum from exposure to a toxic substance to clinically diagnosed disease. The continuum is a time sequence, but the chemical to which the organism is exposed is not necessarily the same chemical in subsequent stages (i.e., metabolites are formed, which can serve as the biomarker). Enzymes produced to enhance metabolism or detoxification can also serve as biomarkers. Susceptibility biomarkers indicate increased vulnerability between the steps. [Adapted from C. F. Bearer, Markers to Detect Drinking during Pregnancy, *Alcohol Research and Health*, 25(3):210–218, 2001.]

Table 4.4 Examples of Biomarkers Following Oral Exposure to Ethanol in Pregnant Women

Exposure/effect step	Biomarker type	Example biomarkers
Internal dose	Alcohol ingestion	Blood ethanol concentration
Biologically effective dose	Ethanol metabolites	Acetaldehyde
		Ethyl glucuronide
		Fatty acid ethyl esters (FAEEs)
		Cocaethylene
Early effects	Enzymes in ethanol metabolic reactions	Cytochrome P450 2E1
		Catalase
		FAEE synthase
Alter function or structure	Target protein alteration	Carbohydrate-deficient transferring
		Serum proteins
		Urinary dolichols
		Sialic acid
	Early target organ damage	γ-Glutamyltransferase
		Aspartate aminotransferase/ alanine aminotransferase
		Mean corpuscular volume
		B-hexosaminidase
Clinical disease	Physiological response, including neurological damage and low birth weight, in newborn baby	Fetal alcohol syndrome

Source: Adapted from C. F. Bearer, Markers to Detect Drinking during Pregnancy, *Alcohol Research and Health,* 25(3):210–218, 2001.

Organic *versus* Inorganic Toxicants

Environmental contaminants fall into two major categories, organic and inorganic. *Organic compounds* are those that have at least one covalent bond between two carbon atoms or between a carbon and a hydrogen atom. Thus, the simplest hydrocarbon, methane (CH_4), has a bond between carbon and each of four hydrogen atoms. Organic compounds are subdivided between aliphatic (chains) and aromatic (rings) compounds. A common group of aliphatic compounds are the chain structures known as alkanes, which are hydrocarbons with the generic formula C_nH_{2n+2}. If these compounds have all of the carbon atoms in a straight line, they are considered "normal" and are known as *n*-alkanes. The simplest aromatic, benzene (C_6H_6), has bonds between carbon atoms and between carbon and hydrogen atoms (see Figure 4.15).

The structure of the compound determines its persistence, toxicity, and ability to accumulate in living tissue. Subtle structural differences can lead to very different environmental behaviors. Even various arrangements with identical chemical formulas (i.e., isomers) can exhibit very different chemical characteristics. For example, the boiling points at 1 atm for *n*-pentane, isopentane, and neopentane (all C_5H_{12}) are 36.1°C, 27.8°C, and 9.5°C, respectively. Among the most important factors are the length of the chains in aliphatic compounds and the number and configurations of the rings in aromatics.

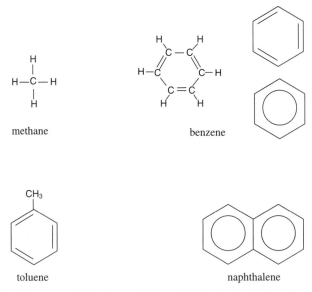

methane   benzene

toluene   naphthalene

Figure 4.15 Organic compound structures. Methane is the simplest aliphatic structure, and benzene is the simplest aromatic structure. Note that the benzene molecule has alternating double and
single bonds between the carbon atoms. The double and single bonds flip (i.e., resonate). This is
why the benzene ring is also shown as the two structures on the right, which are the commonly
used condensed form in aromatic compounds, such as the solvent toluene and the polycyclic
aromatic hydrocarbon, naphthalene.

Arguably, substitutions are even more critical. For example, methane is a gas under
environmental conditions, but it becomes a very toxic and bioaccumulating liquid (carbon
tetrachloride or tetrachloromethane) when chlorine atoms (CCl_4) are substituted for the
hydrogen atoms. Naphthalene, the simplest polycyclic aromatic hydrocarbon ($C_{10}H_8$), is
considered to be a possible human carcinogen, but the data are not sufficient to calculate
a slope factor. However, when an amine group (NH_2) substitutes for a hydrogen atom to
form 2-naphthylamine ($C_{10}H_9N$), the inhalation cancer slope factor is $1.8 \text{ kg} \cdot \text{day mg}^{-1}$.
The formulation of pesticides takes advantage of the dramatic increases in toxicity by
substitution reactions.

CASE STUDY: PESTICIDES AND STERILITY
For many years both Shell Oil and Dow Chemical supplied a pesticide containing dibromochloropropane (DBCP) to Standard Fruit Company for use on
its banana plantations, even though Shell Oil was aware since the 1950s that
DBCP exposure is linked to sterility in laboratory animals. In spite of evidence
that DBCP also causes sterility in humans and was banned in the United
States, Shell continued to market the pesticide in Central America.

$$\text{H}-\overset{\displaystyle \overset{\text{Cl}}{|}}{\underset{\displaystyle \underset{\text{H}}{|}}{\text{C}}}-\overset{\displaystyle \overset{\text{Br}}{|}}{\underset{\displaystyle \underset{\text{H}}{|}}{\text{C}}}-\overset{\displaystyle \overset{\text{Br}}{|}}{\underset{\displaystyle \underset{\text{H}}{|}}{\text{C}}}-\text{H}$$

dibromochloropropane (DBCP)

In 1984, banana plantation workers from several Central American countries filed a class action suit against Shell, claiming that they became sterile and faced a high risk of cancer. In response, Shell claimed that it was inconvenient to continue the case because the workers were in Costa Rica, a claim that was quickly thrown out of court. Shell finally settled out of court with the Costa Rican workers and paid $20 million in damages to the 16,000 claimants. A scientist from Shell is quoted as saying: "Anyway, from what I hear they could use a little birth control down there" (quote from David Weir and Constance Matthiessen, Will the Circle Be Unbroken? *Mother Jones,* June 1989).

Congeners are configurations of a common chemical structure. For example, all polychlorinated biphenyls (PCBs) have two benzene rings bonded together at two carbon atoms. They also have at least one chlorine substitution around the rings, so that there are 209 possible configurations (i.e., 209 PCB congeners). Since the two benzene rings can rotate freely on the connecting bond, for any PCB congener (except decachlorobiphenyl, in which every hydrogen has been substituted by a chlorine), the location of chlorines can differ (e.g., 2,3,4-trichlorobiphenyl is the same as 2',3,4-trichlorobiphenyl and the same as 2,4',6'- trichlorobiphenyl). The location of the chlorine atoms can lead to different physical, chemical, and biological characteristics of molecules, including their toxicity, persistence, and bioaccumulation potential.

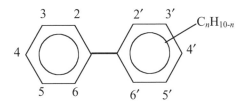

polychlorinated biphenyl structure

Numerous acids are organic, because they contain the C—C and C—H bonds. For example, acetic acid ($HC_2H_3O_2$), benzoic acid ($HC_7H_5O_2$), and cyanoacetic acid ($C_3H_3NO_2$) are organic acids. Like other compounds, organic acids can have substitutions that change their hazard, such as when acetic acid's hydrogen atoms are substituted with chlorines to form trichloroacetic acid ($C_2HCl_3O_2$).

Inorganic compounds are those that do not contain carbon-to-carbon or carbon-to-hydrogen covalent bonds. Thus, even carbon-containing compounds can be inorganic.

For example, the pesticides sodium cyanide (NaCN) and potassium cyanide (KCN) are inorganic compounds, as are the gases carbon monoxide (CO) and carbon dioxide (CO_2), compounds that contain the anions carbonate (CO_3^{2-}) and bicarbonate (HCO_3^-), and inorganic acids, such as carbonic acid (H_2CO_3) and cyanic acid (HCNO).

Metals are particularly important in environmental situations. Like other elements, the compounds formed by metals vary in their toxicity and how rapidly they move and change in the environment. However, certain metals, no matter what their form are hazardous. Unlike carbon, hydrogen, oxygen, and many other elements, which in certain configurations are essential and in others are toxic, heavy metals and metalloids are considered hazardous no matter what the chemical species. For example, any amount of lead or mercury in any form is considered toxic, although some forms are much more toxic than others. And since metals and metalloids are elements, we are not going to be able to "destroy" them as we do organic compounds by using chemical, thermal, and biological processes. Destruction simply means that we are changing compounds into simpler compounds (e.g., hydrocarbons are broken down to CO_2 and H_2O). But metals are already in elemental form. So the engineer must attempt to change the metal or metalloid to make it less toxic and less mobile, and once that is done, to take measures to keep the metal wastes away from people, wildlife, and other receptors.

The oxidation state or *valence* of metals and metalloids is the most important factor in their toxicity and mobility. The outermost electrons determine how readily an element will enter into a chemical reaction and what type of reaction will occur. This is the oxidation number of the element. Most metals contain more than one oxidation state, each with its own toxicity and mobility characteristics. However, in most cleanup situations, all forms of the metal, even those with low toxicity and mobility, must be removed since when environmental conditions change the metals may change to more toxic and mobile forms.

CASE STUDY: JERSEY CITY CHROMIUM

One of the biggest selling points for automobiles in the 1950s and 1960s was the amount of "chrome" displayed. To this day, the metal chromium (Cr) is in high demand since it strongly resists corrosion and oxidation. As such, it is an ingredient of stainless steel and is used to plate other metals.

Jersey City, in Hudson County, New Jersey, was once the chromium-processing capital of the United States, and over the years, 20 million tons of chromate ore processing residue was sold or given away as fill. The city has had at least 120 contaminated sites, including ball fields and basements underlying homes and businesses. It has not been uncommon for brightly colored chromium compounds to crystallize on damp basement walls and to "bloom" on soil surfaces where soil moisture evaporates, creating something like an orange hoar frost of hexavalent chromium, Cr^{6+}. A broken water main in the wintertime resulted in the formation of bright green ice due to the presence of trivalent chromium, Cr^{3+}.

The companies that created the chromium waste problem no longer exist, but liability was inherited by three conglomerates through a series of takeovers. In 1991, Florence Trum, a local resident, successfully sued Maxus Energy, a subsidiary of one of the conglomerates, for the death of her husband, who loaded trucks in a warehouse built directly over a chromium waste disposal site. He developed a hole in the roof of his mouth and cancer of the thorax, and it was determined by autopsy that his death was caused by chromium poisoning. Although the subsidiary company did not produce the chromium contamination, the judge ruled that company managers knew about the hazards of chromium. Such assumed liability is part of the regulations for *potentially responsible parties* (PRPs) under the Superfund law.

The state of New Jersey initially spent $30 million to locate, excavate, and remove some of the contaminated soil. But the extent of the problem was overwhelming and they stopped these efforts. The director of toxic waste cleanup for New Jersey admitted that even if the risks of living or working near chromium were known, the state does not have the money to remove it. Initial estimates for site remediation were well over $1 billion.[45] Citizens of Hudson County were angry and afraid. Those sick with cancer wondered if it could have been prevented. Mrs. Trum perceived the perpetrators as well-dressed business people who were willing to take chances with other peoples' lives. "Big business can do this to the little man . . . ," she said.

The contamination in Jersey City is from industries that used chromium in their processes, including metal plating, leather tanning, and textile manufacturing. The deposition of this chromium residue in dumps has resulted in chromium-contaminated water, soils, and sludge. Chromium is particularly difficult to regulate because of the complexity of its chemical behavior and toxicity, which translates into scientific uncertainty. Uncertainty exacerbates the tendency of regulatory agencies to make conservative and protective assumptions, the tendency of the regulated to question the scientific basis for regulations, and the tendency of potentially exposed citizens to fear potential risk.

Chromium exists in nature primarily in one of two oxidation states: Cr^{3+} and Cr^{6+}. In the reduced form of chromium, Cr^{3+}, there is a tendency to form hydroxides which are relatively insoluble in water at neutral pH values. Cr^{3+} does not appear to be carcinogenic in animal and bioassays. In fact, organically complexed Cr^{3+} has recently become one of the more popular dietary supplements in the United States and can be purchased commercially as chromium picolinate ($C_{18}H_{12}CrN_3O_6$) or with trade names such as Chromalene to help with proper glucose metabolism, to control blood fat concentrations, to aid weight loss and muscle tone, and as essential to gene expression.

When Cr^{3+} oxidized as Cr^{6+}, however, chromium is highly toxic. It is implicated in the development of lung cancer and skin lesions in industrial workers. In contrast to Cr^{3+}, nearly all Cr^{6+} compounds have been shown to be potent mutagens. The U.S. EPA has classified chromium as a human carcinogen by inhalation based on evidence that Cr^{6+} causes lung cancer. However, by ingestion, chromium has not been shown to be carcinogenic.

What confounds the understanding of chromium chemistry is that under certain environmental conditions, Cr^{3+} and Cr^{6+} can interconvert. In soils containing manganese, Cr^{3+} can be oxidized to Cr^{6+}. Given the heterogeneous nature of soils, these redox reactions can occur simultaneously. Although organic matter may serve to reduce Cr^{6+}, it may also complex Cr^{3+} and may make it more soluble—facilitating its transport in groundwater and increasing the likelihood of encountering oxidized manganese present in the soil.

Cleanup limits for chromium are still undecided, but through the controversy there have evolved some useful technologies to aid in resolution of the disputes. For example, analytical tests to measure and distinguish between Cr^{3+} and Cr^{6+} in soils have been developed. Earlier in the history of New Jersey's chromium problem, these assays were not reliable and would have necessitated remediating to soil concentrations based on total chromium. Other technical/scientific advances include *in situ* remediation strategies designed to reduce chemically Cr^{6+} to Cr^{3+} in order to reduce risk without excavation and removal of soil designated as hazardous waste. The establishment of cleanup standards is anticipated, but the proposed endpoint based on contact dermatitis is controversial. Although some perceive contact dermatitis as a legitimate claim to harm, others have jokingly suggested regulatory limits for poison ivy, which also causes contact dermatitis. The methodology by which dermatitis-based soil limits were determined has come under attack by those who question the validity of skin patch tests and the inferences by which patch test results translate into soil Cr^{6+} levels.

The Jersey City community's frustration with slow cleanup and what citizens perceive as double-talk by scientists finally culminated in the unusual step of amending the state constitution to provide funds for hazardous waste cleanups. State environmentalists depicted the constitutional amendment as a referendum on Governor Christine Todd Whitman's (R) environmental record, which they perceived as relaxed enforcement and reduced cleanups. (Whitman was the first administrator of the U.S. Environmental Protection Agency to be named by President George W. Bush.)

Radioisotopes

Different atomic weights of a same element are the result of different numbers of neutrons. The number of electrons and protons of stable atoms must be the same. Elements with differing atomic weights are known as *isotopes*. An element may have numerous isotopes. Stable isotopes do not undergo natural radioactive decay, whereas radioactive isotopes involve spontaneous radioactive decay as their nuclei disintegrate, thus are known as *radioisotopes*. This decay leads to the formation of new isotopes or new elements. The stable product of an element's radioactive decay is known as a *radiogenic isotope*. For example, lead (Pb; atomic number = 82) has four naturally occurring isotopes of different masses (^{204}Pb, ^{206}Pb, ^{207}Pb, ^{208}Pb). Only ^{204}Pb is stable. The isotopes ^{206}Pb and ^{207}Pb are daughter (or progeny) products of the radioactive decay of uranium (U), while ^{208}Pb is a product of thorium (Th) decay. Owing to the radioactive decay, the heavier isotopes of lead will increase in abundance compared to ^{204}Pb. The toxicity of a radioisotope can be twofold (i.e., chemical toxicity and radioactive toxicity). For example, Pb is neurotoxic no matter the atomic weight, but if people are exposed to its unstable isotopes they are also threatened by radiation emitted from decay of the nucleus. The energy of the radioactive decay can alter genetic material and lead to mutations, including cancer.

> **CASE STUDY: RADIATION POISONING IN GOIANIA, BRAZIL**[46]
>
> Sometimes good intentions lead to unfortunate consequences. In the early 1980s a small cancer clinic was opened in Goiana, Brazil, but business was not good and the clinic closed five years later. Left behind in the abandoned building were a radiation therapy machine and some canisters containing waste radioactive material—1400 curies of cesium 137, which has a half-life of 30 years. In 1987 the container of cesium 137 was discovered by local residents and was opened, revealing a luminous blue powder. The material was a local curiosity and children even used it to paint their bodies, which caused them to sparkle. One of the little girls went home for lunch and ate a sandwich without first washing her hands. Six days later she was diagnosed with radiation illness, having received an estimated five to six times the lethal radiation exposure for adults. The ensuing investigation identified the true content of the curious barrel. In all, over 200 persons had been contaminated and 54 were serious enough to be hospitalized, with four people dying from the exposure (including the little girl with the sandwich). Treatment of radiation disease is challenging. The International Atomic Energy Commission characterized the treatment of the Goianian patients as follows:
>
> > . . . the first task was to attempt to rid their bodies of cesium. For this, they administered Prussian blue, an iron compound that bonds with cesium, aiding its excretion. The problem in this case was the substantial delay—at least a week—from initial exposure to treatment. By that time much of the cesium had moved from the bloodstream into the tissues, where it is far more difficult to remove . . . the patients were also treated with antibiotics as needed to combat infections and with cell infusions to prevent bleeding. . . .[47]
>
> By the time the government mobilized the response, the tragic damage was done. A large fraction of the local population had received excessive radiation exposures, and the export of produce from Goiania dropped to zero, creating a severe economic crisis. The incident is now recognized as the second-worst radiation accident in the world, second only to the explosion of the nuclear power plant in Chernobyl.

Factors of Safety

Of the myriad of chemicals in the environment, workplace, and home, relatively few have been associated with chronic diseases such as cancer. However, for those that do, risk seldom is zero. Simple mathematics tells us that if the hazard is zero, the risk must be zero. So only a carcinogen can cause cancer. No matter what the dose, the cancer risk from a noncarcinogen is zero. A prominent hypothesis in carcinogenesis is the *two-hit theory,* suggested by A. G. Knudson[48] in 1971. The theory argues that cancer develops after genetic material [i.e., usually deoxyribonucleic acid (DNA)] is damaged. The first damage is known as *initiation.* This step may, but does not necessarily, lead to cancer. The next step, *promotion,* changes the cell's makeup and nature, such as the loss of normal homeostasis (cellular self-regulation) and the rapid division of clonal tumor cells. Promoters may or may not be carcinogens. So when we say that a noncarcinogen dose cannot lead to cancer, we are talking specifically of compounds that initiate cancer, since exposure to noncarcinogenic promoters, such as excessive dietary fats, can hasten the onset of cancer cells.

The RfD is the principal factor of safety used in assigning hazard to noncarcinogens. The slope factor (SF) is the principal hazard characteristic for carcinogens. Both factors are developed from a mix of mutagenicity studies, animal testing, and epidemiology. Unlike the RfD, which provides a "safe" level of exposure, cancer risk assessments generally assume that there is no threshold. Thus, the thresholds NOAEL and LOAEL are meaningless for cancer risk. Instead, cancer slope factors are used to calculate the estimated probability of increased cancer incidence over a person's lifetime [called the *excess lifetime cancer risk* (ELCR)]. Slope factors are expressed in inverse exposure units since the slope of the dose–response curve is an indication of risk per exposure. Thus, the units are the inverse of mass per mass per time, usually $(\text{mg kg}^{-1} \text{ day}^{-1})^{-1} = \text{kg} \cdot \text{day mg}^{-1}$. This means that the product of the cancer slope factor and exposure (i.e., risk) is dimensionless. This should make sense because risk is a unitless probability of adverse outcomes. The SF values are contaminant- and route-specific. Thus, one must not only know the contaminant, but how a person is exposed (e.g., via inhalation, via ingestion, or through the skin). Inhalation, oral, and dermal cancer slope factors are shown in Table 4.5.

The more potent the carcinogen, the larger the slope factor will be (i.e., the steeper the slope of the dose–response curve). Note, for example, that when inhaled, ingested, or dermally exposed, the slope for the most carcinogenic dioxin tetrachlorodibenzo-*p*-dioxin, is eight orders of magnitude steeper than the slope for aniline. Keep in mind that this is the linear part of the curve. The curve is actually sigmoidal because at higher doses the effect is dampened (i.e., the response is increasing at a decreasing dosage rate). This process is sometimes called the saturation effect. One way to think about this is to consider that if the dose–response curve comes from animal tests of various doses there is a point at which increasing the dose of a chemical adds little to the onset of tumors. The dosage approaches an effective limit and becomes asymptotic. So if chemical A is given to 1000 rats, at increasing dosages an incremental increase in rats with tumors is seen. This is the linear range. Doubling the dose doubles the effect. But at some inflection point, say after 50 rats with tumors, if the dose is doubled, half as many additional rats with tumors are seen. The rate continues to decrease up to a point where even very large doses do not produce many additional tumors. This is one of the challenges of animal experiments and models. Dose is substituted for time; the assumed lifetime of humans is about 70 years, and the doses to carcinogens are usually very small (e.g., parts per billion or trillion). Animal doses may last only a few months and use relatively high doses. We have to extrapolate long-term effects from limited data from short-term studies. The same is somewhat true for human studies, where we try to extrapolate effects from a small number of cases to a much larger population (e.g., a small study comparing cases to controls in one hospital, or a retrospective view of risk factors that may have led to a cluster of cases of cancer).

Table 4.5 Cancer Slope Factors for Some Environmental Contaminants[a]

Contaminant	Inhalation slope factor ($kg \cdot day\ mg^{-1}$)	Oral slope factor ($kg \cdot day\ mg^{-1}$)	Dermal slope factor ($kg \cdot day\ mg^{-1}$)
Acrylonitrile	2.38×10^{-1}	5.40×10^{-1}	6.75×10^{-1}
Aniline	5.70×10^{-3}	5.70×10^{-3}	1.14×10^{-3}
Arsenic	1.51×10^{1}	1.50	1.58×10^{1}
Atrazine	4.44×10^{-1}	2.22×10^{-1}	4.44×10^{-1}
Benzene	2.90×10^{-2}	2.90×10^{-2}	3.22×10^{-2}
Benz[a]anthracene	3.10×10^{-1}	7.30×10^{-1}	1.46
Benzo[a]pyrene	3.10	7.30	1.46×10^{1}
Benzo[b]fluoranthene	3.10×10^{-1}	7.30×10^{-1}	1.46
Bis(2-chloroethyl)ether	1.16	1.16	1.13
Bis(2-chloroisopropyl)ether (DEHP)	3.50×10^{-2}	1.10×10^{-2}	8.75×10^{-2}
Bis(2-ethylhexyl)phthalate	1.40×10^{-2}	7.00×10^{-2}	2.80×10^{-2}
Bromodichloromethane	6.20×10^{-2}	6.20×10^{-2}	6.37×10^{-2}
Bromoform	3.85×10^{-3}	7.90×10^{-3}	1.05×10^{-2}
Cadmium	Not given	6.30	Not given
Chlordane	3.50×10^{-1}	3.50×10^{-1}	4.38×10^{-1}
Chloroethane (ethyl chloride)	2.90×10^{-3}	2.90×10^{-3}	1.28
Chloroform	8.05×10^{-2}	6.10×10^{-3}	6.10×10^{-3}
Chloromethane	3.50×10^{-3}	1.30×10^{-2}	1.63×10^{-2}
Chromium(VI)	3.50×10^{-3}	Not given	Not given
DDD	2.40×10^{-1}	2.40×10^{-1}	3.00×10^{-1}
Dichlorobenzene,1,4-	2.20×10^{-2}	2.40×10^{-2}	2.40×10^{-2}
Dieldrin	1.61×10^{1}	1.61×10^{1}	1.60×10^{1}
Dinitrotoluene, 2,4-	6.80×10^{-1}	6.80×10^{-1}	6.80×10^{-1}
Dioxane, 1,4-	2.20×10^{-2}	1.11×10^{-2}	2.20×10^{-2}
Diphenylhydrazine, 1,2-	7.70×10^{-1}	8.00×10^{-1}	1.60
Ethylene oxide	3.50×10^{-1}	1.02	1.28
Formaldehyde	4.55×10^{-2}	Not given	Not given
Heptachlor epoxide	9.10	9.10	2.28×10^{1}
Hexachlorobenzene	1.61	1.60	2.00
Hexachlorocyclohexane, α	6.30	6.30	6.47
Hexachlorocyclohexane, β	1.80	1.80	1.99
Hexachlorocyclohexane, γ (lindane)	1.30	1.30	1.31
Hexahydro-1,3,5-trinitro-1,3,5-triazine (RDX)	2.22×10^{-1}	1.11×10^{-1}	2.22×10^{-1}
Nitrosodi-n-propylamine, n-	7.00	7.00	1.47×10^{1}
Pentachlorophenol	1.20×10^{-1}	1.20×10^{-1}	2.40×10^{-1}
Polychlorinated biphenyls (Arochlor mixture)	3.50×10^{-1}	2.00	2.35
Tetrachlorodibenzo-p-dioxin, 2,3,7,8-	1.16×10^{5}	1.50×10^{5}	1.68×10^{5}
Tetrachloroethane,1,1,1,2-	2.59×10^{-2}	2.60×10^{-2}	3.25×10^{-2}
Tetrachloroethane,1,1,2,2-	2.03×10^{-1}	2.03×10^{-1}	2.86×10^{-1}
Tetrachloroethene (PCE)	2.00×10^{-3}		5.20×10^{-2}
Tetrachloromethane (carbon tetrachloride)	5.25×10^{-2}	1.30×10^{-1}	1.53×10^{-1}

Table 4.5 (*Continued*)

Contaminant	Inhalation slope factor $(kg \cdot day\ mg^{-1})$	Oral slope factor $(kg \cdot day\ mg^{-1})$	Dermal slope factor $(kg \cdot day\ mg^{-1})$
Toxaphene	1.12	1.10	1.75
Trichloroethane,1,1,2-	5.60×10^{-2}	5.70×10^{-2}	7.04×10^{-2}
Trichloroethene (TCE)	6.00×10^{-3}	1.10×10^{-2}	1.16×10^{-2}
Trichlorophenol, 2,4,6-	1.10×10^{-2}	1.10×10^{-2}	2.20×10^{-2}
Trichloropropane, 1,2,3-	8.75	7.00	8.75
Trinitrotoluene, 2,4,6- (TNT)	6.00×10^{-2}	3.00×10^{-2}	6.00×10^{-2}
Vinyl chloride	3.00×10^{-1}	1.90	2.17

Source: U.S. Environmental Protection Agency, *Integrated Risk Information System,* U.S. EPA, Washington, DC, 2002; U.S. Environmental Protection Agency, *Health Effects Summary Tables,* U.S. EPA, Washington, DC, 1994.

[a]These values are updated periodically. If a carcinogen is not listed in the table, visit http://risk.lsd.ornl.gov/tox/rap_toxp.shtml.

It can be argued that addressing rare and chronic diseases such as cancer, endocrine dysfunction, reproductive disorders, and neurological diseases is an effort in the control of variables to reduce the possibility of an improbable (thankfully!) event. New statistical techniques are being developed to help engineers deal with rare events.

Discussion: Small Changes

Small changes can be very profound in rare events. If you think about it, when you start with very small numbers, a slight change can make a difference. Stockbrokers and retailers use this phenomenon often. For example, a company may be the fastest-growing company in its field this year. Upon investigation, its sales may have been only $5 last year but grew to $5000 this year, a 1000-fold increase. Real estate investors might say that sales grew 100,000% this year, whereas engineers and scientists generally prefer absolute terms and might say that the growth rate was 4.995×10^3 yr^{-1}. Both of these are correct statements. But would you rather invest in a company that had $10 million in sales last year and grew to $20 million this year? That is only a doubling of the income, or only 100% growth. But the absolute growth is 1×10^6 yr^{-1}, or three orders of magnitude greater than that for the small firm. What does this tell us about rare outcomes such as cancer?

In reviewing epidemiological information, are the data given an incidence of disease or prevalence? Disease *incidence* is the number of new cases diagnosed each year, whereas *prevalence* is the number of cases at any given time. We must also be careful to ascertain whether the values are absolute or relative. For example, are the values given a year-over-year change, or are they simply a one-time event? In environmental and public health reports, especially risk assessments, the values are often presented as probabilities in engineering notation; for example, a common target of cleanup of hazardous waste sites is that no more than one additional case of cancer per million population should result from the clean site (i.e., the added risk is less than or equal to 10^{-6}). Like all probabilities, this is simply a fraction and a decimal. However, if the engineer uses it in a public forum, it can be very disarming and not clearly understood. In fact, the entire concept of population risk is foreign to most people. The point is that when the engineer goes about explaining rare events such as cancer, great care must be taken.

The science of toxicology deals with even smaller values and often very limited data. In fact, one of the raging toxicological debates is that of cancer dose–response and where to literally "draw the line." As a matter of scientific policy, in what is known as the *precautionary principle,* many health agencies around the world assume that a single molecule of a carcinogen *can* cause cancer. In other words, there is no threshold under which a dose, no matter how small, would be safe; "one hit potentially leads to a tumor." This approach is commonly known as the *one-hit model.* Most other diseases have such a threshold dose, known as the *no observed adverse effect level* (NOAEL; shown in Figure 4.7). The precautionary principle is in large part due to our lack of understanding of how things work at the molecular level. Toxicological models

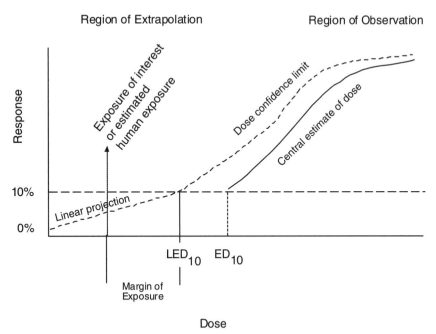

Figure 4.16 Linearized multistage dose–response curve showing the two major regions of data availability. LED_{10} = lower 95% confidence limit on a dose associated with 10% extra risk; ED_{10} = estimate of the dose that would lead to 10% increase in the response (in this case, cancer). (From D. A. Vallero, *Environmental Contaminants: Assessment and Control,* Elsevier Academic Press, Burlington, MA, 2004.)

work better when they use observed data, but at level below this, we are guessing (albeit a very educated guess) as to what is happening (see Figure 4.16). Since risk at very low doses is not directly measurable using animal experiments or from epidemiology, mathematical models are used to extrapolate from high to low doses.

Numerous extrapolation models or procedures may reasonably fit the observed data; however, extremely large differences of risk at low doses can be calculated. Scientists must use different models, depending on the particular chemical compound, as well as use information about how cancer seems to be occurring (i.e., the biological *mechanism of action* at work in the cell).[49] When such biological information is limited, the default is to assume linearity, and since there is no threshold, the curve intersects the *x*-axis and the *y*-axis at 0. For example, the U.S. Environmental Protection Agency usually recommends a linearized multistage procedure as the default model unless sufficient information to the contrary exists. The linearized multistage procedure calls for the fitting of a multistage model to the data. *Multistage models* are exponential models approaching 100% risk at high doses, with a shape at low doses given by a polynomial function. If this is first degree, the model is equivalent to a one-hit model, yielding almost a linear relationship between low dose and cancer risk. An upper bound risk is estimated by applying an appropriate linear term to the statistical bound for the polynomial. At sufficiently small exposures, any higher-order terms in the polynomial are assumed to be negligible, and the graph of the upper bound will appear to be a straight line. The slope of this line is called the *slope factor,* which is a measure of the cancer potency of the compound (i.e., the steeper the slope, the more potent the carcinogen).[50]

The units that we use in engineering can make risk communication unclear. For example, when we treat pollution, we often use a measure of pollutant removal efficiency, such as "percent removal." To see how well an incinerator is destroying a hazardous substance, engineers report the removal efficiency for that compound. In fact, the environmental engineering community uses the *rule of six nines* for extremely hazardous compounds. For example, if the most toxic form of dioxin, tetrachlorodibenzo-*p*-dioxin (TCDD) is in a waste stream, the incinerator must destroy 99.9999% (six nines) of the TCDD. If the incinerator is destroying 99.9998%, theoretically it is out of compliance (of course, this begs the question about the worthiness of our measurement techniques and significant figures, but that is another matter!). Often, however, the removal is reported in units of mass or concentration. If a waste contains a total of 100 mg (mass), or 100 mg L^{-1} (concentration), of TCDD, after treatment in a properly operating incinerator, we are left with 0.0001 mg if we started with 100 mg (100 mg − 0.999999 × 100 mg). If the incinerator increases its efficiency by seven nines (99.99999% removal), we would have 0.00001 mg of TCDD left; that is, the improvement allowed us to remove only 0.00009 mg of TCDD. If you want to make this incinerator improvement look better, you report it as nanograms (ng) removed (10 ng better). If you want to make the difference look insignificant, you report it as grams removed (only 0.00001 g removed). But both removal efficiencies are the same; only the units differ.

A further problem is that removal efficiency is a relative measure of success. If a waste has a large amount of a contaminant, even relatively inefficient operations look good. Taking the TCDD example; if waste A has 100 g of TCDD (scary thought!) and waste B has 100 ng of TCDD and they both comply with the rules of six nines, the waste A incinerator is releasing 0.0001 g or 100 ng of the contaminant to the atmosphere, whereas the waste B incinerator is emitting only 0.0001 ng. That is why environmental laws also set limits on the maximum mass or concentration of a contaminant leaving the stack (or pipe, for water discharges). In addition, the laws require that for some pollutants the ambient concentration not be exceeded. However, for many very toxic compounds that require elaborate and expensive monitoring devices, such ambient monitoring is infrequent and highly localized (e.g., near a known polluter). Regulators often depend on self-reporting by the facilities, with occasional audit (analogous to the IRS accepting a taxpayer's self-reporting, which is verified to some extend by audits of a certain sample of taxpayers).

Statistics and probabilities for extreme and rare events can be perplexing. People want to know about trends and differences in exposures and diseases between their town or neighborhood and those of others. Normal statistical information about central tendencies such as the mean, median, and mode, or ranges and deviations, fail us when we analyze rare events. Normal statistics allows us to characterize the typical behaviors in our data in terms of differences between groups and trends, focusing on the center of the data. *Extreme value theory* (EVT), conversely, lets us focus on the points far out on the tail of our data, with the intent of characterizing a rare event. For example, perhaps we have been collecting health data for 10 years for thousands of workers exposed to a contaminant. What is special about those who have been most highly exposed (e.g., those at the 99th percentile)? What can we expect as the highest exposures over the next 50 years? EVT is one means of answering these questions. The first question can be handled with traditional statistics, but the second is an extrapolation (50 years hence) beyond our data set.

Such extrapolations in EVT are justified by a combination of mathematics and statistics (i.e., probability theory and inference and prediction, respectively). This can be a very powerful analytical tool. However, the challenge may come after the engineer has completed the analysis. The engineer may be confident that the neighborhood does not involve much additional risk based on EVT and traditional methods. But how does the engineer explain how such a conclusion was derived? Many in the audience have not taken a formal course in basic statistics, let alone a course that deviates from the foundations of statistics, such as EVT! Senol Utku, a former colleague at Duke, was fond of saying: "To understand a non-banana, one must first understand a banana." This was in the context of discussing the value of linear relationships in engineering. Everyone recognizes that many engineering and scientific processes and relationships are nonlinear in their behavior, but students must first learn to apply linear mathematics. Our advice is to use the best science possible, but be ready to support your approaches in understandable ways, targeted to the specific audience.

Exposure Estimation

Now consider the second part of the risk equation. An *exposure* is any contact with an agent. For chemical and biological agents, contact can come about from a number of exposure pathways (i.e., routes taken by a substance), from its source to its endpoint (i.e., a target organ such as the liver, or a location short of that, such as in fat tissues). The substances often change to other chemical species as a result of the body's metabolic and detoxification processes. These new substances are known as *degradation products* or *metabolites.*

Physical agents such as electromagnetic radiation, ultraviolet (UV) light, and noise do not follow this pathway exactly. The contact with these sources of energy can elicit a physiological response that may generate endogenous chemical changes that behave somewhat like metabolites. For example, UV light may infiltrate and damage skin cells. The UV light helps to promote skin-tumor promotion by activating the transcription factor complex activator protein-1 (AP-1) and enhancing the expression of the gene that produces the enzyme cyclooxygenase-2 (COX2). Noise (i.e., acoustical energy), can also elicit physiological responses that affect an organism's chemical messaging systems (i.e., endocrine, immune, and neural).

The exposure pathway also includes the manner in which people can come into contact with (i.e., be exposed to) the agent. The pathway has five parts:

1. The source of contamination (e.g., a stack or pipe)
2. An environmental medium and transport mechanism (e.g., the air)
3. A point of exposure (e.g., indoor air)
4. A route of exposure (e.g., inhalation, dietary ingestion, nondietary ingestion, dermal contact, nasal route)
5. A receptor population (those who are actually or potentially exposed)

If all five parts are present, the exposure pathway is known as a *completed exposure pathway.* In addition, the exposure may be short-term, intermediate, or long-term. *Short-term contact* is known as an *acute exposure* [i.e., occurring as a single event or for only a short period of time (up to 14 days)]. An *intermediate exposure* is one that lasts from 14 days to less than one year. *Long-term* or *chronic exposures* are greater than one year in duration.

Determining the exposure for a neighborhood can be complicated. For example, even if we do a good job of identifying all of the contaminants of concern and the possible source of these pollutants (no small task), we may have little idea of the extent to which the receptor population has come into contact with these contaminants (steps 2 through 4). Thus, assessing exposure involves not only the physical sciences but also the social sciences (e.g., psychology and behavioral sciences). People's activities greatly affect the amount and type of exposure. That is why exposure scientists use a number of techniques to establish activity patterns, such as asking potentially exposed individuals to keep diaries, videotaping, using telemetry to monitor vital information (e.g., heart and ventilation rates), and comparing individual records to biomarkers (e.g., cotinine in urine as indication of tobacco smoking).

Ambient measurements, such as air pollution monitoring equipment located throughout cities, are generally not good indicators of actual population exposures. Neither, necessarily, are gross production and release estimates. The metals lead (Pb) and mercury (Hg) and their compounds comprise the greatest mass of toxic substances released into the U.S. environment. This is due largely to the large volume and surface areas involved in metal extraction and refining operations. However, this does not necessarily mean that more people will be exposed at higher concentrations or more frequently to these compounds than to others. The mere fact that a substance is released or even that it is found in the ambient environment is not tantamount to its coming in contact with people. Conversely, even a small amount of a substance under the right circumstances can lead to very high levels of exposure (e.g., in an occupational setting, in certain indoor environments, and through certain pathways, such as nondietary ingestion of paint chips by children).

The Lawrence Berkley National Laboratory recently demonstrated the importance of not simply assuming that the released or even background concentrations are a good indicator of actual exposure.[51] The researchers were interested in how sorption may affect indoor environments, so they set up a room (chamber) made up of typical building materials and furnished with actual furniture such as that found in most residential settings. A number of air pollutants were released into the room and monitored. Figure 4.17 shows an organic solvent, xylene, exhibiting the effects of sorption. With the room initially sealed, the decay observed in vapor-phase concentrations indicates that the compound is adsorbing onto surfaces (walls, furniture, etc.). The adsorption continues for hours, with xylene concentrations reaching a quasi-steady state. At this point the room is flushed with clean air to free all vapor-phase xylene. Shortly after the flush, the xylene concentrations began to rise again, until reaching a new steady state. This rise must be the result of desorption of the previously sorbed xylene, since the initial source is gone.

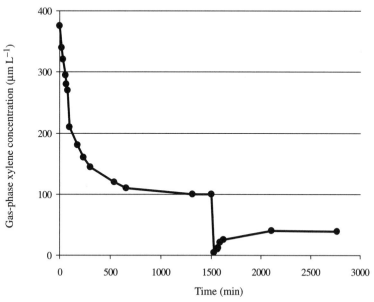

Figure 4.17 Concentrations of xylene measured in its vapor phase in a chamber sealed during adsorption and desorption periods. [Adapted from B. Singer, A Tool to Predict Exposure to Hazardous Air Pollutants, *Environmental Energy Technologies Division News,* 4(4): 5, 2003.]

Sorption is one of the processes that must be considered to account for differences in the temporal pattern of indoor *versus* outdoor concentrations.

Figure 4.18 shows a number of the ways that contaminants can enter and leave an indoor environment. People's activities as they move from one location to another make for unique exposures. For example, people generally spend much more time indoors than outdoors. The simplest quantitative expression of exposure is

$$E = \frac{D}{t} \tag{4.14}$$

where E is the human exposure during the time period t [units of concentration (mass per volume) per time] (mg kg^{-1} day^{-1}), D is the mass of pollutant per body mass (mg kg^{-1}), t is the time (days). Usually, to obtain D, the chemical concentration of a pollutant is measured near the interface of the person and the environment during a specified time period. This measurement is sometimes referred to as the *potential dose* (i.e., the chemical has not yet crossed the boundary into the body, but is present where it may enter the person, such as on the skin, at the mouth, or at the nose).

Exposure is a function of the concentration of the agent and time. It is an expression of the magnitude and duration of the contact. That is, exposure to a contaminant is the concentration of that contact in a medium integrated over the time of contact:

$$E = \int_{t=t_1}^{t=t_2} C(t) \, dt \tag{4.15}$$

where E is the exposure during the time period from t_1 to t_2 and $C(t)$ is the concentration at the interface between the organism and the environment at time t.

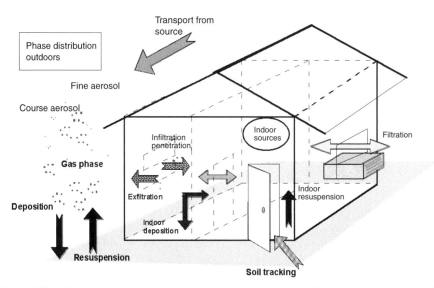

Figure 4.18   The movement and change of a chemical compound (i.e., the mass balance) is a key component of an exposure assessment. (From U.S. Department of Energy, Lawrence Berkeley National Laboratory, http://eetd.lbl.gov/ied/ERA/CalEx/partmatter.html 2003.)

Since the amount of a chemical agent that penetrates from the ambient atmosphere into a building affects the concentration term of the exposure equation, a complete mass balance of the contaminant must be understood and accounted for; otherwise, exposure estimates will be incorrect. The *mass balance* consists of all inputs and outputs as well as chemical changes to the contaminant:

accumulation or loss of contaminant A

$$= \text{mass of A transported in} - \text{mass of A transported out} \pm \text{reactions} \quad (4.16)$$

The reactions may be either those that generate chemical A (i.e., *sources*), or those that destroy chemical A (i.e., *sinks*). Thus, the amount of mass transported in is the *inflow* to the system that includes pollutant discharges, transfer from other control volumes and other media (e.g., if the control volume is soil, the water and air may contribute to the mass of chemical A), and formation of chemical A by abiotic chemistry and biological transformation. Conversely, the *outflow* is the mass transported out of the control volume, which includes uptake by biota, transfer to other compartments (e.g., volatilization to the atmosphere), and abiotic and biological degradation of chemical A. This means that the rate of change of mass in a control volume is equal to the rate of chemical A transported in, minus the rate of chemical A transported out, plus the rate of production from sources, minus the rate of elimination by sinks. Stated as a differential equation, the rate of change for contaminant A is

$$\frac{d[A]}{dt} = -v\,\frac{d[A]}{dx} + \frac{d}{dx}\left(\Gamma\,\frac{d[A]}{dx}\right) + r \quad (4.17)$$

where v is the fluid velocity, Γ is a rate constant specific to the environmental medium, $d[A]/dx$ is the concentration gradient of chemical A, and r represents the internal sinks and sources within the control volume.

Reactive compounds can be particularly difficult to measure. For example, many volatile organic compounds in the air can be measured by collection in stainless steel canisters and analysis in the lab by chromatography. However, some of these compounds, such as the carbonyls (notably, aldehydes such as formaldehyde and acetaldehyde), are prone to react inside the canister, meaning that by the time the sample is analyzed, a portion of the carbonyls is degraded (underreported). Therefore, other methods are used, such as trapping the compounds with dinitrophenyl hydrazine (DNPH)–treated silica gel tubes that are frozen until being extracted for chromatographic analysis. The purpose of the measurement is to see what is in the air, water, soil, sediment, or biota at the time of sampling, so that any reactions before the analysis gives measurement error.

Remember that the chemical that is released may or may not be what the engineer measures. If the chemical released is reactive, some or all of it may have changed into another form (i.e., *speciated*) by the time it is measured. Even relatively nonreactive compounds may speciate between when the sample is collected (e.g., in a water sample, air canister, soil core, or bag) and when the sample is analyzed. In fact, each contaminant has unique characteristics which vary according to the type of medium in which it exists, and extrinsic conditions such as temperature and pressure. Sample preservation and holding times for anions according to EPA Method 300.1 are shown in Table 4.6. These methods vary according to the contaminant of concern and the environmental

Table 4.6   Preservation and Holding Times for Anion Sampling and Analysis

Analyte	Preservation	Holding time
Part A: Common Anions		
Bromide	None required	28 days
Chloride	None required	28 days
Fluoride	None required	28 days
Nitrate-N	Cool to 4°C	48 hours
Nitrite-N	Cool to 4°C	48 hours
o-Phosphate-P	Cool to 4°C	48 hours
Sulfate	Cool to 4°C	28 days
Part B: Inorganic Disinfection By-products		
Bromate	50 mg L^{-1} EDA	28 days
Bromide	None required	28 days
Chlorate	50 mg L^{-1} EDA	28 days
Chlorite	50 mg L^{-1} EDA, cool to 4°C	14 days

Source: U.S. Environmental Protection Agency, *EPA Method 300.1: Determination of Inorganic Anions in Drinking Water by Ion Chromatography,* Revision 1.0., U.S. EPA, Washington, DC, 1997.

medium from which it is collected, so engineers need to find and follow the correct methods.

The general exposure equation (4.15) is rewritten to address each route of exposure, accounting for chemical concentration and the activities that affect the time of contact. The exposure calculated from these equations is actually the chemical intake (*I*) in units of concentration (mass per volume or mass per mass) per time, such as mg kg^{-1} day^{-1}:

$$I = \frac{C \cdot CR \cdot EF \cdot ED \cdot AF}{BW \cdot AT} \tag{4.18}$$

where *C* is the chemical concentration of contaminant (mass per volume), *CR* is the contact rate (mass per time), *EF* is the exposure frequency (number of events, dimensionless), and *ED* is the exposure duration (time). These factors are further specified for each route of exposure, such as the lifetime average daily dose (LADD) as shown in Table 4.7. The LADD is obviously based on chronic, long-term exposure.

Acute and subchronic exposures require different equations, since the exposure duration (ED) is much shorter. For example, instead of LADD, acute exposures to noncarcinogens may use the maximum daily dose (MDD) to calculate exposure. However, even these exposures follow the general model given in equation (4.18).

Table 4.7   Equations for Calculating Lifetime Average Daily Dose (LADD) for Various Routes of Exposure

Route of exposure	Equation: LADD (in mg kg^{-1} day^{-1}) =	Definition
Inhaling aerosols (particulate matter)	$$\frac{C \cdot PC \cdot IR \cdot RF \cdot EL \cdot AF \cdot ED \cdot 10^{-6}}{BW \cdot TL}$$	C = concentration of the contaminant on the aerosol/particle (mg kg^{-1}) PC = particle concentration in air (g m^{-3}) IR = inhalation rate (m^{-3} h^{-1}) RF = respirable fraction of total particulates (dimensionless, usually determined by aerodynamic diameters, e.g., 2.5 μm) EL = exposure length (h day^{-1}) ED = duration of exposure (days) AF = absorption factor (dimensionless) BW = body weight (kg) TL = typical lifetime (days) 10^{-6} is a conversion factor (kg to mg)
Inhaling vapor-phase contaminants	$$\frac{C \cdot IR \cdot EL \cdot AF \cdot ED}{BW \cdot TL}$$	C = concentration of the contaminant in the gas phase (mg m^{-3}) Other variables the same as above
Drinking water	$$\frac{C \cdot CR \cdot ED \cdot AF}{BW \cdot TL}$$	C = concentration of the contaminant in the drinking water (mg L^{-1}) CR = rate of water consumption (L day^{-1}) ED = duration of exposure (days) AF = portion (fraction) of the ingested contaminant that is physiologically absorbed (dimensionless) Other variables the same as above
Contact with soil-borne contaminants	$$\frac{C \cdot SA \cdot BF \cdot FC \cdot SDF \cdot ED \cdot 10^{-6}}{BW \cdot TL}$$	C = concentration of the contaminant in the soil (mg kg^{-1}) SA = skin surface area exposed (cm^{-2}) BF = bioavailability (percent of contaminant absorbed per day) FC = fraction of total soil from contaminated source (dimensionless) SDF = soil deposition, the mass of soil deposited per unit area of skin surface (mg cm^{-1} day^{-1}) Other variables the same as above

Source: M. Derelanko, Risk Assessment, in *CRC Handbook of Toxicology,* M. J. Derelanko and M. A. Hollinger, (Eds.), CRC Press, Boca Raton, FL, 1999.

Example:

Exposure Calculation

Over an 18-year period, a polymer manufacturer has contaminated the soil on its property with vinyl chloride. The plant closed two years ago but vinyl chloride vapors continue to reach the neighborhood surrounding the plant at an average concentration of 1 mg m⁻³. Assume that people are breathing at a ventilation rate of 0.5 m³ h⁻¹ (about the average of adult males and females over 18 years of age[52]). The legal settlement allows neighboring residents to evacuate and sell their homes to the company. However, they may also stay. The neighbors have asked for advice on whether to stay or leave, since they have already been exposed for 20 years.

Vinyl chloride is highly volatile, so its phase distribution will be mainly in the gas phase rather than the aerosol phase. Although some of the vinyl chloride may be sorbed to particles, we will use only the vapor-phase LADD equation, since the particle phase is likely to be relatively small. Also, we will assume that outdoor concentrations are the exposure concentrations. This is unlikely, however, since people spend very little time outdoors, so this may provide an additional factor of safety. To determine how much vinyl chloride penetrates living quarters, indoor air studies would have to be conducted. For a scientist to compare exposures, indoor air measurements should be taken.

   Find the appropriate equation in Table 4.7 and insert values for each variable. Absorption rates are published by the EPA and the Oak Ridge National Laboratory (http://risk.lsd.ornl.gov/cgi-bin/tox/TOX_select?select=nrad). Vinyl chloride is well absorbed, so we can assume that $AF = 1$. We will also assume that a person who stays in the neighborhood is exposed to the average concentration 24 hours a day ($EL = 24$) and that a person lives the remainder of an entire typical lifetime exposed at the measured concentration.

   Although the ambient concentrations of vinyl chloride may have been higher when the plant was operating, the only measurements we have are those taken recently. Thus, this is an area of uncertainty that must be discussed with clients. The common default value for a lifetime is 70 years, so we can assume that the longest exposure would be 70 years (25,550 days). Table 4.8 gives some of the commonly used default values in exposure assessments. If the person is now 20 years of age, has already been exposed for that time, and lives a remaining 50 years exposed at 1 mg m⁻³:

$$\text{LADD} = \frac{C \cdot IR \cdot EL \cdot AF \cdot ED}{BW \cdot TL}$$

$$= \frac{1 \cdot 0.5 \cdot 24 \cdot 1 \cdot 25{,}550}{70 \cdot 25{,}550}$$

$$= 0.2 \text{ mg kg}^{-1} \text{ day}^{-1}$$

If the 20-year-old leaves today, the exposure duration would be for the 20 years that the person lived in the neighborhood. Thus, only the ED term would change: from 25,550 days to 7300 days (i.e., 20 years).

Table 4.8 Commonly Used Human Exposure Factors[a]

Exposure factor	Adult male	Adult female	Child (3–12 years of age)[b]
Body weight (kg)	70	60	15–40
Total fluids ingested (L day^{-1})	2	1.4	1.0
Surface area of skin, without clothing (m^2)	1.8	1.6	0.9
Surface area of skin, wearing clothes (m^2)	0.1–0.3	0.1–0.3	0.05–0.15
Respiration/ventilation rate (L min^{-1}) Resting	7.5	6.0	5.0
Light activity	20	19	13
Volume of air breathed (m^3 day^{-1})	23	21	15
Typical lifetime (yr)	70	70	N.A.[c]
National upper-bound time (90th percentile) at one residence (yr)	30	30	N.A.
National median time (50th percentile) at one residence (yr)	9	9	N.A.

Source: U.S. Environmental Protection Agency, *Exposure Factor Handbook,* U.S. EPA, Washington, DC, 2003; and Agency for Toxic Substances and Disease Registry, *ATSDR Public Health Assessment Guidance Manual,* ATSDR, Washington, DC, 2003.

[a] These factors are updated periodically by the U.S. EPA in the *Exposure Factors Handbook* at www.epa.gov/ncea/exposfac.htm.

[b] The definition of *child* is highly variable in risk assessment. The *Exposure Factors Handbook* uses these values for children between the ages of 3 and 12 years.

N.A., not applicable.

> Therefore, the LADD falls to $\frac{2}{7}$ of its value:
> $$\text{LADD} = 0.05 \text{ mg kg}^{-1} \text{ day}^{-1}$$

Once the hazard and exposure calculations are done, we are able to characterize the risk quantitatively. There are two general ways that such risk characterizations are used in environmental problem solving: direct risk assessments and risk-based cleanup standards.

Direct Risk Calculations

In its simplest form, risk is the product of the hazard and the probability of exposure to that hazard, but assumptions can greatly affect risk estimates. For example, cancer risk can be defined as the theoretical probability of contracting cancer when exposed continually for a lifetime (e.g., 70 years) to a given concentration of a substance (carcinogen). The probability is usually calculated as an upper confidence limit. The maximum estimated risk may be presented as the number of chances in a million of contracting cancer.

Two measures of risk are commonly reported. One is the *individual risk*, the probability of a person developing an adverse effect (e.g., cancer) due to the exposure. This is often reported as a *residual* or increased probability above background. For example, if we want to characterize the contribution of all U.S. power plants to increased cancer incidence, the risk above background would be reported. The second way that risk is reported is *population risk*, the annual excess number of cancers in an exposed population. The maximum individual risk might be calculated from exposure estimates based on a *maximum exposed individual* (MEI). The hypothetical MEI lives an entire lifetime outdoors at the point where pollutant concentrations are highest. Assumptions about exposure will greatly affect the risk estimates. For example, the cancer risk from U.S. power plants has been estimated to be 100- to 1000-fold lower for an average exposed person than that calculated for the MEI.[53]

For cancer risk assessments, the hazard is generally assumed to be the slope factor, and the long-term exposure is the lifetime average daily dose:

$$\text{cancer risk} = SF \times \text{LADD} \tag{4.19}$$

Cancer Risk Calculation

Applying the lifetime average daily dose value from the vinyl chloride exposure calculation earlier, estimate the direct risk to the people living near the abandoned polymer plant. What advice would you give the neighbors?

Insert the calculated LADD values and the vinyl chloride inhalation slope factor of 3.00×10^{-1} from Table 3.5. For the two LADD values under consideration, the cancer risk to the neighborhood exposed for an entire lifetime (exposure duration = 70 years) gives us 0.2 mg kg^{-1} day^{-1} $\times$ 0.3 (mg kg^{-1} day^{-1})$^{-1}$ = 0.06. This is an incredibly high risk! The threshold for concern is often 1 in a million (0.000001), but this is a probability of 6%.

Even at the shorter-duration period (20 years of exposure instead of 70 years), the risk is calculated as $0.05 \times 0.3 = 0.017$, nearly a 2% risk. The combination of a very steep slope factor and very high lifetime exposures leads to a very high risk. Vinyl chloride is a liver carcinogen, so unless corrective actions significantly lower the ambient concentrations of vinyl chloride, the prudent course of action is that the neighbors accept the buyout and leave the area.

Incidentally, vinyl chloride has a relatively high water solubility and can be absorbed to soil particles, so ingestion of drinking water (e.g., people on private wells drawing water from groundwater that has been contaminated) and dermal exposures (e.g., children playing in the soil) are also conceivable. The total risk from a single contaminant such as vinyl chloride is equal to the sum of risks from all pathways (e.g., vinyl chloride in the air, water, and soil):

$$\text{total risk} = \sum \text{risks from all exposure pathways} \tag{4.20}$$

Requirements and measures of success are seldom, if ever, as straightforward as the vinyl chloride example. In fact, the engineer would be ethically remiss if the only advice given is to the local community (i.e., whether or not to accept the buyout). Of course, one of the canons is to be a "faithful agent" to the clientele. However, the first engineering canon is to hold paramount the health and safety of the public. Thus, the engineer must balance any proprietary information that the client wants to protect with the need to protect public health. In this case, the engineer must tell the client and prime contractors, for example, that the regulatory agencies need to know that even though the neighbors are moving, a threat continues for others, including future populations. In other words, just because one's clients are taken out of harm's way does not obviate the need for remediation to reduce the vinyl chloride concentrations to acceptable levels.

The risk of adverse outcome other than cancer ("noncancer risk") is generally called the *hazard quotient* (HQ). It is calculated by dividing the maximum daily dose (MDD) by the acceptable daily intake (ADI):

$$\text{noncancer risk} = \text{HQ} = \frac{\text{MDD}}{\text{ADI}} = \frac{\text{exposure}}{\text{RfD}} \tag{4.21}$$

Note that this is an index, not a probability, so it is really an indication of relative risk. If the noncancer risk is greater than 1, the potential risk may be significant, and if the noncancer risk is less than 1, the noncancer risk may be considered to be insignificant. As shown in equation (4.21), the reference dose, RfD, is one type of ADI.

Example:

> **Noncancer Risk Calculation**
>
> *Chromic acid (Cr^{6+}) mist has a dermal chronic RfD of 6.00×10^{-3} mg kg^{-1} day^{-1}. If the actual dermal exposure of people living near a metal processing plant is calculated (e.g., by intake or LADD) to be 4.00×10^{-3} mg kg^{-1} day^{-1}, calculate the hazard quotient for the noncancer risk of chromic acid mist to the neighborhood near the plant and interpret the meanings.*
>
> From equation (4.21),
>
> $$\frac{\text{exposure}}{\text{RfD}} = \frac{4.00 \times 10^{-3}}{6.00 \times 10^{-3}} = 0.67$$
>
> Since this is less than 1, one would not expect people chronically exposed at this level to show adverse effects from skin contact. However, at this same chronic exposure (i.e., 4.00×10^{-3} mg kg^{-1} day^{-1}) to hexavalent chromic acid mists via oral route, the RfD is 3.00×10^{-3} mg kg^{-1} day^{-1}, meaning the HQ = 4/3 or 1.3. The value is greater than 1, so we cannot rule out adverse noncancer effects.

If a population is exposed to more than one contaminant, the hazard index (HI) can be used to express the level of cumulative noncancer risk from pollutants 1 through n:

$$HI = \sum_1^n HQ \qquad (4.22)$$

The HI is useful in comparing risks at various locations (e.g., benzene risks in St. Louis, Cleveland, and Los Angeles). It can also give the cumulative (additive risk) in a single population exposed to more than one contaminant. For example, if the HQ for benzene is 0.2 (not significant), toluene is 0.5 (not significant), and tetrachloromethane is 0.4 (not significant), the cumulative risk of the three contaminants is 1.1 (potentially significant).

It is desirable to have realistic estimates of the hazard and exposures in such calculations. However, precaution is the watchword for risk. Estimations of both hazard (toxicity) and exposure are often worst-case scenarios, because the risk calculations can have large uncertainties. Models usually assume that effects occur even at very low doses. Human data are usually gathered from epidemiological studies, which no matter how well they are designed, are fraught with error and variability (science must be balanced with the rights and respect of subjects, populations change, activities may be missed, and confounding variables are ever present). Uncertainties exist in every phase of risk assessment, from the quality of data, to limitations and assumptions in models, to natural variability in environments and populations.

Risk-Based Cleanup Standards

Environmental protection for most of the second half of the twentieth century was based on two types of controls: technology-based and quality-based. Technology-based controls are set according to what is "achievable" from the current state of the science and engineering. These are feasibility-based standards. The Clean Air Act has called for best achievable control technologies (BACT), and more recently, for maximally achievable control technologies (MACT). Both standards reflect the reality that even though from an air quality standpoint it would be best to have extremely low levels of pollutants, technologies are not available or are not sufficiently reliable to reach these levels. Requiring unproven or unreliable technologies can even exacerbate the pollution, such as in the early days of wet scrubbers on coal-fired power plants. Theoretically, the removal of sulfur dioxide could be accomplished by venting the power plant flue through a slurry of carbonate, but technology at the time was unproven and unreliable, allowing all-too-frequent releases of untreated emissions while the slurry systems were being repaired. Selecting a new technology over older proven techniques is unwise if the trade-off of the benefit of improved treatment over older methods is outweighed by numerous failures (i.e., no treatment).

Wastewater treatment, groundwater remediation, soil cleaning, sediment reclamation, drinking water supply, air emission controls, and hazardous waste site cleanup all are in part determined by availability and feasibility of control technologies.

Quality-based controls are those that are required to ensure that an environmental resource is in good enough condition to support a particular use. For example, a stream may need to be improved so that people can swim in it and so that it can be a source of water supply. Certain streams may need higher levels of protection than others, such as the so-called "wild and scenic rivers." The parameters will vary but usually include

Biographical Sketch: Earle Phelps

Earle Phelps (1876-1953) graduated from MIT with a degree in chemistry. He was a student of William Sedgwick, who also mentored other notable early sanitary engineers and public health scientists. After graduation, Phelps worked for a while with the Massachusetts Board of Health at the Lawrence Experiment Station and was closely involved in the development of new treatment technology. He moved on first to be a faculty member at MIT and then to the U.S. Public Health Service, where he did his most influential work on stream pollution, authoring the classic text *Stream Pollution,* which for decades was considered the definitive text on the subject.

Phelps was immensely practical. He recognized that pollution would always exist, but the objective would be to reduce the effect to some reasonable level that can be attained economically using available technology. As he stated: "It is wasteful and therefore inexpedient to require a nearer approach to [the optimal] than is readily obtainable under current engineering practices and at justifiable costs." From this reasoned approach to pollution was born the *Principle of Expediency.*

Phelps argued that the objective of regulatory science is to couple the ethics of societal protection with the science of regulation. He defined public health practice as "the application of the science of preventive medicine, through government, for social ends."

minimum levels of dissolved oxygen and maximum levels of contaminants. The same goes for air quality, where ambient air quality must be achieved, with the goal that concentrations of contaminants listed as National Ambient Air Quality Standards, as well as certain toxic pollutants, are below levels established to protect health and welfare.

A third type of standard has recently emerged, one based on risk. Although numerous federal agencies were involved, environmental protection in the United States was spearheaded by the U.S. Environmental Protection Agency, created in 1970 and led during its formative years by William Ruckelshaus. After returning for his second term, Ruckelshaus saw the need for "risk-based" environmental standards and recognized that such standards would receive public support. Risk-based approaches to environmental protection, especially contaminant target concentrations, are designed to require engineering controls and preventive measures to ensure that risks are not exceeded. The risk-based approach actually embodies elements of both technology-based and quality-based standards. The technology assessment helps determine how realistic it will be to meet certain contaminant concentrations, while the quality of the environment sets the goals and means to achieve cleanup. Engineers are often asked: How clean is clean? When do we know that we have done a sufficient job of cleaning up a spill or hazardous waste site? It is often not possible to have nondetectable concentrations of a pollutant. Commonly, the threshold for cancer risk to a population is 1 in a million excess cancers. However, one may find that the contaminant is so difficult to remove that we almost give up on dealing with the contamination and put in measures to prevent exposures (i.e., fencing an area in and prohibiting access). This is often done as a first step in remediation but is unsatisfying and controversial (and usually, politically and legally unacceptable). Thus, even if costs are high and technology unreliable, the engineer must find suitable and creative ways to clean up the mess and meet risk-based standards.

Biographical Sketch: William Ruckelshaus

The strength and legitimacy of the U.S. Environmental Protection Agency owes much to the leadership of its first administrator, William D. Ruckelshaus (born 1932). Ruckelshaus is a graduate of Princeton University with a law degree from Harvard. After graduation he was a deputy attorney general in Indiana and was then elected to the Indiana House of Representatives. In 1970 he was asked by President Nixon to head the nascent U.S. EPA. During the EPA's formative years he was able to blend the various federal agencies that oversaw pollution and environmental health into one cohesive structure, took action against the severely polluted cities and industrial polluters, oversaw the setting of health-based standards for both air and water pollution, and developed the first regulations controlling emissions from automobiles (amid general anguish from the automobile and petroleum industries, which claimed that it could not be done). He worked with the states to develop both water quality standards and ambient air quality plans, and he worked to ban the use of some pesticides, such as DDT. Almost all of the environmental legislation we presently enjoy in the United States was guided through the Congress during the years William Ruckelshaus was head of the U.S. EPA.

In 1973 he stepped down from the directorship to become at first the acting director of the FBI and then briefly as deputy attorney general in the Justice Department. He distinguished himself in this post by refusing to fire the special prosecutor investigating the Watergate break-in; instead, resigning his post.

Following the disastrous tenure of Ann Gorsuch as the administrator of the U.S. EPA during the first Reagan administration, in which she was apparently charged with scuttling the agency (a popular move to numerous politicians at the time), William Ruckelshaus was once again asked to take over. He worked to rekindle both the work and the morale of the agency employees and developed widely accepted principles of risk-based decision-making in environmental controls. In his first all-hands speech, that was piped into the EPA offices around the nation, he received great applause when he ensured the employees that the agency would uphold the law and work toward its mission to protect the environment. His work in restoring and protecting water quality in the Chesapeake Bay, in developing processes for cleaning up hazardous waste sites, and in the banning of many chlorinated pesticides were significant accomplishments during his second tenure as the chief of the U.S. EPA.

Risk-based target concentrations can be calculated by solving for the target contaminant concentration in the exposure and risk equations. Since risk is the hazard (e.g., slope factor) times the exposure (e.g., LADD), a cancer risk–based cleanup standard can be found by enumerating the exposure equation (4.18) within the risk equation (in this instance, the drinking water equation from Table 4.7) gives

$$\text{risk} = \frac{C \cdot CR \cdot EF \cdot ED \cdot AF \cdot SF}{BW \cdot AT} \tag{4.23}$$

and solving for C, we have

$$C = \frac{\text{risk} \cdot BW \cdot AT}{CR \cdot EF \cdot ED \cdot AF \cdot SF} \tag{4.24}$$

This is the target concentration for each contaminant needed to protect the population from the specified risk (e.g., 10^{-6}). In other words, this is the concentration that must not be exceeded to protect a population having an average body weight and over a specified averaging time from an exposure of certain duration and frequency that leads to a risk of 1 in a million. Although 1-in-a-million added risk is a commonly used benchmark, cleanup may not always be required to achieve this level. For example, if a site is considered to be a "removal" action (i.e., the principal objective is to get rid of a sufficient amount of contaminated soil to reduce possible exposures), the risk reduction target may be as high as one additional cancer per 10,000 (i.e., 10^{-4}). This is an area of risk management, which is discussed in greater detail in Chapter 5.

Example:

Risk-Based Contaminant Cleanup

A well is the principal water supply for the town of Apple Chill. A study has found that the well contains 80 mg L^{-1} tetrachloromethane (CCl_4). Assuming that the average adult in the town drinks 2 L day^{-1} of water from the well and lives in the town for an entire lifetime, what is the lifetime cancer risk to the population if no treatment is added? What concentration is needed to ensure that the population cancer risk is below 10^{-6}?

The lifetime cancer risk added to Apple Chill's population can be estimated using the LADD and slope factor for CCl_4. In addition to the assumptions given, we will use default values from Table 4.8. We will also assume that people live in the town for their entire lifetimes and that their exposure duration is equal to their typical lifetime. Thus, the ED and TL terms cancel, leaving the abbreviated

$$\text{LADD} = \frac{C \cdot CR \cdot AF}{BW}$$

Since we have not specified male or female adults, we will use the average body weight, assuming that there are about the same number of males as females. We look up the absorption factor for CCl_4 and find that it is 0.85, so the adult lifetime exposure is

$$\text{LADD} = \frac{80 \cdot 2 \cdot 0.85}{65} = 4.2 \text{ mg kg}^{-1} \text{ day}^{-1}$$

Using the midpoint value between the default values $[(15 + 40)/2 = 27.5 \text{ kg}]$ for body weight and default CR values (1 L day^{-1}), the lifetime exposure for children is

$$\text{LADD} = \frac{80 \cdot 1 \cdot 0.85}{27.5} = 2.5 \text{ mg kg}^{-1} \text{ day}^{-1}$$

for the first 13 years, and the adult exposure of $4.2 \text{ mg kg}^{-1} \text{ day}^{-1}$ thereafter. The oral SF for CCl_4 is $1.30 \times 10^{-1} \text{ kg day}^{-1}$, so the added adult lifetime risk from drinking the water is

$$4.2 \times (1.30 \times 10^{-1}) = 5.5 \times 10^{-1}$$

and the added risk to children is

$$2.5 \times (1.30 \times 10^{-1}) = 3.3 \times 10^{-1}$$

However, for children, environmental and public health agencies recommend an additional factor of safety beyond what would be used to calculate risks for adults. This is known as the *10× rule*: that is, children need to be protected 10 times more than adults because they are more vulnerable, have longer life expectancies (so latency periods for cancer need to be accounted for), and their tissue is developing prolifically and changing. So in this case, with the added risk, our reported risk would be 3.3. Although this is statistically impossible (i.e., one cannot have a probability greater than 1 because it would mean that the outcome is more than 100% likely, which of course is impossible!) However, what this tells us is that the combination of a very high slope of the dose–response curve and a very high LADD leads to much needed protection, and removal of either the contaminants from the water or the provision of a new water supply. The city engineer or health department should mandate bottled water immediately.

The cleanup of the water supply to achieve risks below 1 in a million can also be calculated from the same information and reordering the risk equation to solve for C:

$$\text{risk} = \text{LADD} \times SF$$

$$\text{risk} = \frac{C \cdot CR \cdot AF \cdot SF}{BW}$$

$$C = \frac{BW}{CR \cdot AF \cdot SF \cdot \text{risk}}$$

Based on adult LADD, the well water must be treated so that the tetrachloromethane concentrations are below

$$C = \frac{65 \cdot 10^{-6}}{2 \cdot 0.85 \cdot 0.13} = 2.9 \times 10^{-4} \text{ mg L}^{-1} = 290 \text{ ng L}^{-1}$$

Based on the children's LADD, and the additional $10\times$, the well water must be treated so that the tetrachloromethane concentrations are below

$$C = \frac{27.5 \cdot 10^{-7}}{1 \cdot 0.85 \cdot 0.13} = 2.5 \times 10^{-5} \text{ mg L}^{-1} = 25 \text{ ng L}^{-1}$$

The town will have to remove the contaminant, so that the concentration of CCl_4 in the finished water must be treated to a level six orders of magnitude less than the untreated well water (i.e., lowered from 80 mg L^{-1} to 25 ng L^{-1}).

Cleanup standards are part of the arsenal needed to manage risk. However, other considerations needed to be given to a contaminated site, such as how to monitor progress in lowering pollutant levels and how to ensure that the community stays engaged and is participating in the cleanup actions, where appropriate. Even when the engineering solutions are working well, the engineer must allot sufficient time and effort to these other activities; otherwise, skepticism and distrust can arise.

CASE STUDY: THE DRAKE CHEMICAL COMPANY SUPERFUND SITE[54]

One of the downsides of industrial development is a legacy of harmful chemicals. The Drake Chemical Company of Lock Haven, Pennsylvania, was a major producer of chemicals during World War II and continued to provide employment opportunities to the economically depressed town after the war. Among the chemicals that the company disposed of in an open pit was β-naphthylamine (also known as 2-naphthylamine), a compound used as a dye.

β-naphthylamine

Unfortunately, β-naphthylamine is also a potent carcinogen (inhalation and oral cancer slope factor = 1.8),[55] having been found to be a known human carcinogen based on sufficient evidence of carcinogenicity in humans. Epidemiological studies have shown that occupational exposure to β-naphthylamine alone or when present as an impurity in other compounds is causally associated with bladder cancer in workers.[56]

In 1962, the state of Pennsylvania banned the production of this chemical, but the damage to the groundwater had already been done with the disposal of β-naphthylamine into the uncontrolled pit. An order from the state caused Drake to stop manufacturing β-naphthylamine, but the company continued to produce other chemicals, seemingly without much concern for the environment or the health of the people in Lock Haven. Finally, in 1981, the U.S. EPA closed the company site and took control of the property.

Cleanup crews discovered several unlined lagoons and hundreds of often unmarked barrels of chemicals stored in makeshift buildings. After removing the drums and draining the lagoons, the crews discovered that the β-naphthylamine had seeped into nearby property and into creeks, creating a serious health hazard. The EPA's attempts to clean the soil and the water were, however, met with public opposition. Much of the public blamed the EPA for forcing Drake Chemical, a major local employer, to close the plant. In addition, the best way to treat the contaminated soil was to burn it in an incinerator, and the EPA made plans to bring in a portable combustion unit. The public, not at all happy with EPA being there in the first place, became concerned with the emissions from the incinerator. After many studies and the involvement of the U.S. Army Corps of Engineers, the incinerator was finally allowed to burn the soil, which after treatment was spread out and covered with 3.5 feet of topsoil. The groundwater was pumped and treated, and this continued until the levels of β-naphthylamine reached background concentrations. The project was not completed until 1999. Ironicaly, part of the cleanup cost included the EPA paying the legal fees of the lawyers who argued against the cleanup.

Some general principles have been adopted almost universally by regulatory agencies, especially those concerned with cancer risks from environmental exposures (see Table 4.9).

Zero risk can occur only when either the hazard (e.g., toxicity) does not exist or the exposure to the hazard is zero. A substance found to be associated with cancers based on animal testing or observations of human populations can be further characterized to improve the certainty of linking exposure to cancer. Association of two factors, such as the level of exposure to a compound and the occurrence of a disease, does not necessarily mean that one necessarily "causes" the other. Often, after study, a third variable explains the relationship. However, it is important for science to do what it can to link causes with effects. Otherwise, corrective and preventive actions cannot be identified. So strength of association is a beginning step toward cause and effect (see the biographical sketch of Sir Bradford Hill later in the chapter). A major consideration in the strength of association is the application of sound technical judgment of the weight of evidence. For example, characterizing the weight of evidence for carcinogenicity in humans consists of three major steps:[57]

1. Characterization of the evidence from human studies and from animal studies individually

2. Combination of the characterizations of these two types of data to show the overall weight of evidence for human carcinogenicity

3. Evaluation of all supporting information to determine if the overall weight of evidence should be changed

Note that none of these steps is absolutely certain.

Students are rightfully warned in their introductory statistics courses not to confuse association with causality. One can have some very strong statistical associations that are not causal. For example, if one were to observe ice cream eating in Kansas City and

Table 4.9   General Principles Applied to Health and Environmental Risk Assessments in the United States

Principle	Explanation
Human data are preferable to animal data.	For purposes of hazard identification and dose–response evaluation, epidemiological and other human data better predict health effects than do animal models.
Animal data can be used in lieu of sufficient, meaningful human data.	Although epidemiological data are preferred, agencies are allowed to extrapolate hazards and to generate dose–response curves from animal models.
Animal studies can be used as a basis for risk assessment.	Risk assessments can be based on data from the most highly sensitive animal studies.
The route of exposure in animal study should be analogous to human routes.	Animal studies are best if based on the same route of exposure as in humans (e.g., inhalation, dermal, or ingestion routes). For example, if an air pollutant is being studied in rats, inhalation is a better indicator of effect than if the rats are dosed on the skin or if the exposure is dietary.
A threshold is assumed for noncarcinogens.	For noncancer effects (e.g., neurotoxicity, endocrine dysfunction, and immunosuppression), there is assumed to be a safe level under which no effect would occur [e.g., no observed adverse effect level (NOAEL), which is preferred, but also lowest observed adverse effect level (LOAEL)].
The threshold is calculated as a reference dose or reference concentration (air).	Reference dose (RfD) or concentration (RfC) is the quotient of the threshold (NOAEL) divided by factors of safety (uncertainty factors and modifying factors; each usually multiples of 10): $$\text{RfD} = \frac{\text{NOAEL}}{\text{UF} \times \text{MF}}$$
Sources of uncertainty must be identified.	Uncertainty factors (UFs) address: • Interindividual variability in testing • Interspecies extrapolation • LOAEL-to-NOAEL extrapolation • Subchronic-to-chronic extrapolation • Route-to-route extrapolation • Data quality (precision, accuracy, completeness, and representativeness) Modifying factors (MFs) address uncertainties that are less explicit than the UFs.
Factors of safety can be generalized.	The uncertainty and modifying factors should follow certain protocols: e.g., 10 = for extrapolation from a sensitive individual to a population; 10 = rat-to-human extrapolation, 10 = subchronic-to-chronic data extrapolation), and 10 = LOAEL used instead of NOAEL.
No threshold is assumed for carcinogens.	No safe level of exposure is assumed for cancer-causing agents.
Precautionary principle is applied to the cancer model.	A linear, no-threshold dose–response model is used to estimate cancer effects at low doses [i.e., to draw the unknown part of the dose–response curve from the region of observation (where data are available) to the region of extrapolation].
Precautionary principle is applied to cancer exposure assessment.	The most highly exposed person is generally used in the risk assessment (upper-bound exposure assumptions). Agencies are reconsidering this worst-case policy and considering more realistic exposure scenarios.

Source: U.S. Environmental Protection Agency, *General Principles for Performing Aggregate Exposure and Risk Assessment,* Office of Pesticides Programs, U.S. EPA, Washington, DC, 2001.

counted the number of people wearing shorts, one would find a strong association between shorts-wearing and ice cream–eating. Does wearing shorts cause more people to eat more ice cream? In fact, both findings are caused by a third variable, ambient temperature. Hotter temperatures drive more people to wear shorts *and* to eat more ice cream.

People have a keen sense of observation, especially when it has to do with the health and safety of their families and neighborhoods. They can "put 2 and 2 together." Sometimes, it seems to them that we engineers are telling them that 2 + 2 does *not* equal 4. That cluster of cancers in town may have nothing to do with the green gunk that is flowing out of the abandoned building's outfall. But in their minds, the linkage is obvious.

The challenge is to present information in a meaningful way without violating or overextending the interpretation of the data. If we assign causality when none really exists, we may suggest erroneous solutions. But if all we can say is that the variables are associated, the public is going to want to know more about what may be contributing an adverse affect (e.g., learning disabilities and blood lead levels). This was particularly problematic in early cancer research. Possible causes of cancer were being explored and major research efforts were being directed at myriad physical, chemical, and biological agents. So there needed to be some manner of sorting through findings to see what might be causal and what is more likely to be spurious results. Sir Austin Bradford Hill is credited with articulating key criteria that need to be satisfied to attribute cause and effect in medical research.[58] His recommended factors to be considered in determining whether exposure to an agent elicits an effect are as follows:

Criterion 1: Strength of association. For an exposure to an agent to cause an effect, the exposure must be associated with that effect. Strong associations provide more certain evidence of causality than is provided by weak associations. Common epidemiological metrics used in associations include risk ratio, odds ratio, and standardized mortality ratio.

Criterion 2: Consistency. If the exposure is associated with an effect consistently under different studies using diverse methods of study of assorted populations under varying circumstances by different investigators, the link to causality is stronger. For example, if carcinogenic effects of chemical X are found in mutagenicity studies, mouse and Rhesus monkey experiments, and human epidemiological studies, there is greater consistency between chemical X and cancer than if only one of these studies showed the effect. Consistency is one of the important factors in risk models. For example, if the animal and human data do not agree, an increased uncertainty factor is added to the reference dose (RfD).

Criterion 3: Specificity. The specificity criterion holds that the cause should lead to only one disease and that the disease should result from this single cause only. This criterion appears to be based in the germ theory of microbiology, where a specific strain of bacteria and viruses elicits a specific disease. This is rarely the case in studying most chronic environmental diseases, since a chemical can be associated with cancers in numerous organs, and the same chemical may elicit cancer, hormonal, immunological, and neural dysfunctions.

Criterion 4: Temporality. Timing of exposure is critical to causality. This criterion requires that exposure to the chemical must precede the effect. For example, in a retrospective study, the researcher must be certain that the manifestation of a disease was not already present before exposure to the chemical. If the disease were present prior to the exposure, it may not mean that the chemical in question is not a cause, but it does mean that it is not the sole cause of the disease (see criterion 3 above). This can be challenging, for example, for diseases with extended latency periods and large sub-clinical periods before being diagnosed.

Criterion 5: Biologic gradient. This is another essential criterion for environmental risks. Gradient is a familiar concept to engineers (e.g., it is central to Fick's law concentration change with distance). In risk assessment, the biological gradient is known as the *dose–response step* in risk assessment. If the level, intensity, duration, or total level of chemical exposure is increased a concomitant, progressive increase should occur in the toxic effect.

Criterion 6: Plausibility. Generally, an association needs to follow a well-defined explanation based on a known biological system. However, "paradigm shifts" in the understanding of key scientific concepts do occur. A noteworthy example is the change in the latter part of the twentieth century in the understanding of how the endocrine, immune, and neural systems function: from the view that these are exclusive systems, to today's perspective that in many ways they constitute an integrated chemical and electrical set of signals in an organism. For example, Candace Pert, a pioneer in endorphin research, has espoused the concept of mind/body, with all the systems interconnected, rather than separate and independent systems.

Criterion 7: Coherence. The criterion of coherence suggests that all available evidence concerning the natural history and biology of a disease should "stick together" (cohere) to form a cohesive whole. By that, the proposed causal relationship should not conflict or contradict information from experimental, laboratory, epidemiologic, theory, or other knowledge sources. For some time, for example, human studies linked arsenic exposure to cancer, but these were not replicated in animal studies. Eventually the animal studies also showed the linkage, but until then, it was a quandary for risk assessors. More often, animal studies first show the link and it is the human data that are more uncertain. See Criterion 2 (Consistency).

Criterion 8: Experimentation. Experimental evidence in support of a causal hypothesis may come in the form of community and clinical trials, *in vitro* laboratory experiments, animal models, and natural experiments.

Criterion 9: Analogy. The term *analogy* implies a similarity in some respects among things that are otherwise different. It is thus considered one of the weaker forms of evidence.

In assessing and managing environmental risks, some of Hill's criteria are more important than others. Risk assessments rely heavily on strength of association (e.g., to establish dose–response relationships). Coherence is also very important. Animal and human data should be extensions of one another and should not disagree. Biological gradient is crucial, since this is the basis for the dose–response relationship (the more dose, the greater the biological response).

Temporality is crucial to all scientific research (i.e., the cause must precede the effect). However, this is sometimes difficult to see in some instances, such as when the exposures to suspected agents have been continuous for decades and the health data are only recently available.

Biographical Sketch: Sir Bradford Hill

In 1965, Austin Bradford Hill (1897–1991) published his famous paper, "The Environment and Disease: Association or Causation?" which included the nine guidelines for establishing the relationship between environmental exposure and effect.[59] Hill meant for the guidelines to be just that—guidelines, not an absolute test for causality. A situation does not have to meet Hill's nine criteria to be shown to be causally related. In the introduction to his paper, Hill acknowledges this by suggesting that there will be circumstances where not all of the nine criteria need to be met before action is taken. He recommended that action may need to be taken when the circumstances warrant. In his opinion, in some cases "the whole chain may have to be unraveled" or in other situations "a few links may suffice." The case of the 1853 cholera epidemic in London, concluded by John Snow to be waterborne and controlled by the removal of the pump handle, is a classic example in which only a few links were understood.

Biographical Sketch: Sir William Richard Shoboe Doll

Richard Doll (1912–2005) graduated from medical school and then served as a physician in the Royal Army Medical Corps during World War II. After the war he returned to England to conduct epidemiological research, concentrating on the relationship between radiation and leukemia and the effect of stress on the formation of peptic ulcers.

In 1950, Bradford Hill and Richard Doll initiated a study on the environmental cause of lung cancer, using the then-held hypothesis that automobile exhaust was the causative agent. They soon discovered, through statistical evaluations of large-scale trials, that the only positive correlation existed between cigarette smoking and lung cancer. This was the first time that an unequivocal connection had been made between cigarette smoking and lung cancer. All subsequent studies in this field have been based on this groundbreaking work. Their famous article in the *British Medical Journal* concluded "The risk of developing the disease increases in proportion to the amount smoked. It may be 50 times as great among those who smoke 25 or more cigarettes a day as among non-smokers."

Sir Richard Doll was knighted by Queen Elizabeth for his outstanding contribution to the field of epidemiology.

The key is that sound engineering and scientific judgment, based on the best available and most reliable data, should always be used when estimating risk. Linking cause and effect is often difficult in environmental matters. The best we can do is to be upfront and clear about the uncertainties and the approaches we use.

Environmental risk by nature addresses probable impossibilities. From a statistical perspective, it is extremely likely that cancer will not be eliminated during our lifetimes. But the efforts to date have shown great progress toward reducing risks from several forms of cancer. This risk reduction can be attributed to a number of factors, including changes in behavior (smoking cessation, dietary changes, and improved lifestyles), source controls (fewer environmental releases of cancer-causing agents), and the reformulation of products (substitution of chemicals in manufacturing processes).

RISK ASSESSMENT: MERELY THE FIRST STEP

We have covered a wide array of elements needed to access environmental risks. These are crucial because engineers must first understand the science before being able to intervene to make things better. These elements must be pulled together. *Risk characterization* is the stage where the engineer summarizes the necessary assumptions, describes the scientific uncertainties, and determines the strengths and limitations of the analyses. The risks begin to be understood by integrating the analytical results, interpreting adverse outcomes, and describing the uncertainties and weights of evidence. This can be very important for many minority communities, because much of their culture and livelihood

Biographical Sketch: Daniel A. Okun

Dan Okun (born 1917) has viewed his career as a mission— a mission to help others live longer and healthier lives through better sanitation and public health. His efforts have been focused in three directions: research, education, and outreach. In research, he was one of the principal developers of the original oxygen probe, a galvanic device that made measurement of dissolved oxygen in the field practical and convenient. In education, he led the development of an outstanding environmental engineering and science program at the University of North Carolina, recognizing early on the value of interaction among air pollution control, industrial hygiene, epidemiology, radiological health, hazardous waste management, and what was then known as sanitary engineering. But his greatest contribution was in outreach efforts, particularly in Central America and in the Pacific. He spearheaded the development of many environmental engineering programs at such universities as San Paulo in Guatemala, and brought many Central American and South American engineers to UNC to study alongside students from the United States. His legacy as an innovator and mentor will be felt for a very long time in the New World.

is linked directly to ecosystems, such as Native American subsistence agriculture, silviculture, and fishing, African American communities in or near riparian and littoral habitats, and Hispanic American families exposure to agricultural chemicals.

A reliable risk assessment is the groundwork for determining whether risks are disproportionate in a given neighborhood or region; as such, it is a first step in achieving environmental justice. Exposures to hazards can be disproportionate, which leads to disproportionate risk. There are also situations where certain groups of people are more sensitive to the effects of pollutants. Such things are difficult to quantify, but need to be addressed, as we discuss in Chapter 5. Risk assessment is a process distinct from risk management, where actions are taken to address and reduce the risks. But the two are deeply interrelated and require continuous feedback with each other. Engineers are key players in both efforts. In addition, risk communication between the engineer and the client further complicate the implementation of the risk assessment and management processes. What really sets risk assessment apart from the actual management and policy decisions is that the risk assessment must follow the prototypical rigors of scientific investigation and interpretation that we outlined in this chapter. As we see in the next chapters, risk management draws upon the technical risk assessment, but must also factor in other social considerations. The challenge is to maintain the rigors of science and engineering *and* incorporate the societal needs of the community.

REFERENCES AND NOTES

1. The segregation of risk assessment (science) from risk management (feasibility) is a recent approach. In fact, one of its advocates was William Ruckelshaus (see his biographical sketch later in this chapter).
2. C. Mitcham and R. S. Duval, Responsibility in Engineering, in *Engineering Ethics,* Prentice Hall, Upper Saddle River, NJ, 2000.
3. Although presented within the context of how risk is a key aspect of environmental justice, the information in this chapter is based on two principal sources: D. A. Vallero, *Environmental Contaminants: Assessment and Control,* Elsevier Academic Press, Burlington, MA, 2004; and D. A. Vallero, *Paradigms Lost: Learning from Environmental Mistakes, Mishaps, and Misdeeds,* Butterworth-Heinemann, Burlington, MA, 2006.
4. T. M. Apostol, *Calculus,* Vol. II, 2nd ed., Wiley, New York, 1969.
5. This discussion and examples are based on United Nations Environmental Program, International Labor Organisation, International Programme on Chemical Safety, Environmental Health Criteria 214, *Human Exposure Assessment,* UNEP Geneva, 2000.
6. C. B. Fleddermann, Safety and Risk, *Engineering Ethics,* Prentice Hall, Upper Saddle River, NJ, 1999.
7. R. Beyth-Marom, B. Fischhoff, M. Jacobs-Quadrel, and L. Furby, Teaching Decision Making in Adolescents: A Critical Review, in *Teaching Decision Making to Adolescents,* J. Baron and R. V. Brown (Eds.), Lawrence Erlbaum Associates, Hillsdale, NJ, 1991, pp. 19–60.
8. K. Smith, Hazards in the Environment, in *Environmental Hazards: Assessing Risk and Reducing Disaster,* Routledge, London, 1992.
9. This calls to mind Jesus's parable of the lost sheep (Matthew 18). In the story, the shepherd leaves (abandons, really) 99 sheep to find a single lost sheep. Some might say that if we as professionals behaved like that shepherd, we would be acting irresponsibly. However, it is actually how most of us act. We must give our full attention to one patient or client at a time.

There are a number of ways to interpret the parable, but one is that there is an individual value to each member of society, and the value of society's members is not mathematically divisible. In other words, a person in a population of 1 million is not one-millionth of the population's value. The individual value is not predicated on the group's value. This can be a difficult concept for those of us who are analytical by nature, but it is important to keep in mind when estimating risk.

10. See, for example, D. Dockery, Epidemiologic Evidence of Cardiovascular Effects of Particulate Air Pollution, *Environmental Health Perspectives,* 109(Supplement), 2001, pp. 483–486.

11. State University of New York–Stony Brook, http://www.matscieng.sunysb.edu/disaster/, accessed November 6, 2004.

12. P. Sandman's advice is found in S. Rampton and J. Stauber, *Trust Us, We're Experts: How Industry Manipulates Science and Gambles with Your Future,* Jeffrey B. Tarcher/Putnam, New York, 2001.

13. According to the Internet Encyclopedia of Philosophy (http://www.iep.utm.edu/): "A red herring is a smelly fish that would distract even a bloodhound. It is also a digression that leads the reasoner off the track of considering only relevant information."

14. American Council on Science and Health, *America's War on "Carcinogens": Reassessing the Use of Animal Tests to Predict Human Cancer Risk,* ACSH, New York, 2005.

15. H. W. Lewis, The Assessment of Risk, in *Technological Risk,* W.W. Norton, New York, 1990.

16. The need for this section grew from discussions between Vallero and John Ahearne, a renowned risk assessment expert.

17. C. E. Wormuth, "Homeland Security Risk Assessments: Key Issues and Challenges," Testimony before the Subcommittee on Intelligence, Information Sharing and Terrorism Risk Assessment Committee on Homeland Security, United States House of Representatives, November 17, 2005.

18. E. Cameron and G. G. Peloso, Risk Management and the Precautionary Principle: A Fuzzy Logic Model, *Risk Analysis,* Vol. 25, No. 4, 2005.

19. National Research Council, *Improving Risk Communication,* National Academy Press, Washington, DC, 1989, pp. 53 and 321.

20. National Research Council, P. C. Stern and H. V. Fineberg (eds.), *Understanding Risk: Informing Decisions in a Democratic Society,* National Academy Press, Washington, DC, 1996, pp. 215–216.

21. N. C. Rasmussen, The Application of Probabilistic Risk Assessment Techniques, *Annual Review of Energy,* Vol. 6, pp. 123–138, 1981.

22. U.S. Environmental Protection Agency, *Exposure Factors Handbook,* EPA/600/8-89/043, U.S. EPA, Washington, DC, 1990.

23. C. Tesar, POPs: What They Are; How They Are Used; How They Are Transported, *Northern Perspectives,* 26(1):2–5, 2000.

24. J. D. Graham and J. B. Wiener, Confronting Risk Tradeoffs, in *Risk Versus Risk: Tradeoffs in Protecting Health and the Environment,* J. D. Graham and J. B. Wiener, Eds., Harvard University Press, Cambridge, MA, 1995.

25. This goes well beyond mere translation. For example, it is very common in emergency departments of hospitals to hear the physician or nurse conversing in Spanish themselves or via a translator with a patient or family member. Spanish has many dialects and idioms that, if translated incorrectly, can lead to a miscommunication of symptoms and risks. Other tools are used, such as graphical devices (e.g., amount of pain ranging from 0 to 10 is shown as a smiling face and an obviously painful grimace, respectively, or the need for preventing the spread of germs with cartoons of people wearing masks and lathering soap). One cannot help concluding, however, that many of the more subtle and complex medical information is lost on the subjects and their attending medical staffs. For example, does nodding by a person

from Japan mean the same as that of a person from Ecuador or Mozambique (or East LA *versus* Beverly Hills, for that matter)?

26. Typically, researchers withheld treatment to observe the course of the disease (an excruciatingly painful process).

27. J. K. Hammitt, E. S. Belsky, J. I. Levy, and J. D. Graham, Residential Building Codes, Affordability, and Health Protection: A Risk–Tradeoff Approach, *Risk Analysis,* 19(6):1037–1058, 1999.

28. At least on its face, this runs contrary to some of the measures calling for improved environmental risk-based science proposed in the 1980s, especially the separation of risk assessment and risk management. This advice actually was an attempt to make risk science more objective and empirical, so that the science does not become "contaminated" by vested interests in the findings (such as political and economic considerations).

29. The phenomenon where children eat such nonfood material as paint chips is known as *pica.* For young children, especially those in poorer homes with older residences, this type of ingestion was (and still is, in some places) a major lead exposure pathway in children. Other pathways include soil ingestion (also pica), inhalation of lead on dust particles (which can be very high when older homes are renovated), and lead in food [such as lead leaching into food from glazes on cooking and dining ware, common in some ethnic groups (e.g., Mexican, Mexican American).]

30. Hydraulics and hydrology provide very interesting case studies in the failure domains and ranges, particularly how absolute and universal measures of success and failure are almost impossible. For example, a levee or dam breach, such as the recent catastrophic failures in New Orleans during and in the wake of Hurricane Katrina, experienced failure when flow rates reached cubic meters per second. Conversely, a hazardous waste landfill failure may be reached when flow across a barrier exceeds a few cubic centimeters per decade.

31. U.S. Environmental Protection Agency, *Guidelines for Carcinogen Risk Assessment,* EPA/630/R-00/004, *Federal Register* 51(185):33992–34003, U.S. EPA, Washington, DC, 1986; and R. I. Larsen, An Air Quality Data Analysis System for Interrelating Effects, Standards, and Needed Source Reductions, Part 13: Applying the EPA *Proposed Guidelines for Carcinogen Risk Assessment* to a Set of Asbestos Lung Cancer Mortality Data, *Journal of the Air and Waste Management Association,* 53:1326–1339, 2003.

32. J. Duffus and H. Worth, Training program: The Science of Chemical Safety: Essential Toxicology, Part 4: Hazard and Risk, *IUPAC Educators' Resource Material,* International Union of Pure and Applied Chemistry, Research Triangle Park, NC, 2001.

33. Actually, another curve could be shown for essential compounds such as vitamins and certain metallic compounds. In such a curve, the left-hand side (low dose or low exposure) of the curve would represent deficiency and the right-hand side (high dose or exposure) would represent toxicity, with an optimal, healthy range between these two adverse responses (see Fig. 4.19). Note that the two responses will differ at the low and high doses. For example, anemia and its related effects may occur at the low end, with neurotoxicity at the high end of exposures. Ideally, the optimal range has neither effect. Like the other curves, the safe levels of both effects would be calculated and appropriate factors of safety applied.

34. For an excellent summary of the theory and practical applications of the Ames test, see K. Mortelmans and E. Zeiger, The Ames *Salmonella*/Microsome Mutagenicity Assay, *Mutation Research,* 455:29–60, 2000.

35. The source for much of the technical information in the rest of the chapter is D. A. Vallero, *Paradigms Lost: Learning from Environmental Mistakes, Mishaps, and Misdeeds,* Butterworth-Heinemann, Burlington, MA, 2006.

36. The source for the physicochemical properties of DDT and its metabolites is United Nations Environmental Programme, Chemicals: North American Regional Report, Regionally Based

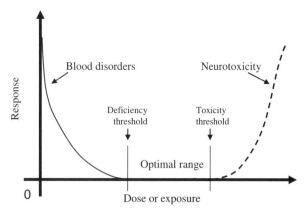

Figure 4.19   Hypothetical dose–response curves for essential substances.

Assessment of Persistent Toxic Substances, Global Environment Facility, UNEP, New York, 2002.

37. The two principal isomers of DDD are *p,p'*-2,2-bis(4-chlorophenyl)-1,1-dichloroethane and *o,p'*-1-(2-chlorophenyl)-1-(4-chlorophenyl)-2,2-dichloroethane. The principal isomer of DDE is *p,p'*-1,1'-(2,2-dichloroethenylidene)-bis[4-chlorobenzene].

38. The source for the first part of this discussion is the U.S. Commission on Civil Rights report, *Not in My Backyard.*

39. Chatham College, Leaders of Cancer Alley, http://www.chatham.edu/rci/well/women21-30/canceralley.html, accessed April 10, 2003.

40. Elizabeth Teel, deputy director, Environmental Law Clinic, Tulane Law School, testimony before the U.S. Commission on Civil Rights, hearing, Washington, DC, January 11, 2002, official transcript, p. 117.

41. German Federal Ministry for Economic Cooperation and Development, Environmental Handbook: Documentation on Monitoring and Evaluating Environmental Impacts, Vol. III, Compendium of Environmental Standards, http://www.gtz.de/uvp/publika/English/vol369.htm, accessed November 29, 2004.

42. Mossville Environmental Action Network, Breathing Poison: The Toxic Costs of Industries in Calcasieu Parish, Louisiana, http://www.mapCruzin.com/mossville/reportondioxin.htm, 2000.

43. State of Georgia, *Watershed Protection Plan Development Guidebook,* The State, Atlanta, GA, 2003.

44. National Research Council, *Biologic Markers in Reproductive Toxicology,* National Academies Press, Washington, DC, 1989.

45. Associated Press, Jersey to Ship Out Chromium, *New York Times,* December 15, 1989.

46. A general source of information for this case is NBC-Med, http://www.nbc-med.org/SiteContent/MedRef/OnlineRef/CaseStudies/csGoiania.html, accessed December 3, 2004.

47. M. Sun, Radiation Accident Grips Goiania, *Science,*. 238:1028–1031, 1987.

48. A. G. Knudson, Hereditary Cancer, Oncogenes, and Antioncogenes, *Cancer Research,* 45(4): 1437–1443, 1985.

49. E. E. McConnell, H. A. Solleveld, J. A. Swenberg, and G. A. Boorman, Guidelines for Combining Neoplasms for Evaluation of Rodent Carcinogenesis Studies, *Journal of the National Cancer Institute,* 76(2):283–289, 1986.

50. U.S. Environmental Protection Agency, *EPA Approach for Assessing the Risks Associated with Chronic Exposures to Carcinogens: Integrated Risk Information System,* Background Document 2, U.S. EPA, Washington, DC, 1992.

51. B. Singer, A Tool to Predict Exposure to Hazardous Air Pollutants, *Environmental Energy Technologies Division News,* 4(4):5, 2003.

52. U.S. Environmental Protection Agency, *Exposure Factors Handbook,* EPA/600/P-95/002Fa, U.S. EPA, Washington, DC, 1997.

53. U.S. Environmental Protection Agency, *Study of Hazardous Air Pollutant Emissions from Electric Steam Generating Units—Final Report to Congress,* EPA-453/R-98-004a, U.S. EPA, Washington, DC, 1998; and G. M. Gray, Forget Chemical Use, Let's Report Risk—*Risk in Perspective,* 5:277, 1997.

54. L. D. Budnick, D. C. Sokal, H. Falk, J. N. Logue, and J. M. Fox, Cancer and Birth Defects near the Drake Superfund Site, *Pennsylvania Archives of Environmental Health,* 39:409–413, 1984.

55. California Office of Environmental Health Hazard Assessment, California Cancer Potency Values, 2002, http://www.oehha.ca.gov/risk/chemicalDB/index.asp, accessed November 23, 2004.

56. International Agency for Research on Cancer, *IARC Monographs on the Evaluation of the Carcinogenic Risk of Chemicals to Man: Some Aromatic Amines, Hydrazine and Related Substances, N-Nitroso Compounds and Miscellaneous Alkylating Agents,* Vol. 4, 1974, *IARC Monographs on the Evaluation of the Carcinogenic Risk of Chemicals to Humans: Chemicals and Industrial Processes Associated with Cancer in Humans,* Supplement 1, 1979; *IARC Monographs on the Evaluation of the Carcinogenic Risk of Chemicals to Humans: Chemicals, Industrial Processes and Industries Associated with Cancer in Humans,* Supplement 4, 1982; *IARC Monographs on the Evaluation of Carcinogenic Risks to Humans: Overall Evaluations of Carcinogenicity,* Supplement 7, 1987; IARC, Lyon, France.

57. U.S. Environmental Protection Agency, *Guidelines for Carcinogen Risk Assessment,* EPA/630/R-00/004, *Federal Register,* 51(185):33992–34003, U.S. EPA, Washington, DC, 1986.

58. A. Bradford-Hill, The Environment and Disease: Association or Causation? Proceedings of the Royal Society of Medicine, *Occupational Medicine,* 58:295, 1965; and A. Bradford-Hill, The Environment and Disease: Association or Causation? President's Address, *Proceedings of the Royal Society of Medicine,* 9:295–300, 1965.

59. M. S. Legator and D. L. Morris, What Did Bradford Hill Really Say?, *Archives of Environmental Health,* 58(11):718–720.

5

Indirect Risks to Human Health and Welfare

The Roman goddess of the hearth, Vestal, was worshiped in a temple that contained a perpetual fire, tended by six virgins. The Vestal Virgins were highly respected and took on demigod status within the community. Mortals were not allowed even to touch them. Occasionally, one of the Vestal Virgins fell from grace and had a liaison. When this violation was discovered, the offending Virgin was led to a deep underground cell, given some bread and water, and left to die. This punishment seemed to be a compromise between two strict rules: Do not touch a Vestal Virgin, and each Vestal Virgin must remain chaste or be put to death. By leaving her in an the underground cell to die, she was not touched by anyone and yet she paid for her transgression.

The lore of the Vestal Virgins illustrates that there are two ways to kill, be it a human being or any other living thing: by harming it directly, or killing it indirectly by removing the sustaining environment. In this chapter we argue that just engineers must consider the protection of our environment as a means for protecting human health and welfare, for present as well as future people.

RISK TO THE ENVIRONMENT

As a cynic might put it, the Corps has hired technical experts to turn mission-oriented sows' ears into ecological silk purses.

Lynton K. Caldwell, during one of the early National Environmental Policy Act court proceedings, as best remembered by Caldwell's former student, Timothy Kubiak[1]

You have to hand it to the bureaucrats! Some seem to stop at nothing when trying to implement the objectives of particular programs. This can be a good thing if you agree with the mission, but in many instances, any effort not seen to be directly associated with the bureaucracy's purposes will be ignored or may even become the target of hostilities. There are parallels between Caldwell's quote above and what has occurred since issuance of an Executive Order mandating federal government agencies to embrace and to encourage environmental justice. Both the National Environmental Policy Act of 1970 (NEPA) and the Executive Order were bold statements that federal "business as usual" was falling short. NEPA was a recognition that many major projects were causing significant adverse environmental impacts; and the EJ Executive Order recognized that federal agencies had allowed disproportionate harm to be done to certain segments of society and had not adequately involved people from these communities in decisions that could have major effects on their quality of life.

The parallel can be drawn further. People tend to be *xenophobic;* they fear that which is different. Scientists and engineers are not exempt. In fact, we may be some of the worst and most difficult to convince that something needs change. Thomas S. Kuhn's thesis in his groundbreaking work *The Structure of Scientific Revolutions*[2] was that scientific paradigms, such as the move from Newtonian physics to relativity, come only after great consternation (even "violence") within the scientific community. In other words, scientists can be some of the most closed-minded people when it comes to carrying out their missions. So when we are asked to consider whether the systems that we have supported may somehow be working unfairly, we may be tempted to put some spin on our response. We are in effect trying to turn an unjust "sow's ear" into an environmental "silk purse." This is doubly wrong for scientists, since we are expected to be objective. We are called to work from factually sound premises.

Further, since we are trained for many years in our disciplines, we may be tempted to ignore or at least not to give proper credence to the views of the lay public. We may be insensitive, unsympathetic, and certainly not empathetic to the situations being encountered by the people affected by our designs. It is perilous to perceive that a decision is "just too complicated" for the general public. It is our responsibility to disclose the facts and to describe completely the various alternatives available. Again, this attention to details and full disclosure of the potential adverse effects from each alternative grew out of the environmental impact statement. The law and its regulations did not excuse federal agencies from fully disclosing and ensuring that the consequences, actual and potential, were completely understood by the affected public. In other words, the courts held that agency "experts" could not condescend. They must consider the people who could be affected as peers (or at least as fellow sojourners).

Although profit or corporate mission may come to mind first as examples of deluding the public and poorly communicating risks, the physical and environmental sciences have their share of intellectual deception. At a recent workshop of the National Academy of Engineering, John Ahearne, who served for seven years as executive director of Sigma Xi, the scientific research society, warned that scientists should never be arguing about the facts. Facts simply exist. We certainly must debate their meanings and give appropriate attention to their various interpretations, but we must not redefine facts simply to fit our paradigms. We may not like what the facts are telling us, but we must be objective if we are to follow the scientific method. We have seen students debate the ethics of full disclosure. In one recent debate, a number of environmental science students at Duke questioned the ethics of the findings of a journal article regarding sea turtles. Some students sympathized with Stephen Schneider, a Stanford University climatologist and former government advisor, who argued some years ago that scientists ought to "offer up scary scenarios, make simplified dramatic statements, and make little mention of any doubts we may have. Each of us has to decide what the right balance is between being effective and being honest."[3] Ahearne argues, and we agree, that a distorted version of "utilitarian science" is dangerous. It violates a first canon of science, honest inquiry and reporting. The English philosopher of science C. P. Snow articulated this principle some

50 years ago: "The only ethical principle which has made science possible is that the truth shall be told all the time. If we do not penalise false statements made in error, we open up the way, don't you see, for false statements by intention. And of course a false statement of fact made deliberately, is the most serious crime a scientist can commit."[4]

Not every scientist and environmental professional buys this. The recent debate that precipitated the Duke student's quandary was actually over the seemingly innocuous field of biological taxonomy. The dialogues exposed the acceptance by some of the justification of using morally unacceptable means to achieve the greater good.[5] The journal *Conservation Biology* published a number of scientific, philosophical, and ethical perspectives on whether to misuse science to promote the larger goal (conservation) to protect the black sea turtle (since the black and green turtles together would have sufficient numbers so as no longer to be considered "endangered"). Even though the taxonomy is scientifically incorrect (i.e., the black sea turtle is not a unique species), some writers called for a "geopolitical taxonomy."[6] The analogy of war was invoked as a justification, with one writer declaring that "it is acceptable to tell lies to deceive the enemy." The debate moderators asked a telling question: Should legitimate scientific results then be withheld, modified, or spun to serve conservation goals? Continuing with the war analogy, some scientists likened the deceptive taxonomy to propaganda needed to prevent advances by the enemy. The problem is that, as Snow would put it, once you stop telling the truth, you have lost credibility as scientists, even if the deception is for a noble cause.[7] Two writers, Kristin Shrader-Frechette and Earl D. McCoy, emphasized that credible science requires that ". . . in virtually all cases in professional ethics, the public has the right to know the truth when human or environmental welfare is at issue."[8]

Another example of late is the debate about global climate change, discussed later in this chapter. A particularly troublesome topic for many scientists is the role of nuclear power generation. For decades, many in the scientific and engineering communities have argued vociferously against the use of nuclear energy for weapons and its use as a source of energy. Nuclear weapons often boils down to differences of opinion of geopolitical issues, such as proliferation and the appropriateness of mutual destruction. Nuclear power generation is a different matter. Some scientists and engineers have resisted while others have supported the nuclear power plants. The public seems to support existing plants and are neutral to wary of new plants.[9] These technological optimists argue that nuclear power is a reliable energy source and, with sufficient fail-safe measures, can provide an alternative to the dirty and unsafe fossil fuel sources.

In the past decade or so, with the increasing links (intellectual, if not necessarily scientific) between anthropogenic sources of greenhouse gases, especially carbon dioxide (CO_2) and increasing mean global temperatures, many in the scientific community have argued for curtailing the use of fossil fuels, especially in developed nations. However, pro-nuclear scientists see these findings as a requisite to revisit the need for nuclear power, since fission produces no CO_2 (there is no combustion, i.e., oxidation of a hydrocarbon). As in the case of the sea turtle taxonomy debate, scientists all too often revert to an advocacy position rather than one of objectivity. They know that nuclear power will produce long-lived radioactive by-products that will need to be stored for thousands of years. They also remember Chernobyl and fear similar accidents. But do these justify a full and open debate on whether nuclear power is a viable means of reducing the emission of greenhouse gases? It is a challenge even to get some scientists to consider the issue, not as a geopolitical decision, but as an engineering, science, or mathematical problem.

If it is the responsibility of government to protect the lives of its citizens against foreign invasion or criminal assault, it is equally responsible for protecting the health and lives of its citizens from other potential dangers, such as falling bridges and toxic air pollutants. Government has a limited budget, however, and we expect that this money is distributed so as to achieve the greatest benefits to health and safety. If two chemicals provide the same public benefit, but are placing people at risk, it is rational that funds and effort be expended to eliminate the chemical that results in the greatest risk.

But is this what we really want? Suppose, for example, it is cost-effective to spend more money and resources to make coal mines safer than it is to conduct heroic rescue missions if accidents occur. It might be more risk-effective to put the money we have into safety, eliminate all rescue squads, and simply accept the few accidents that will still inevitably occur. But since there would no longer be rescue teams, the trapped miners would then be left on their own. The net effect would be, however, that overall, fewer coal miners' lives would be lost.

Even though this conclusion would be risk-effective, we would find it unacceptable. Human life is considered sacred. This does not mean that infinite resources have to be directed at saving lives, but rather that one of the sacred rituals of our society is the attempt to save people in acute or critical need, such as crash victims and trapped coal miners. Thus, purely rational calculations, such as the coal miners example above, might not lead us to conclusions that we find acceptable.

In all such risk analyses, the benefits are usually to humans only, and they are short-term benefits. Similarly, the costs determined in the cost-effectiveness analysis are real budgetary costs, money that comes directly out of the pocket of the agency. Costs related to environmental degradation and long-term costs that are very difficult to quantify are not easily included in these calculations. When long-term and environmental costs can nowhere be readily considered in these analyses, coupled with the blatant abuse of benefit–cost analysis by governmental agencies, it is necessary to bring into action another decision-making tool: the analysis of environmental impact.

ANALYZING RISK TO THE ENVIRONMENT

Projects receiving funding or regulatory action from the U.S. government require compliance with the National Environmental Policy Act (NEPA)[10] of 1969 and the Council on Environmental Quality (CEQ) Guidelines.[11] NEPA serves as the nation's basic environmental protection charter. It is to ensure that federal agencies consider the environmental consequences of their actions and decisions as they conduct their respective missions. For "major federal actions significantly affecting the quality of the human environment," the federal agency must prepare a detailed environmental impact statement (EIS) that assesses the proposed action and all reasonable alternatives.[12] EISs must be broad in scope to address the full range of potential effects of the proposed action on public health and the environment. Regulations established by both the CEQ and the EPA require that socioeconomic impacts associated with significant physical environmental impacts be addressed in the EIS.

The first step toward compliance is to determine whether an action is categorically excluded from NEPA. Compliance with NEPA requires public participation throughout the process, starting with initial meetings, public meetings, and presentations during the assessment process, and final public hearing(s). The process begins with an initial session involving interested federal, state, and local agencies. The environmental assessment follows, with the public involved. The key output of the assessment is a determination of whether or not the federal action will significantly affect the quality of the human environment. If the results indicate that the action has no significant impact or the impacts can be mitigated, a *finding of no significant impact* (FONSI) is issued. This is akin to a "negative declaration." The public has 30 days to review the FONSI. There is a public hearing during the review process, and comments are received and formal responses are prepared. If the FONSI is still deemed the appropriate action after the hearings and the comments, the lead agency formally approves the document, allowing the action (or project) to go forward.

If the FONSI is appropriate, this does not mean that the project will not be scrutinized for environmental impact. It will probably have to go through some form of *environmental assessment* (EA) for the state. Environmental assessments are most common for treatment plant upgrades and other modifications.

If the FONSI is not appropriate, which is the case for almost all wastewater treatment plants, an *environmental impact statement* must be prepared. This process starts with the filing of a notice of intent that is published by U.S. EPA in the *Federal Register*. Although there are not strict guidelines on how the EIS is to be prepared, the following discussion is a description of several alternatives within a general framework. The EIS should be in three parts: *inventory, assessment,* and *evaluation.*

The first duty in the writing of any EIS is the gathering of data, such as hydrological, meteorological, and biological information. A list of the species of plants and animals in the area of concern, for example, is included in the inventory. There are no decisions made at this stage, since everything properly belongs in the inventory.

The second stage is analysis, commonly called the *assessment.* This is the mechanical part of the EIS in that the data gathered in the inventory are fed to the assessment mechanism and the numbers are crunched accordingly. Numerous assessment methodologies have been suggested, only some of which are discussed here.

The *quantified checklist* is possibly the simplest quantitative method of comparing alternatives. It involves first listing those areas of the environment that might be affected by the proposed project, and then an estimation of the *importance* of the impact, the *magnitude* of the impact, and the *nature* of the impact (whether negative or positive). Commonly, the importance is given numbers such as 0 to 5, where 0 means no importance whatever and 5 implies extreme importance. A similar scale is used for magnitude. The nature of the impact is expressed simply as -1, negative (or adverse) impact, or $+1$, positive (or beneficial) impact. The environmental impact (EI) is then calculated as

$$\text{EI} = \sum_{i=1}^{n} I_i M_i N_i \tag{5.1}$$

where I_i is the importance of the ith impact; M_i is the magnitude of the ith impact; N_i is the nature of the ith impact, so that $N = +1$ if beneficial and $N = -1$ if detrimental;

and n is the total number of areas of concern. The environmental impact, which can be negative (detrimental) or positive (beneficial), is calculated for each alternative and tabulated. The following example illustrates use of the quantitative checklist.

Example:

Analyzing Environmental Impact I

A community has two alternatives: increase the refuse collection frequency from once to twice a week, or allow the burning of rubbish on premises. Analyze these two alternatives using a quantified checklist.

First, the areas of environmental impact are listed. In the interest of brevity, only five areas are shown below, while recognizing that a thorough assessment would include many other concerns. Following this, values for importance and magnitude are assigned (0 to 5) and the nature of the impact ($+/-$) is indicated. The three columns are then multiplied.

Alternative 1: Increasing collection frequency

Area of concern	Importance (I)	Magnitude (M)	Nature (N)	Total (I × M × N)
Air pollution (trucks)	4	2	−1	−8
Noise	3	3	−1	−9
Litter in streets	2	2	−1	−4
Odor	2	3	−1	−5
Traffic congestion	3	3	−1	−9
Groundwater pollution	4	0	−1	0
(*Note:* No new refuse will be landfilled)				
				EI = −35

Alternative 2: Burning on premises

Air pollution (burning)	4	4	−1	−16
Noise	0	0	−1	0
Litter (present system)	2	1	+1	+2
(*Note:* Present system causes litter)				
Odor	2	4	−1	−8
Traffic congestion	0	0	−1	0
Groundwater pollution	4	1	+1	+4
(*Note:* Less refuse will be landfilled)				
				EI = −18

On the basis of this analysis, burning the refuse would result in the lowest adverse effect.

For simple projects, the quantified checklist is an adequate assessment technique, but it gets progressively unwieldy for larger projects, such as the construction of a dam, where there are many smaller actions all combining to produce the overall final product. The effect of each of these smaller actions should be judged separately with respect to impact. Such an interaction between the individual actions and areas of concern gives rise to the *interaction matrix,*[13] where once again the importance and the magnitude of the interaction are judged (such as by the 0 to 5 scale used previously). There seems to be no agreement on what calculation should be made to produce the final numerical quantity. In some cases, the interaction value is multiplied by the magnitude, and the products summed as before, whereas other procedures are simply to add all the numbers on the table. In the example below, the products are summed into a grand sum shown on the lower right corner of the matrix.

Example:

Analyzing Environmental Impact II

Continuing the preceding example: Use the interaction matrix assessment technique to decide on the alternatives presented.

Figure 5.1 shows the calculations using the matrix technique. Note again that these are incomplete lists used only for illustrative purposes. The results indicate that once again it makes more sense to burn the paper.

Before moving on to the next technique, we want to emphasize again that the method illustrated in the example above can have many variations and modifications, none of

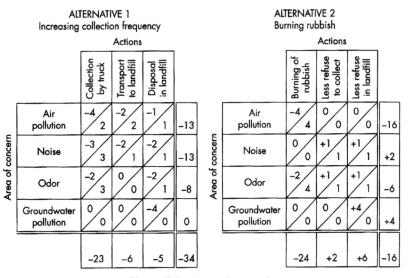

Figure 5.1   Interaction matrix.

which are "right" or "wrong" but which depend on the type of analysis conducted, for whom the report is prepared, and what is being analyzed. Individual initiative is often a most valuable component in the development of a useful EIS. However, whichever approach is used, it must be fully described and supported technically.

The *common parameter checklist* is a third technique for environmental impact assessment. It differs from the quantified checklist technique only in that instead of using arbitrary numbers for importance and magnitude, the importance term is called the effect (*E*) and is calculated from actual environmental data (or predicted quantitative values), and the magnitude is expressed as weighting factors, *W*.

The basic objective of this technique is to reduce all data to a common parameter, thus allowing the values to be added. The data (actual or predicted) are translated to the effect term by means of a function that describes the relationship between the variation of the measurable value and the effect of that variation. This function is commonly drawn for each interaction on a case-by-case basis. Three typical functions are illustrated in Figure 5.2. The curves show that as the value of the measured quantity increases, the effect on the environment (*E*) also increases, but that this relationship can take several forms. The value of *E* ranges from 0 to ± 1.0, with the positive sign implying beneficial impact and the negative detrimental.

Consider, for example, the presence of a toxic waste on the health and survival of a certain aquatic organism. The concentration of the toxin in the stream is the quantity measured, and the health of the aquatic organism is the effect. The effect (detrimental) increases as the concentration increases. A very low concentration has no detrimental effect, whereas a very high concentration can be disastrous. But what type of function (curve) makes the most sense for this interaction? The straight-line function (linear portion of Figure 5.3A) implies that as the concentration of the toxin increases from zero, the detrimental effects are felt immediately. This is seldom true. At very low concentrations, most toxins do not show a linear relationship with effect, and thus this function does not appear to be useful. For the same reasons, Figure 5.3B, the next curve, is also clearly incorrect. But Figure 5.3 seems much more reasonable, since it shows that a number of substances are actually beneficial (Figure 5.3A) at certain concentrations (e.g., vitamins and essential metals) and also implies that the effect of the toxin is very small at lower concentrations, but when it reaches a threshold level, it becomes very toxic quickly (Figure 5.3B). As the level increases above the toxic threshold, there can be no further damage since the organisms are all dead and the effect levels off at 1.0 (i.e., 100%).

The conditions in Figure 5.3, where the safe dose is also the efficacious dose, is ideal for drugs and food additives, for example. Often, however, real-world exposures are more like that depicted in Figure 5.4, where the efficacious dose carries with it commensurate risk, with the positive and harmful effects increasing together (e.g., vaccines, cancer treatments, lipophilic vitamins, and metals). If the adverse effect (Figure 5.4B) is a relatively innocuous side effect (e.g., skin rash) compared to a lifesaving beneficial effect (hemophilia treatment), the risk–benefit conclusions are easy. However, when the benefits are dramatic but the harm is also substantial (e.g., applying DDT to eliminate malaria-bearing mosquitoes, with the possibility of causing endocrine disorders), the risk–benefit relationship becomes quite complicated.

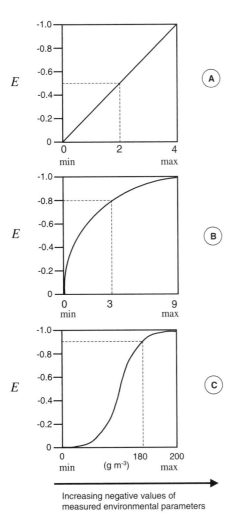

Figure 5.2 Three specific functions of environmental effect.

Once the effect (E) terms are estimated for each characteristic, they are multiplied by weighting factors (W) which are distributed among the various effects according to the importance of the effect. Typically, the weighting terms add up to 100, but this is not important as long as an equal number of weighting terms are used for each alternative considered.

The final impact is then calculated by adding up the products of the effect terms (E) and weighing factors (W). Thus, for each alternative considered,

$$\text{EI} = \sum_{i=1}^{n} E_i \times W_i \tag{5.2}$$

Remember that E terms can be negative (detrimental) or positive (beneficial).

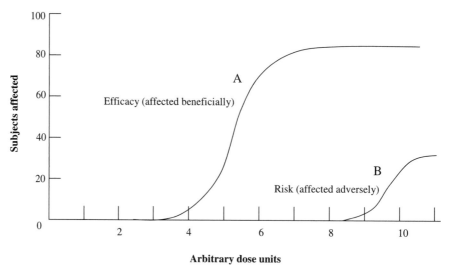

Figure 5.3 Hypothetical dose–response curve comparing exposure to a substance that can be either beneficial or adverse, depending on the dose. In this case, risk assessment is straightforward since the efficacious dose (about 7 units) is well below the unsafe level (greater than 8 units). This demonstrates a substance where the greatest efficacy is realized at the lowest dose. (Adapted from W. W. Lowrance, *Of Acceptable Risk: Science and the Determination of Safety,* William Kaufmann, Los Altos, CA, 1976, p. 97.)

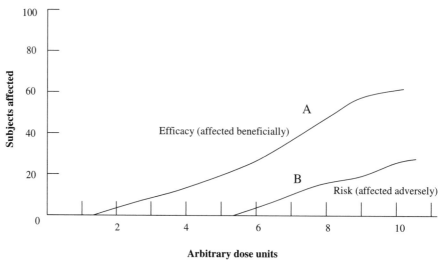

Figure 5.4 Hypothetical dose–response curve comparing exposure to a substance that can be either beneficial or adverse, depending on the dose. In this case, risk assessment is complicated since the dose–response relationship has a large area with both efficacy and risk. This demonstrates a substance where the risk–benefit determination will rest on the weight of the gain *versus* that of the harm. (Adapted from W. W. Lowrance, *Of Acceptable Risk: Science and the Determination of Safety,* William Kaufmann, Los Altos, CA, 1976, p. 98.)

Example:

Analyzing Environmental Impact III

Continuing the preceding example: Considering only litter, odor, and airborne particulates as an example of air quality, calculate the environmental impact of rubbish burning using the common parameter–weighted checklist. (This is only a very small part of the total impact assessment that would be necessary when using this technique.)

Assume that the three curves shown in Figure 5.2 are the proper functions relating the environmental characteristics of litter, odor, and airborne particulates, respectively. Assume that it has been estimated that the burning of rubbish will result in a litter level of 2 on a scale of 0 to 4, an odor number of 3 on a scale of 0 to 9, and a airborne particulate increase to 180 g m^{-3} on a concentration scale. Entering Figure 5.2A, B, and C at 2, 3, and 180, respectively, the effects (E) are read off as -0.5, -0.8, and -0.9. (*Note:* We are assuming that the litter function looks like the curve in Figure 5.2A, the odor function looks like the curve in Figure 5.2B, and the air pollution function looks like the curve in Figure 5.2C; however, the actual functions could all be the same curve or three functions other than those shown in Figure 5.2.) It is now necessary to assign weighting factors, and out of a total of 10 it is decided to assign 2, 3 and 5, respectively, implying that the most important effect is the air quality, and the least important is the litter. Then

$$EI = (0.5 \cdot 2) + (0.8 \cdot 3) + (-0.9 \cdot 5) = -7.9$$

A similar calculation would be performed for other alternatives, and the EI values would be compared.

Clearly, this technique is wide open to individual modifications and interpretations, and the example above should not be considered in any way a standard method. It does, however, provide a numerical answer to the question of environmental impact, and when this is done for several alternatives, the numbers can be compared. This process of comparison and evaluation represents the third part of an EIS.

The comparison of the results of the assessment procedure and the development of the final conclusions is all covered under evaluation. It's important to recognize that the previous two steps, inventory and assessment, are simple and straightforward procedures compared to the final step, which requires judgment and common sense. During this step in writing the EIS, the conclusions are drawn up and presented. Often, the reader of the EIS sees only the conclusions and never bothers to review all the assumptions that went into the assessment calculations, so it is important to include in the evaluation the flavor of these calculations and to emphasize the level of uncertainty in the assessment step.

But even when the EIS is as complete as possible, and the data have been gathered and evaluated as carefully as possible, conclusions concerning the use of the analysis are open to severe differences. For example, the EIS written for the Alaska oil pipeline, when all the pages in the volumes are stacked, represents 14 feet of work. At the end of all that effort, good people on both sides drew diametrically opposite conclusions on the effect of the pipeline. The trouble was, they were arguing over the *wrong thing.* They may have been arguing about how many caribou would be affected by the pipeline, whereas their disagreement was actually how deeply they cared that the caribou were affected by the pipeline. For a person who does not care one twit about the caribou, the impact is zero, whereas those who are concerned about the herds and the long-range effects on the sensitive tundra ecology care very much. What, then, is the solution? How can engineering decisions be made in the face of conflicting values? Such decisions require engineers to make value judgments, and this is where the concepts of justice enter the analysis of environmental impact.

Once the draft EIS has been published in the *Federal Register,* a formal comment period begins that leads to preparation of the final environmental impact statement, another comment period, and finally, a record of decision. This process is time consuming, particularly if the project is controversial or the environmental documentation is challenged. This needs to be factored into the project implementation schedule. Also, many permits are contingent on having an approved environmental document (e.g., an NPDES permit); this can affect the entire project time line because the permits themselves are time intensive.

EISs are all about balance and fairness. Various aspects of the human and natural environment must be protected. Similarly, the environmental justice goal of "fair treatment" is not to shift risks among populations, but rather, to identify potential disproportionately elevated and harmful effects and to identify alternative approaches that may ameliorate and mitigate these effects. This makes the EIS an ideal environmental justice mechanism. EISs have been used as tools for three decades, striving for the best option among many, just as environmental justice concerns should engage in serious considerations of credible alternatives, including "no action" and modified designs to consider how a publicly funded project may affect minority and disadvantaged communities disproportionately. The designers of NEPA called for community education and participation, much as environmental fairness calls for involvement. This is not to say that the guidance and existing approaches for community input are sufficient. Although many of the NEPA methods for public involvement and education have been honed for three decades, special situations brought on by historical disenfranchisement will undoubtedly need new tools, such as that shown in Figure 5.5. Note that neither the need for credible science nor the need for continuous involvement with those potentially affected by the project is sacrificed. There are key points in the critical path where the needs of the public and those of the professional intersect.

The preparation of an EIS or an environmental assessment of any type provides an opportunity to incorporate environmental justice into project design. Determining the distributional dimensions of environmental impacts on particular populations is entirely consistent with the NEPA process since socioeconomics and other "human" factors are components of the environment. Credible assessments heighten awareness of environmental justice issues within NEPA analyses and aid in considering the full potential for disproportionately adverse human health or environmental effects on minority and low-income populations. Assessments also identify situations early in the design phase where environmental justice issues may be encountered, laying out options for addressing disproportionately high and adverse effects. The assessment process also provides approaches for communicating with the affected population throughout the design, siting, building, and operation of projects. Thus, the application of many of the same tools currently intrinsic to the NEPA process applies directly to environmental justice.[14]

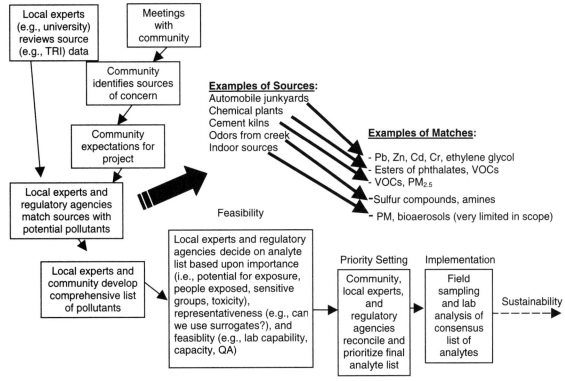

Figure 5.5 Possible process for incorporating community input into an environmental assessment process. (Adapted from D. A. Vallero, *Environmental Contaminants: Assessment and Control,* Elsevier Academic Press, Burlington, MA, 2004.)

MAJOR AREAS OF ENVIRONMENTAL CONCERN

Although there can be an almost infinite number of areas of concern on a regional or global basis for the effect of contaminants on environmental quality, in this section we concentrate on but a few, not so much because they pose the greatest actual or potential risks to the environment or health, but because they vividly illustrate the complexity of the science, technology, and social perspectives. They illustrate the difference in risk endpoints. For example, smog can affect ecosystems but is addressed principally as a human health issue. Acid rain is predominantly a threat to ecosystems. Global climate change would affect ecosystems directly, such as shifting biomes and loss of species, but indirectly would affect human health and welfare (e.g., a greater number of tropical diseases shifting to presently temperate climates). These global- and regional-scale problems provide a template for addressing complex problems by using a systematic approach (i.e., gathering reliable information, understanding and explaining the scientific processes, and characterizing potential adverse outcomes *before* recommending and designing solutions to environmental problems). Environmental systems from the molecular (popularly known as "nano") to the planetary scales are seldom linear and are often chaotic

Biographical Sketch: Eliza Sanderson

Sometimes people become heroes by standing up for what is right, losing, and then being remembered for generations for their unwillingness to submit to defeat. The name of Eliza Sanderson is not well known to the public, but in the world of personal rights law, her fame lives on.

In 1868 when Eliza Sanderson bought land in a town now known as Green Ridge, PA, near Scranton, PA, all she wanted to do was build a house and run a small farm. She and her husband, J. Gardner Sanderson, had by all accounts a successful and comfortable home. Running through their property was a small stream, Meadow Brook, and the brook was one of the reasons she bought the land. The water was sufficient for their rural use and was of high quality.

In 1870, the powerful Pennsylvania Coal Company sunk a deep vertical shaft at their mine north of the Sandersons' property, but this filled with water, and they started to pump it out. Unfortunately, the only place this water could flow was into Meadow Brook, making the water in the brook so foul that it could not be used for human consumption or for watering livestock.

Since Pennsylvania was a riparian law state, the property owner had every right to use the water that flowed through their property, and upstream property owners could not use or contaminate the water. It had to flow into their property "unimpaired in quality and undiminished in quantity", according to the law. Mrs. Sanderson thought she had a pretty good case and went to court.

But she did not count on the power of the Pennsylvania coal industry. The judge presiding on the case decided that the needs of the coal company exceeded the rights of the Sandersons to have their creek back. The judgment was that since "the coal could only be mined where it was, the contamination of the water was part of the mining operation." In other words, the mine could not move, but the Sandersons could. Personal rights be damned.

The willingness of Eliza Sanderson to stand up to the Pennsylvania Coal Company, even though she lost, will long be remembered. She will continue to be a hero of the environmental justice movement.

More information of the case can be found in *The Pennsylvania Coal Company v. Sanderson and Wife,* Supreme Court of Pennsylvania, No. 389, 113 Pa. 126; 6 A. 453; 1886 Pa. LEXIS 343

and ill posed. Their solutions must be matched to these conditions. We certainly recognize that experience and scientific knowledge will provide new and unanticipated problems in the future, but firmly believe the systematic approach will be needed to address them.

Photochemical Smog

The components of automobile exhaust are particularly important in the formation of secondary pollutants. The well-known and much discussed Los Angeles smog is a case of secondary pollutant formation. Photochemical smog starts with high-temperature, high-pressure combustion such as occurs within the cylinders of an internal combustion (gasoline) engine. The first reaction is

$$N_2 + O_2 - \text{high temperature and pressure} \rightarrow 2NO \tag{5.3}$$

which immediately oxidizes to NO_2, as

$$2NO + O_2 \rightarrow 2NO_2 \tag{5.4}$$

Nitrogen dioxide, NO_2, is very active photochemically and breaks apart as the first reaction in the sequence of photochemical smog formation:

$$NO_2 + h\nu \rightarrow NO + O \tag{5.5}$$

where h is Planck's constant [6.62×10^{-34} joule-second (J-s)] and ν is the frequency (s^{-1}). The atomic oxygen molecule, being quite unstable, reacts as

$$O + O_2 + M \rightarrow O_3 + M \tag{5.6}$$

where M is some radical that catalyzes the reaction. The ozone thus created is a strong oxidant, and it will react with whatever it finds to oxidize. From the reaction (5.6), NO is available, and thus

$$O_3 + NO \rightarrow NO_2 + O_2 \tag{5.7}$$

and we are back where we started, with NO_2. Since all of these reactions are fast, there is no way that high levels of ozone could build up in the atmosphere. This puzzled early air pollution researchers, until Arie Haagen-Smit came up with the answer. The answer is that one of the other constituents of polluted urban air is hydrocarbons, and some of these can oxidize the NO to NO_2, leaving ozone in excess. For example,

$$HCO_3{}^0 + NO \rightarrow HCO_2{}^0 + NO_2 \tag{5.8}$$

This explanation allowed for the possibility of ozone buildup as well as the formation of other components of photochemical smog.

Ozone is an eye irritant and can cause severe damage to plants and to materials such as rubber. The peroxyacetyl nitrates (PANs) are also effective in damaging crops and materials. Other hydrocarbons can cause breathing problems and also damage plants. Finally, NO_2 itself is an irritant and has a brownish orange color. Anyone who has ever flown into the Los Angeles airport on a sunny day would have seen the dramatic contrast between the blue sky above and the orange crud below into which the plane was heading.

Photochemical smog is possible if a series of conditions exits:

- High-temperature, high-pressure combustion of fossil fuels such as gasoline
- Plentiful sunlight (source of photo energy)
- Stable atmospheric conditions, which would allow time for the reactions to occur
- Hydrocarbon emissions, which would scavenge the NO from the mix and allow for the buildup of ozone

Cities in the United States that have these conditions include some of the largest urban areas, such as Los Angeles, Denver, and Houston, and even some rapidly growing areas such as Raleigh–Durham, North Carolina.

Reducing photochemical smog has been a challenge to environmental scientists and managers for many years. Many different schemes for reducing the smog in Los Angeles, for example, have been suggested, such as boring a huge tunnel through the surrounding mountains and blowing the polluted air into the valleys on the other side. Never mind what the people on the other side of the mountain might say. Also, rough calculations show that the tunnel large enough to be able to exchange the air over Los Angeles needs to be about 2 miles in diameter! So other solutions have been sought, the most reasonable being not producing NO_2 in the first place by reducing the number of cars in the city and/or clearing up their emissions. The cars exhaust is already very clean compared to earlier models, and the people of LA are not keen on giving up their driving privilege, so this solution is not useful. Another proposed solution is to replace all the gasoline-powered vehicles with electric cars, also a nonstarter, at least for the next few decades, although hybrid gasoline-electric cars are having some impact. As of this writing, no clear solution is in sight.

Acid Rain

Normal, uncontaminated rain has a pH of about 5.6. This low pH can be explained by its adsorption of carbon dioxide, CO_2. As the water droplets fall through the air, the CO_2 in the atmosphere becomes dissolved in the water, setting up an equilibrium condition:

$$CO_2 \text{ (gas in air)} \leftrightarrow CO_2 \text{ (dissolved in the water)} \tag{5.9}$$

The CO_2 in the water reacts to produce hydrogen ions:

$$CO_2 + H_2O \leftrightarrow H_2CO_3 \leftrightarrow H^+ + HCO_3^- \tag{5.10}$$

$$HCO_3^- \leftrightarrow 2H^+ + CO_3^{2-} \tag{5.11}$$

Thus, the more CO_2 that is dissolved in the rain, the lower the pH will be. But this reaction will, at equilibrium, produce rainwater with a pH of 5.6, and no lower. How is it, then, that some rain can have a pH value less than 2 (i.e., 100,000 times more acidic than neutral water).

This pH reduction occurs as the result of interactions with air contaminants. For example, sulfur oxides produced in the burning of fossil fuels (especially coal) is a major contributor to low pH in rain. In its simplest terms, SO_2 is emitted from the combustion of fuels containing sulfur, the reaction being

$$S + O_2 \xrightarrow{\text{heat}} SO_2 \tag{5.12}$$

$$SO_2 + O \xrightarrow{\text{sunlight}} SO_3 \tag{5.13}$$

$$SO_3 + H_2O \rightarrow H_2SO_4 \rightarrow 2H^+ + SO_4^{2-} \tag{5.14}$$

H_2SO_4 is, of course, sulfuric acid. Sulfur oxides do not literally produce sulfuric acid in the clouds, but the concept and environmental impact are the same.[15] The precipitation from air containing high concentrations of sulfur oxides is poorly buffered and readily drops its pH.

Biographical Sketch: Arie Haagen-Smit

In the 1940s, as the number of clear days in southern California became fewer and fewer, concern was being expressed as to what the cause of this bad visibility might be. But during World War II, little was done until a butadiene plant was built in downtown Los Angeles. When the plant went online, it had severe upsets and produced noxious fumes that caused office buildings to be evacuated. Although we now know that the butadiene plant had little to do with the poor visibility, the event caused a public outcry and a demand to do something about the air pollution. Experts were sent for, arriving with their instruments for measuring SO_2 and smoke, but these were worthless since the levels of sulfur oxides were very low. In 1947 the County of Los Angeles got the police power to do something about the air pollution, and the LA County Air Pollution District was formed. Early efforts to reduce air pollution in Los Angeles centered on the reduction of SO_2 from such sources as backyard burning, but to no avail. Finally, it was decided that some research was necessary.

Arie Haagen-Smit, a Dutch-born biologist working at Cal Tech on the fumes emitted by pineapples, decided to distill the contents of the Los Angeles air and discovered peroxyorganic substances, which were no doubt the source of the eye irritation. The source, if this were true, had to have been the gasoline-powered automobile, and the publication of this research set off a firestorm of protest by these industries. Scientists at the Stanford Research Institute, which had been doing smog research on behalf of the transportation industry, presented a paper at Cal Tech accusing Haagen-Smit of bad science. This made him so angry that he abandoned his pineapple work and began to work on the smog problem.

Some of his previous research had been on plant damage caused by ozone, and he discovered that the effect of automobile exhaust on plants produced a similar injury, suggesting that the exhaust contained ozone. But ozone was not emitted by automobiles, so where did it come from? By mixing automobile exhaust with hydrocarbons in a large air chamber, and subjecting the mixture to strong light, he was able to demonstrate that ozone is formed by reactions in the atmosphere. The ozone in the Los Angeles smog was created by reactions that began with the oxides of nitrogen in automobile exhaust.

The powerful automobile and gasoline industries denied that this could be the cause of the smog, and suggested instead that the ozone had to be in the smog because it somehow descended from the stratosphere, which of course was nonsense. Haagen-Smit's courageous work, taking on the most powerful political forces in California, paved the way for an eventual (but not yet present) solution to the smog problem in Los Angeles.

Nitrogen oxides, emitted mostly from automobile exhaust but also from any other high-temperature combustion, contribute to the acid mix in the atmosphere. The chemical reactions that apparently occur with nitrogen are

$$N_2 + O_2 \rightarrow 2NO \tag{5.15}$$

$$NO + O_3 \rightarrow NO_2 + O_2 \tag{5.16}$$

$$2NO_2 + O_3 + H_2O \rightarrow 2HNO_3 + O_2 \rightarrow 2H^+ + 2NO_3^- + O_2 \tag{5.17}$$

where HNO_3 is nitric acid. However, like everything else in environmental science, acid rain formation and its effects are far more complicated than these reactions. A major complication is buffering. If soils have a large buffering capacity (i.e., they keep pH from changing significantly even when acidic water is added), the lakes are not severely impacted. Thus, in the United States, acid rain is not nearly as much a problem west of the Mississippi River than it is to the east (see Figure 5.6). A major reason for this difference in buffering capacity is the availability of carbonates in the soil from parent rock materials. The carbonic acid in rain will release some bicarbonate into the water (see Figure 5.7). This is why soils underlain by limestone ($CaCO_3$) and dolomite [$CaMg(CO_3)_2$] resist pH change much more than the granite and basaltic (noncarbonated) based soils in the eastern part of the United States and much of Scandinavia.

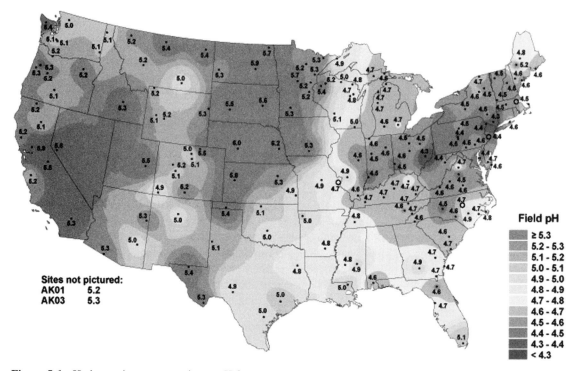

Figure 5.6   Hydrogen ion concentration as pH from measurements made at field laboratories, 2003. (Source: National Atmospheric Deposition Program/National trends Network, http://nadp.sws.uiuc.edu.)

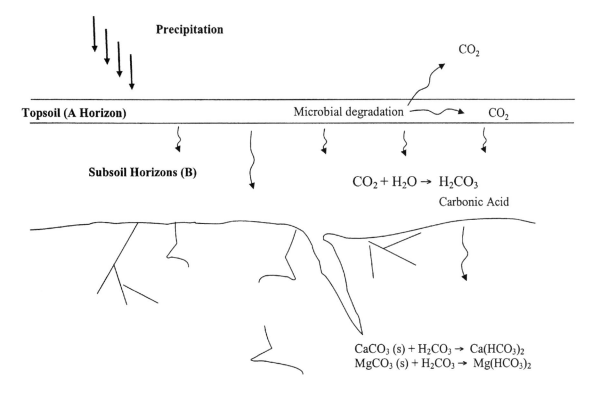

Figure 5.7   Process of releasing carbonates and providing natural soil buffers against acid rain.

The effect of acid rain has been devastating. Hundreds of lakes in North America and Scandinavia have become so acidic that they no longer can support fish life. In a recent study of Norwegian lakes, more than 70% of the lakes having pH less than 4.5 contained no fish, and nearly all lakes with pH 5.5 and above contained fish. The low pH not only affects fish directly, but contributes to the release of potentially toxic metals such as aluminum, thus magnifying the problem. For example, much of the fish toxicity can be attributed to the release of aluminum into the water. Aluminum does not readily become dissolved in neutral waters. However, the trivalent ion (Al^{3+}), which is highly toxic to most aquatic fauna, is released when the pH of water drops.

In North America, acid rain has already wiped out all fish and many plants in 50% of the high mountain lakes in the Adirondacks. The pH in many of these lakes has reached such levels of acidity as to replace the trout and native plants with acid-tolerant mats of algae.

The deposition of atmospheric acid on freshwater aquatic systems prompted the U.S. EPA to suggest a limit of from 10 to 20 kg SO_4^{2-} per hectare per year. If Newton's law of air pollution is used (what goes up must come down), it is easy to see that the amount of sulfuric and nitric oxides emitted is vastly greater than this limit. For example, for the state of Ohio alone, the total annual emissions are 2.4×10^6 metric tons of gaseous sulfur dioxide (SO_2) per year. If all of this is converted to SO_4^{2-} and is deposited on the state of Ohio, the total would be 360 kg per hectare per year.[16]

But not all of this sulfur falls on the folks in Ohio, and much of it is exported by the atmosphere to places far away. Similar calculations for the sulfur emissions for the northeastern United States indicates that the rate of sulfur emission is four to five times greater than the rate of deposition. Where does it all go?

The Canadians have a ready and compelling answer. They have for many years blamed the United States for the formation of most of the acid rain that invades across the border. Similarly, much of the problem in Scandinavia can be traced to the use of tall stacks in Great Britain and the lowland countries of continental Europe. For years British industry simply built taller and taller stacks as a method of air pollution control, reducing the local, immediate ground-level concentration, but emitting the same pollutants into the higher atmosphere. The air quality in the United Kingdom improved, but at the expense of acid rain in other parts of Europe.

CASE STUDY: THE MYSTERY OF THE DISAPPEARING FISH

In the 1960s, the mountain lakes of Norway were full of cold-water game fish, but the populations of these fish seemed to be decreasing. Studies conducted by the Norwegian Institute for Water Research showed a steady decline in fish. The lakes were in remote areas, far away from sources of pollution, and various nonpollution causes were considered for the declining fish populations. Using water quality samples, the institute found that the pH in the lakes had been steadily decreasing. More important, rain gauges set up around the lakes showed a wide variation in the pH of the rain. But where was the acid rain coming from? The mystery was solved by tracking the paths of storms that produced particularly low-pH rainwater and determining where these storms had been prior to dumping the rainwater in Norway. Knowing what was happening to the fish in Norwegian lakes was one thing, of course, and quite another was trying to do something about it.

Acid rain is an indicator of another, larger societal problem. Pollution across political boundaries is a particularly difficult regulatory problem. The big stick of police power is no longer available. Why *should* the UK worry about acid rain in Scandinavia? Why *should* the Germans clean up the Rhine before it flows through The Netherlands? Why *should* Israel stop taking water out of the Dead Sea, which it shares with Jordan? Laws are no longer useful, and threats of retaliation are unlikely. What forces are there to encourage these countries to do the right thing? Is there such a thing as *international ethics*?

One of the difficulties in dealing with acid rain and global climate change is that they in essence are asking that society cut back on combustion; and to many, combustion is tantamount to progress and development. This a but another example of risk trade-offs.

Discussion: *Where There's Fire, There's Smoke*

Fire may not be one of the first things to come to mind when thinking about environmental justice. After all, humans have been around fire throughout our existence. We need it, yet we fear it. Combustion is a relatively simple phenomenon: oxidation of a substance in the presence of heat. Chemically, efficient combustion is

$$a(CH)_x + bO_2 \rightarrow cCO_2 + dH_2O \tag{5.18}$$

where a, b, c, and d are stoichiometric constants depending on the hydrocarbon combusted.

Following are additional complex combustion reactions, balanced combustion reactions for selected organic compounds:

Chlorobenzene	$C_6H_5Cl + 7O_2 \rightarrow 6CO_2 + HCl + 2H_2O$
Tetrachloroethene (TCE)	$C_2Cl_4 + O_2 + 2H_2O \rightarrow 2CO_2 + HCl$
Hexachloroethane (HCE)	$C_2Cl_6 + \frac{1}{2}O_2 + 3H_2O \rightarrow 2CO_2 + 6HCl$
Postchlorinated polyvinyl chloride (CPVC)	$C_4H_5Cl_3 + 4\frac{1}{2}O_2 \rightarrow 4CO_2 \cdot 3HCl + H_2O$
Natural gas fuel (methane)	$CH_4 + 2O_2 \rightarrow CO_2 + 2H_2O$
PTFE teflon	$C_2F_4 + O_2 \rightarrow CO_2 + 4HF$
Butyl rubber	$C_9H_{16} + 13O_2 \rightarrow 9CO_2 + 8H_2O$
Polyethylene	$C_2H_4 + 3O_2 \rightarrow 2CO_2 + 2H_2O$

Wood is considered to have the composition $C_{6.9}H_{10.6}O_{3.5}$. Therefore, the combustion reactions are simple carbon and hydrogen combustion:

$$C + O_2 \rightarrow CO_2$$

$$H + 0.25O_2 \rightarrow 0.5H_2O$$

Most fires, however, do not reach complete combustion of all of the compounds in the fuel being oxidized. They are usually oxygen-limited, so that a large variety of new compounds, many that are toxic, is released. Decomposition of a substance in the absence of oxygen is known as *pyrolysis*. So a fire consists of both combustion and pyrolytic processes; and the fire itself is not homogeneous, with temperatures varying in both space and time. Plastic fires, for example, can release over 450 different organic compounds.[17] The relative amount of combustion and pyrolysis in a fire affects the actual amounts and types of compounds released. Temperature is also important, but there is no direct relationship between temperature and pollutants released. As evidence, Figure 5.8 shows that in a plastics fire (i.e., low-density polyethylene pyrolysis), some compounds are generated at lower temperatures, whereas for others the optimal range is at higher temperatures. However, the aliphatic compounds in this fire (i.e., 1-dodecene, 9-nonadecane, and 1-hexacosene) are generated in higher concentrations at lower temperatures (about 800°C), while the aromatics

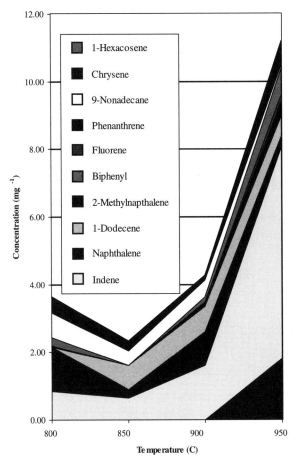

Figure 5.8 Selected hydrocarbon compounds generated in a low-density polyethylene fire (pyrolysis) in four temperature regions. (Data from R. A. Hawley-Fedder, M. L. Parsons, and F. W. Karasek, Products Obtained During Combustion of Polymers Under Simulated Incinerator Conditions, *Journal of Chromatography,* 314:263–272, 1984.)

need higher temperatures (see Figure 5.9). This may be because with increasing temperature, the chains are being transformed to aromatic rings. Also, there appears to be a general trend toward heavier molecular weight compounds with increasing temperature (probably because the increased heat energy allows for more synthesis or combination chemical reactions).

All societies depend on fire. The way that a nation addresses burning is an important measure of how advanced it is, not only in dealing with pollution, but in the level of sophistication of its economic systems. For example, many poorer nations are confronted with the choice of saving sensitive habitat or allowing large-scale biomass burns. And, the combustion processes in developing countries are usually much less restrictive than those in more developed nations. For example, it may be surprising that in Latin America, some of the

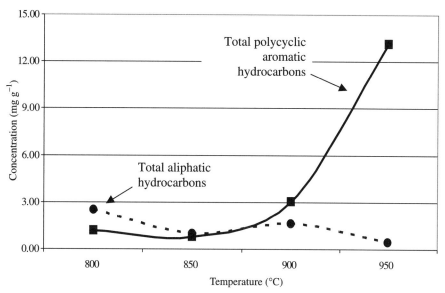

Figure 5.9 Total aliphatic (chain) hydrocarbons *versus* polycyclic aromatic hydrocarbons (PAHs) generated in a low-density polyethylene fire (pyrolysis) in four temperature regions. (Data from R. A. Hawley-Fedder, M. L. Parsons, and F. W. Karasek, Products Obtained During Combustion of Polymers Under Simulated Incinerator Conditions, *Journal of Chromatography,* 314:263–272, 1984.)

largest sources of chlorinated dioxins are in brickmaking. Refractories in developing nations are often small-scale neighborhood operations (see Figure 5.10). Often, the heat source used to reach refractory temperatures are furnaces with scrapped materials as fuel, especially petroleum-derived substances such as automobile tires.

Even within developed countries, there are still remnants of such processes. El Paso, Texas, and Ciudad Juarez, Mexico, with a combined population of 2 million, are located in a deep canyon between two mountain ranges, which can contribute to thermal inversions in the atmosphere (see Figure 5.11). The air quality has been characterized by the U.S. EPA as seriously polluted, with brickmaking on the Mexican side identified as a major source.[18]

The workers who make bricks are called *ladrilleros*. Nearly 400 of them live in unregulated shantytowns known as *colonias* on the outskirts of Ciudad Juarez. The kilns, which are of the same design as that used in Egypt thousands of years ago, are located within these neighborhoods, next to the small houses. The *ladrilleros* are not particular about the fuel they use, burning anything with caloric value, including scrap wood and old tires as well as more conventional fuels such as methane and butane. The dirtier fuels (e.g., tires), release large black plumes of smoke that contain a myriad of contaminants.

Children are at an elevated risk of health problems when exposed to these plumes, since their lungs and other organs are undergoing prolific tissue

Figure 5.10   Large urban city in Mexico showing the effects (loss of visibility and smog) of air pollution.

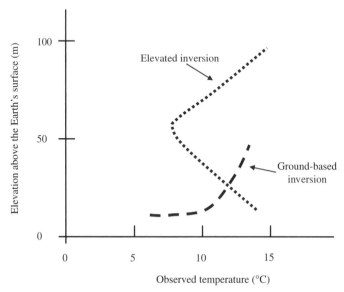

Figure 5.11   Two types of thermal inversions that contribute to air pollution.

growth. Thus, the *ladrilleros'* families are at particularly elevated risks, due to their frequent and high-dose exposures. "The health impact is not only of concern to the worker but also the entire family, especially pregnant women and children who, because of their socioeconomic status, tend to be undernourished," according to Beatriz Vera, project coordinator for the U.S.–Mexico Border Environment and Health Projects. She adds that "many times the entire family participates in the process. Sometimes children are put directly into the area where the kiln is fired."

The two nations' governments are at least somewhat cognizant of the problem, as are numerous *nongovernmental organizations* (known as NGOs). These have included Environmental Defense, Physicians for Social Responsibility, the Federación Mexicana de Asociaciónes Privadas de Salud y Desarrollo Comunitario (FEMAP), and El Paso Natural Gas (EPNG). FEMAP and EPNG, for example, offer courses to the *ladrilleros* from throughout the region on ways to use higher-quality fuel, together with improved safety and business practices. Often, however, even if the brickmakers know about cleaner fuels, they cannot afford them. For example, they have used butane, but in 1994 the Mexican government started to phase out its subsidy, and at about the same time, the peso was devalued, leading to a sharp increase in butane costs. The *ladrilleros* were forced to return to using cheaper fuels. In the meantime the Mexican government banned the burning of tires, so much of the more recent tire burning has been done at night surreptitiously.

A number of solutions to the problem have been proposed, including more efficient kilns. However, arguably the best approach is to prevent the combustion in the first place. In fact, many of the traditional villages where bricks are now used had previously been constructed with adobe. A return to such a noncombustion approach could hold the key. The lesson here is that often in developing countries, the simpler, "low-tech" solutions are the most sustainable.

In the mid-1990s, the U.S. EPA and the Texas Natural Resource Conservation Commission conducted a study in the Rio Grande valley region to address concerns about the potential health impact of local air pollutants, especially since little air quality information was available at the time. There are numerous "cottage industries," known as *maquiladoras,*[19] along both sides of the Texas–Mexico border. In particular, the study addressed the potential for air pollution to move across the U.S.–Mexican border into the southern part of Texas. Air pollution and weather data were collected for a year at three fixed sites near the border in and near Brownsville, Texas. The study found overall levels of air pollution to be similar to or even lower than in other urban and rural areas in Texas and elsewhere and that transport of air pollution across the border did not appear to impact air quality adversely across the U.S. border. Although these "technical" findings may not be particularly surprising, the study provided some interesting results in how to work with local communities, particularly those with much cultural diversity and richness.

Scientists, engineers, and planners must be sensitive to specific challenges in addressing EJ issues. The National Human Exposure Assessment Survey and Lower Rio Grande Valley Transboundary Air Pollution Project evaluated total human exposure to multiple chemicals on community and regional scales from 1995 to 1997. Lessons learned from this research[20] apply to most EJ studies:

1. Develop focused study hypotheses and objectives and develop the linkages between the objectives, data requirements to meet the objectives, and available resources.

2. Ensure that questionnaires address study objectives and apply experiences of previous studies. Wording must apply to the study participants and reflect differences in demographics (e.g., language, race, nationality, gender, age).

3. Develop materials and scripts to address potential concerns that participants may have about the study (e.g., time burden, purpose and value of study, collection of biological samples, legitimacy of the study).

Before recruiting, build knowledge of the study and support in the community, including support, sponsorship, and advertising from a range of organizations.

1. Plan and conduct follow-up visits with community leaders and stakeholders for timely dissemination of information learned in the study.

2. Integrate recruitment staff into the research team for continuity with participants and to build on experience.

3. Limit field teams to two people per visit (about 1 to 1.5 hours) and have the same staff member work with the study participant at all visits.

4. Ensure that laboratories have the expertise and capacity for planned analyses.

5. Establish a format for reporting data before any field samples are collected.

Also, the way the information is conveyed to the public must be understandable, with a minimum possibility for misinterpretation. Obviously, when dealing with a largely Latino population, technical information must be presented in Spanish (see Figure 5.12). Ongoing EJ research is extending this knowledge. For example, it is becoming painfully obvious that without a readily understandable "benchmark" of environmental measurements, citizens in EJ communities can be left with possible misunderstandings, ranging from a failure to grasp a real environmental problem that exists (a false negative) to perceiving a problem even when values are better than or do not differ significantly from those of the general population (a false positive).

Not too long ago in the United States, the standard means of getting rid of household trash was the daily burn. Each evening, people in rural areas, small towns, and even larger cities made a trip into the backyard, dumped the trash they had accumulated into a barrel,[21] and burned the contents. Also, burning was a standard practice elsewhere, such as intentional fires to remove brush,

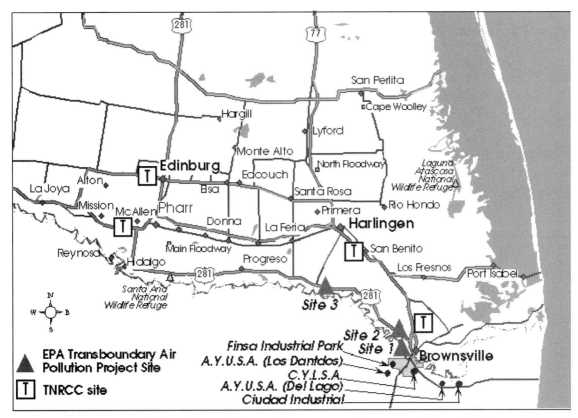

Figure 5.12   Sitios 1, 2, y 3 de Monitoreo de Aire usados en este estudio y en otros sitios establecidos por laTNRCC en el Valle Bajo del Río Grande. Los parques de maquiladoras industriales listados e identificados se identifican con flechas para mostrar sus lugares con relación a los sitios de monitoreo

Figure 5.12   Map annotated in Spanish used to help to explain the Lower Grande Study to community members. English translation: Air monitoring sites 1, 2, and 3 used in this study and other sites established by TNRCC in the Lower Rio Grande Valley. Maquiladora industrial parks listed and identified by arrows to show location relative to monitoring sites. (From S. Mukerjee, D. Shadwick, and K. Dean, *Lower Rio Grande Valley Transboundary Air Pollution Project,* EPA/600/R-99/047, U.S. EPA, Washington, DC, April 1999.)

and even in cottage industries such as backyard smelters and metal recycling operations. Beginning in the 1960s and 1970s, the public acceptance and tolerance for open burning was waning. Local governments began to restrict and eventually to ban many fires. Often, these restrictions had multiple rationales, especially public safety (fires becoming out of control, especially during dry seasons) and public health (increasing awareness of the association between particulate matter in the air and diseases such as asthma and lung cancer).

This new intolerance for burning was a type of paradigm shift. What one did in one's own yard was no longer the sole primacy of the homeowner; the action had an effect on a larger area (i.e., wherever the plume migrated). It also had a cumulative effect. A fire or two may not cause considerable harm, but when large numbers of fires occur, health thresholds could easily be crossed. In fact, the paradigm was new to environmental regulation—but not to ethics. Immanuel Kant is credited with the deontological (i.e., duty-based) view of ethics. A metric used by Kant to determine the morality of an action or decision is whether that action or decision, if universalized, would lead to a better or worse society. This is known as the *categorical imperative.* Somewhere in the 1960s and early 1970s, environmental science and policy shifted toward the categorical imperative. Slogans like "Think globally and act locally," and "We all live in Spaceship Earth" are steeped in the categorical imperative. Our duty and obligation to our fellow human beings, now and in the future, must drive our day-to-day decisions.

So, then, what is so bad about open burning? Vallero recalls as a teenager working with his Uncle Joe in his salvage business. Uncle Joe was truly amazing in his ability to identify sources of profit. He would drive by residences and see yards in need of cleaning, offer for a fee to clean up the site, load the "junk" into his 2-ton truck, and haul it to his "recycling facility," which for some reason not clear to Vallero, he called "the Ponderosa." When they arrived at the site, they would separate the various items by their material makeup. The motors, such as those from washing machines and tools, would be placed in one pile, wires in another, cast iron and bulk metals in another, and "trash" in another. The trash would be burned.

It is instructive to understand the vertical and horizontal integration of the salvage business to its possible environmental benefits and problems. In addition to cleaning up sites, Uncle Joe would visit local junkyards. East St. Louis, Illinois is a rail center and, as such, has been a prime location for rail-side salvage businesses. Often, the salvages had more unclaimed and dumped cargo than they could handle internally, so they often "contracted out" with wildcatters like Vallero's uncle. So we would visit a number of junkyards and buy motors, armatures, and wire at a bulk rate, take them back to the Ponderosa, and separate the valuable materials (e.g., No. 1 and 2 copper from the armatures) from the bulk materials (the cast iron housings from the armatures). Profit was the difference between the few dollars that Uncle Joe paid per hundreds of pounds of bulk material and the 60 cents per pound he received from the same salvage yards (or another if they paid more) for the clean copper. Unfortunately, not all of the copper that was removed and handled was bare wire. Much of it contained insulation, usually made of compounds and polymers similar to polyvinyl chloride (PVC). At the time, the only way available to remove the sheathing insulation *en masse* was by burning the wire. In addition, some of the materials contained dielectric fluids, probably with high concentrations of polychlorinated biphenyls (PCBs). Let us consider what happens when both PVC and PCBs are burned.

Like polyethylene, polyvinyl chloride is a polymer. However, rather than a series of ethylenes in the backbone chain, a chlorine atom replaces the hydrogen on each of the ethylene groups by free-radical polymerization of vinyl chloride (see Figure 5.13). The first thing that can happen when PVC is heated

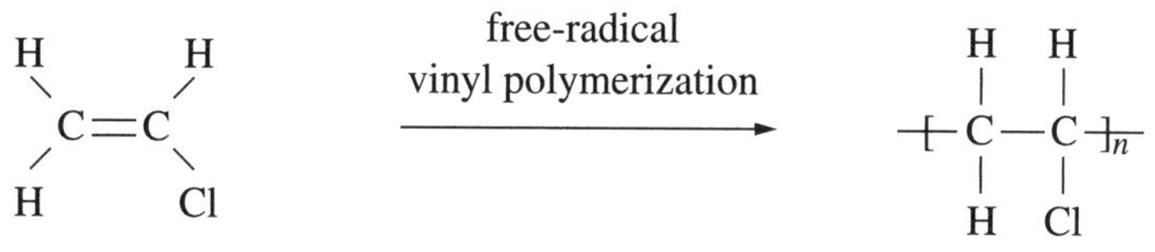

Figure 5.13   Free-radical polymerization of vinyl chloride to form polyvinyl chloride.

is that the polymers become unhinged and chlorine is released. Also, dioxins and furans can be generated from the thermal breakdown and molecular rearrangement of PVC in a heterogeneous process [i.e., the reaction occurs in more than one phase (in this case, in the solid and gas phases)]. The active sorption sites on the particles allow for the chemical reactions, which are catalyzed by the presence of inorganic chloride compounds and ions sorbed to the particle surface. The process occurs within the temperature range 250 to 450°C, so most of the dioxin formation under the precursor mechanism occurs away from the high temperatures of the fire, where the gases and smoke derived from combustion of the organic materials have cooled. Dioxins and furans may also form de novo, wherein dioxins are formed from moieties different from those of the molecular structure of dioxins, furans, or precursor compounds. The process needs a chlorine donor (a molecule that "donates" a chlorine atom to the precursor molecule). This leads to the formation and chlorination of a chemical intermediate that is a precursor to dioxin.

In addition, PVC is seldom in a pure form. In fact, most wires have to be pliable and flexible. On its own, PVC is rigid, so plasticizers must be added, especially phthalates. These compounds have been associated with chronic effects in humans, including endocrine disruption. Also, since PVC catalyzes its own decomposition, metal stabilizers have been added to PVC products. These have included lead, cadmium, and tin (e.g., butylated forms). Another very common class of toxic compounds released when plastics are burned are the polycyclic aromatic hydrocarbons (PAHs).

The change in the attitude and the acceptance of pollution has been dramatic, but since it has occurred incrementally, it may be easy to forget just how much the baseline has changed. For example, like the open burning that was the norm, emissions from industrial stacks were pervasive and contained myriad toxic components. Vallero recalls his despair as a child walking by a chemical plant in Washington Park, Illinois and riding in a car near the coke ovens in Granite City, Illinois. He believes that the plant in Washington Park processed zinc and released compounds that apparently displaced the oxygen in the air downwind. He remembers coughing and gasping for air as he neared the plant. Vallero also recalls holding his nose when passing the Granite City coke ovens, which had a constant plume of obnoxious smelling compounds, probably metallic and sulfur compounds that volatilized during the conversion of coal to coke needed for steel manufacturing. He also remembers, even as a young child, wondering how people could live so close (within a few meters) of these facilities their entire lives. They had grown familiar with the depressed oxygen or odors. While these areas continue to be industrialized, such ambient air quality as that in the 1960s is no longer tolerated.

Coke remains an important component in steel making around the world. It is produced by blending and heating bituminous coals in coke ovens to 1000 to 1400°C in the absence of oxygen.[22] Lightweight oils and tars are distilled from the coal, generating various gases during the heating process. Every half hour or so, the flows of gas, air, and waste gas are reversed to maintain uniform temperature distribution across the wall. In most modern coking systems, nearly half of the total coke oven gas produced from coking is returned to the heating flues for burning after having passed through various cleaning and co-product recovery processes. Coke oven emissions are the benzene-soluble fraction of the particulate matter generated during coke production. What Vallero intuited as a child has since been borne out by scientific research. Coke oven emissions are known to contain human carcinogens. These emissions are truly an awful concoction.

Coke oven emissions are actually complex mixtures of gas, liquid, and solid phases, usually including a range of about 40 PAHs, as well as other products of incomplete combustion: notably, formaldehyde, acrolein, aliphatic aldehydes, ammonia, carbon monoxide, nitrogen oxides, phenol, cadmium, arsenic, and mercury. More than 60 organic compounds have been collected near coke plants. A metric ton of coal yields up to 635 kg of coke, up to 90 kg of coke breeze (large coke particulates), 7 to 9 kg of ammonium sulfate, 27.5 to 34 L of coke-oven gas tar, 55 to 135 L of ammonia liquor, and 8 to 12.5 L of light oil. Up to 35% of the initial coal charge is emitted as gases and vapors. Most of these gases and vapors are collected during by-product coke production. Coke oven gas is comprised of hydrogen, methane, ethane, carbon monoxide, carbon dioxide, ethylene, propylene, butylene, acetylene, hydrogen sulfide, ammonia, oxygen, and nitrogen. Coke-oven gas tar includes pyridine, tar acids, naphthalene, creosote oil, and coal-tar pitch. Benzene, xylene, toluene, and solvent naphthas may be extracted from the light oil fraction. Coke production in the United States increased steadily between 1880 and the early 1950s, peaking at 65 million metric tons in 1951. In 1976, the United States was second in the world, with 48 million metric tons of coke, 14.4% of the world production. By 1990, the United States produced 24 million metric tons, falling to fourth in the world. A gradual decline in production has continued; production has decreased from 20 million metric tons in 1997 to 15.2 million metric tons in 2002. Demand for blast furnace coke has also declined in recent years because technological improvements have reduced the amount of coke consumed per amount of steel produced by as much as 25%.

The junkyards of some decades ago had their own emissions as well. In fact, the combination of fires, wet muck (comprised of soil, battery acid, radiator fluids, motor oil, corroded metal, and water), and oxidizing metals created a rather unique odor around the yards. Neurophysiologists have linked the olfactory center to the memory center of the human brain (olfactory bulb in the cerebellum; see Figure 5.14). In other words, when we smell something, it evokes a strong memory response. This was the case when Vallero was involved in air sampling in lower Manhattan shortly after the attacks on the World Trade Center on September 11, 2001. The smell from the burning plastics, oxidizing metal, semivolatile and volatile organic compounds, and particulate matter was very similar to what Vallero smelled as a teenager in the East St. Louis junkyards, and it brought back 35-year-old memories.

Odors have often been associated with environmental nuisances. In addition to the link between memory and olfactory centers, however, the nasal–neural connection is important to environmental exposure. This goes beyond nuisance and is an indication of potential adverse health effects. For example, nitric oxide (NO) is a neurotoxic gas released from many sources, such as confined animal feeding operations, breakdown of fertilizers after they are applied to soil and crops, and emissions from vehicles. In addition to being inhaled

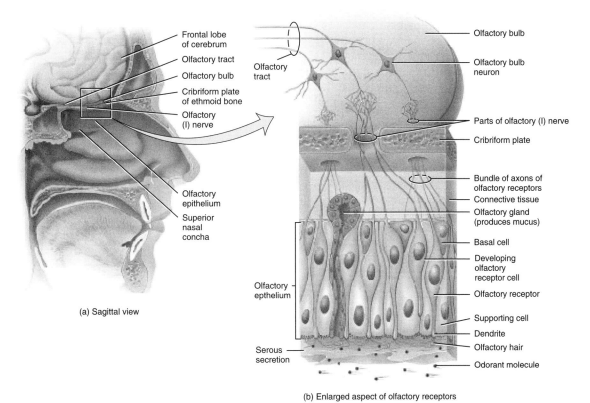

(a) Sagittal view

(b) Enlarged aspect of olfactory receptors

Figure 5.14 Human smell is initiated by sniffing, which transports air with concentrations of odorant molecules past curved bony structures (i.e., turbinates) in the nose. The turbinates generate turbulent airflow that mixes volatile compounds and particles and carries them to a thin mucus layer that coats the olfactory epithelium. The olfactory epithelium contains odor-sensing cells. (From S. G. Turtora and B. Derrickson, *Principles of Anatomy and Phyisology*, 11th edition, copyright 2006, courtesy of John Wiley & Sons.)

into the lungs, NO can reach the brain directly. That is, without having to travel through the lungs and circulatory system, the gas can pass through a thin membrane via the nose to the brain.

The nasal exposure is a different paradigm from that commonly used to calculate exposure. In fact, most sources do not have a means for calculating exposures other than dermal, inhalation, and ingestion. Research at Duke University, for example, linked people's emotional states (i.e., moods) to odors around confined animal feeding operations in North Carolina.[23] People who live near swine facilities are negatively affected when they smell odors from the facility. This is consistent with other research, which has found that people experience adverse health symptoms more frequently when exposed to livestock odors. These symptoms include eye, nose, and throat irritation, headache, nausea, diarrhea, hoarseness, sore throat, cough, chest tightness, nasal congestion, palpitations, shortness of breath, stress, and drowsiness. There is quite a bit of diversity in response, with some people being highly sensitive even to low concentrations of odorant compounds, whereas others are relatively unfazed even at much higher concentrations. Actually, response to odors can be triggered by three different mechanisms. In the first mechanism, symptoms can be induced by exposure to odorant compounds at sufficiently high concentrations to cause irritation or other toxicological effects. The irritation, not the odor, evokes the health symptoms. The odor sensation is merely an exposure indicator. In the second mechanism, symptoms of adverse effects result from odorant concentrations lower than those eliciting irritation. This can be due to genetic predisposition or conditioned aversion. In the third mechanism, symptoms can result from a coexisting pollutant (e.g., an endotoxin), which is a component of the odorant mixture.

Therefore, to address this new paradigm, better technologies will be needed. For example, the Duke and other findings have encouraged some innovative research, including the development of "artificial noses."[24]

During the aftermath of the World Trade Center attacks, odors played a key role in response. People were reminded daily of the traumatic episode, even when the smoke from the fire was not visible; and the odors were evidence of exposure to potentially toxic substances. In this instance, it was not NO, but volatile, semivolatile organic, and metallic compounds. The presence of the odors was one of many factors that kept New Yorkers on edge, and until the ordors subsided significantly, they continued to be a source of anxiety. Analysis of the samples had confirmed the brain's olfactory-memory connection. The plume from the World Trade Center fires contained elevated concentrations of PAHs, dioxins, furans, volatile organic compounds (e.g., benzene), and particles containing metals (see Figures 5.15 to 5.17).

Obviously, combustion still goes on in the United States, but the *open burning* paradigm has thankfully shifted quite far toward cleaner processes and better control technologies in recent decades. In fact, the past few decades have clearly shifted toward an appreciation for the common value of air sheds, notwithstanding Garrett Hardin's laments in "The Tragedy of the Commons."[25] This is not to say that Hardin was wrong (i.e., people *do* tend to favor their

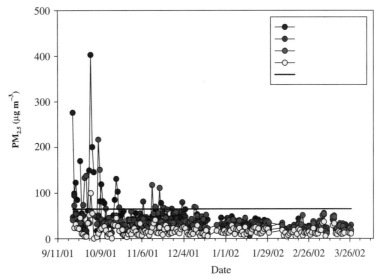

Figure 5.15 Mean daily particulate matter concentrations (aerodynamic diameter ≤2.5 micrometers = $PM_{2.5}$) at the World Trade Center site in 2001 and 2002.

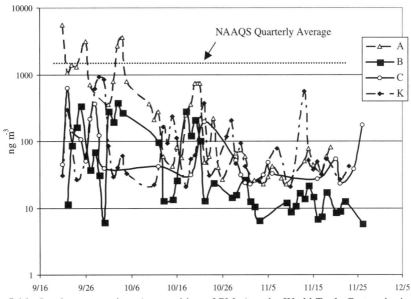

Figure 5.16 Lead concentrations (composition of $PM_{2.5}$) at the World Trade Center site in 2001.

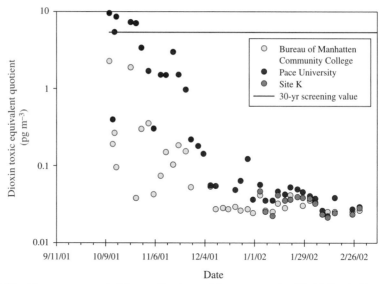

Figure 5.17 Measurements of airborne dioxins and furans near the World Trade Center site in 2001 and 2002.

personal utilities). However, once society's members saw the problems being wrought by unchecked combustion, they willingly accepted restrictions and regulations, as long as they were effective and perceived to be fair. In this way, the open-burning paradigm shift can be seen as a codicil to Thomas Hobbes' social contract; that is, had we continued on the path of unbridled combustion and emissions, we would have reached a societal "state of nature" where the collective of "brutes" with fires would damage the entire population, including our personal and familial well-being.

Ozone Depletion

The atmosphere surrounding the Earth is conveniently divided into identifiable layers. More than 80% of the airmass is in the lowest layer, the *troposphere,* which is between 10 and 12 km deep. The fastest airlines fly at around 38,000 feet, which is about 11.6 km. The mass of air in the troposphere varies with distance from the equator. At the equator, the troposphere is about 18 km deep, whereas at the poles it is only about 6 km deep. The temperature decreases with elevation, and the air within the troposphere is in motion with significant mixing.

Above the troposphere is the *stratosphere,* which is stable and reaches to about 50 km. There is little movement in or out of the layer, and anything that gets to that layer can be expected to stay there for a long time. Taken together, the troposphere and the stratosphere account for about 99.9% of all the air, and everything above this remaining 0.1% layer is considered "space."

One of the major gases in the stratosphere is ozone. Ozone (O_3) is a pollutant, but is also beneficial. It is a matter of where the O_3 is. The O_3 in the stratosphere acts as an ultraviolet radiation shield, and there is concern that this protective shield is being destroyed, leaving Earth and its inhabitants vulnerable to the effects of ultraviolet radiation, which can result in an increased risk of skin cancer as well as producing changes in the global ecology in unpredictable ways.

Ozone in the stratosphere is created by the reaction of oxygen with light energy, first splitting the oxygen molecule into free oxygen atoms:

$$O_2 + h\nu \rightarrow O + O \qquad (5.19)$$

where h is Planck's constant [6.62×10^{-34} joule-second (J-s)] and ν is the frequency (s^{-1}). Think of the combination of h and ν as light energy. The important point is that this reaction can take place only if the light energy is in the ultraviolet wavelength region, below about 400 mn.

Because the oxygen atoms are highly unstable, they will seek out oxygen molecules to form ozone,

$$O + O_2 + M \rightarrow O_3 + M \qquad (5.20)$$

where M is some third body used to carry away the heat generated.

Ozone is in turn destroyed by light energy as ultraviolet radiation:

$$O_3 + h\nu \rightarrow O_2 + O \qquad (5.21)$$

and the atomic oxygen can then once again react with molecular oxygen to form ozone. The light energy responsible for this reaction has a wavelength between 200 and 320 nm, which is smack in the middle of the ultraviolet part of the spectrum, thus adsorbing the ultraviolet energy and preventing it from reaching Earth's surface. If the stratosphere does not contain ozone, a much larger percentage of the ultraviolet energy will pass through the stratosphere, causing damage to living tissues. In the stratosphere there is a balance with oxygen being broken down, forming ozone, and the ozone being destroyed returning to oxygen molecules, all the time using up the energy from the ultraviolet radiation. The presence of the ozone at some steady-state concentration has become immensely important in protecting all life on Earth from damaging ultraviolet radiation.

The problem with the depletion of upper atmospheric ozone is due to the manufacture and discharge of certain compounds, especially a class of chemicals called *chlorofluorocarbons* (CFCs). These compounds have been widely used as propellants and in refrigeration systems, but may also responsible for both global warming as well as the depletion of the protective ozone layer in the stratosphere. Before the synthesis of CFCs, refrigeration units used CO_2, isobutene, methyl chloride, or sulfur dioxide, all of which are toxic, flammable, or inefficient. CFCs are nontoxic, not water soluble, nonreactive, nonbiodegradable, and are easily liquefied under pressure.[26] When they evaporate they produce very cold temperatures, making them ideal refrigerants. Two of the most important CFCs are trichlorofluoromethane, $CFCl_3$ (industrial designation CFC-11) and dichlorodifluoromethane, CF_2Cl_2 [industrial designation CFC-12, also known as Freon (trade name of DuPont)]. Other chlorinated refrigerants include hydrochlorofluorocarbons (HCFCs) such as CHF_2Cl (industrial designation HCFC-22) and hydrofluorocarbons (HFCs), which do not contain chlorine. One HFC of importance is CH_2FCF_3 (HFC 134a) which is becoming widely used in automobile air conditioners. Both the HCFCs and HFCs are still potentially ozone-depleting gases, but they can be broken down by sunlight and thus their lifetime in the atmosphere is significantly shorter than that of the CFCs and they don't end up in the stratosphere, whirling around for hundreds of years.

The problematic CFCs are no longer manufactured, although many air-conditioning systems still have CFCs, and thus the potential for damage by CFCs remains high. The problem is that the coolants can escape from the refrigeration units and enter the atmosphere, where they are inert and non–water soluble and do not wash out. They drift into the upper atmosphere and are eventually destroyed by shortwave solar radiation, releasing chlorine, which can react with ozone.

The effect of the CFCs on this delicate balance can be devastating. Because the CFCs are so nonreactive and nonsoluble, once they have drifted into the atmosphere, they will stay there for a long time. Eventually, they will be broken down by ultraviolet radiation, forming other chlorinated fluorocarbons and releasing atomic chlorine. For example, CFC-12, a leading refrigerant, breaks down as

$$CF_2Cl_2 + h\nu \rightarrow CF_2Cl + Cl \tag{5.22}$$

The atomic chlorine acts as a catalyst in breaking down ozone:

$$Cl + O_3 \rightarrow ClO + O_2 \tag{5.23}$$

$$ClO + O \rightarrow Cl + O_2 \tag{5.24}$$

with atomic chlorine again forming to continue to promote more destruction of ozone. The atomic oxygen, meanwhile, can also help in destroying ozone:

$$O + O_3 \rightarrow 2O_2 \tag{5.25}$$

Thus, a single Cl atom can make thousands of loops until it eventually reacts with something like methane and is tied up chemically. The reaction with methane produces HCl and ClO. The ClO then can react with nitrogen dioxide (another air pollutant):

$$ClO + NO_2 \rightarrow ClONO_2 \text{ (chlorine nitrate)} \tag{5.26}$$

The chlorine in HCl and $ClONO_2$ is trapped and can no longer enter into the foregoing reactions. The problem is that under certain conditions, both of these chemicals, HCl and $ClONO_2$, can break up and re-release the chlorine so that it can continue to do damage. These reactions occur when surfaces are available for the reaction to take place, and these surfaces are provided by molecules such as sulfate aerosols. The breakup of HCl and $ClONO_2$ plays an important role in understanding the "ozone hole" over the Antarctic.

During the winter months there exists a polar vortex, formed as a whirling mass of very cold air. This vortex is so strong that it isolates air above the South Pole from the rest of the atmosphere. When spring comes, this exceedingly cold ($-90°C$) air forms clouds of ice crystals in the stratosphere. On the surfaces of these crystals, the "inert" chlorinated compounds react as follows:

$$ClONO_2 + H_2O \rightarrow HOCl + HNO_3 \tag{5.27}$$

$$HOCl + HCl \rightarrow Cl_2 + H_2O \tag{5.28}$$

$$ClONO_2 + HCl \rightarrow Cl_2 + HNO_3 \tag{5.29}$$

These reactions tie up the chlorine on the ice crystals so that it cannot enter into the reactions to break up ozone. But when spring comes, this large quantity of stored chlorine suddenly becomes available, and light energy breaks it apart:

$$Cl_2 + h\nu \rightarrow 2Cl \tag{5.30}$$

The sudden dumping of all that chlorine into the stratosphere results in a high rate of ozone destruction, producing the "hole" in the ozone layer. In the southern hemisphere's later spring (November) air from other parts of the globe rushes in, replenishing the ozone and reducing the size of the hole. The effect of the annual depletion is strongly felt in countries such as Australia and New Zealand, which are close to the Antarctic continent.

CASE STUDY: DISCOVERY OF THE OZONE HOLE

Ozone levels in the stratosphere had been measured since the 1970s using satellites. In addition to satellite data, a British team of scientist in the Antarctic began to measure the concentration of ozone in the stratosphere using sophisticated instruments at the base near the South Pole. In 1985 the team discovered that there had been a dramatic drop in the ozone levels in the stratosphere, creating a large hole (i.e., O_3 concentration lower than the rest of the strastosphere) over the Pole. The satellite date did not show such a hole, however, and the scientists concluded that their instruments must be faulty. They had a new set of instruments flown in, and even with careful calibration, they had to conclude that this huge hole had indeed developed. The team studied the satellite data and found that the software being used was designed to ignore very low levels of ozone and reported this simply as "no data." The satellite had seen this hole for many years, but the data reduction did not recognize the presence of the hole. When the scientists adjusted the program to report low levels of ozone, the satellite data agreed with the data obtained on the ground. There indeed was this huge hole over the middle of the South Pole.

The size of the ozone layer has been increasing over the years, prompting increasing concern about the effect on the Earth of ultraviolet radiation. Calculations show that a 1% increase in UV radiation can result in a 0.5% increase in melanoma, a particularly aggressive form of skin cancer, and a 2.5% increase in other forms of skin cancer. Epidemiological statistics show that melanoma is, in fact, increasing at a rate of 2 to 3% annually in the United States. It must be pointed out, however, that much of this increase might be explained by a longer life span as well as more outdoor activities and recreation.

UV radiation can also cause eye damage and can suppress the immune system in humans. A very troublesome effect of UV radiation is the suppression of photosynthesis in aquatic plants. A slower rate of photosynthesis would result in a higher atmospheric concentration of CO_2, exacerbating the problems associated with global warming.

The story with the ozone layer and CFCs is one of two environmental researchers who published a courageous paper in 1974 suggesting that the depletion of ozone was possible as a direct result of escaped refrigerants. Industrial interests at first contested these findings, and it might never have become a global concern were it not for the discovery of the Antarctic ozone hole. From that point on, international concern forced action, culminating in the *Montreal Protocol on Substances That Deplete the Ozone Layer,* with 23 nations, including the United States, agreeing to cut the use of CFCs by 50% by 1999, and eventually, to cease production totally. In 1988, DuPont, the largest manufacturer of CFCs, stopped making them, prompting a huge underground industry based on smuggling these refrigerants into the United States, since so many cooling systems depended on them. In 1990, the availability of new refrigerants, and the growing concern with the ozone hole in the Antarctic, prompted a revised schedule, with a new timetable calling for a complete phase-out of CFCs by 2000, with a few exceptional chemicals needed for medical and other uses to be phased out by 2010. A convention in Copenhagen in 1992 again accelerated the phase-out, with the production and importation of CFCs banned as of 1994. The U.S. EPA responded to these protocols by banning nonessential use of CFCs, such as in noise horns, requiring automobile mechanics to save and reuse CFC in air conditioners, and mandating the removal of CFCs from cars headed for demolition.

The Montreal Protocol was mainly the product of both UK and U.S. leadership, and was possible because the two largest manufacturers of refrigerants, DuPont in the United States and ICI in the United Kingdom, already had developed alternative coolants. The Protocol was therefore welcomed by these industries since this represented a way of increasing their market share in these chemicals.

Global Warming

The first question that comes to mind in a discussion global warming is: Is it happening? Is the Earth really getting warmer? Quite obviously, the average temperature of the earth is difficult to measure, but the measurements agree that the Earth is getting warmer. This is a very small overall change that would not be detectable to humans based solely on short-term and regional variations.

The second question, provided that we have convinced ourselves that the Earth is in fact getting warmer, is what might be causing this increase in temperature. One explanation would be that we are simply seeing a natural cycle of temperature fluctuations, such as has occurred on Earth for hundreds of thousands of years. Studies of ice cores in Russia have shown that, amazingly enough, the mean temperature of the Earth has not changed in over 200,000 years. There have been wide fluctuations, such as the ice ages, but on balance, the mean temperature has remained constant, prompting some scientists to speculate some whimsical causes for such consistency.

There is another explanation for the sudden (relative to geological time) increase we are presently experiencing, and that is that the presence of certain gases in the atmosphere are causing the Earth to become less efficient in returning the heat energy from the sun back into space. The Earth acts as a reflector of the sun's rays, receiving radiation from the sun, reflecting some of it into space (called *albedo*), and adsorbing the rest, only to reradiate this into space as heat. In effect, the Earth acts as a wave converter, receiving high-energy, high-frequency radiation from the sun and converting most of it into low-

Biographical Sketches: F. Sherwood Rowland and Mario J. Molina

Mario Molina was born and raised in Mexico. Following his university graduation with a degree in physical chemistry, he went to Berkeley, where he did research on photochemical reactions. In 1973 he joined a research group headed by Sherwood Rowland at the University of California at Irvine. Rowland, who had come to Irvine after an undergraduate degree from Ohio Wesleyan and a Ph.D. in chemistry from the University of Chicago, offered several research opportunities to Molina, who chose the little-known problem of understanding the fate of chlorofluorocarbons in the atmosphere. In 1972, Rowland had heard a talk by the British scientist James Lovelock about detecting these chemicals in the atmosphere but not understanding their fate. At first the study was only one of scientific curiosity, but soon they began to realize that the presence of CFCs in the stratosphere would have profound environmental consequences. Other scientists had recognized the effect of chlorine on stratospheric ozone, but none had shown that the CFCs would have a dramatic impact on the ozone concentration. Molina and Rowland published their findings in *Nature* in 1974 and immediately became the targets of severe scientific criticism from industrial interests. The two scientists persevered, however, and went to great lengths to publicize their results and to testify at congressional hearings. Finally, the scientists at DuPont, the largest manufacturer of CFCs, acknowledged that Molina and Rowland were correct, and pledged to cease the manufacture of these compounds. (The fact that they had already developed an alternative refrigerant no doubt played a role in their decision.)

In 1974, Rowland and Molina, along with Paul Crutzen of the Max-Planck Institute, were awarded the Nobel Prize in Chemistry for their work in understanding the chemical processes involved in ozone depletion.

energy, low-frequency heat to be radiated back into space. In this manner, the Earth maintains a balance of temperature.

To better understand this balance, light and heat energy have to be defined in terms of their radiation patterns, as shown in Figure 5.18. The incoming radiation (light) wavelength has a maximum at around 0.5 nm, and almost all of it is less than 3 nm. The heat energy spectrum, the energy reflected back into space, has the maximum at about 10 nm, almost all of it at a wavelength higher than 3 nm.

As both light and heat energy pass through the Earth's atmosphere, they encounter the aerosols and gases surrounding the Earth. These can either allow the energy to pass through, or they can interrupt it by scattering or absorption. If the atoms in the gas molecules vibrate at the same frequency as the light energy, they will absorb the energy

Biographical Sketch: Charles Keeling

Charles Keeling thought that atmospheric chemists should study the atmosphere from the field, and not from the laboratory, and he took his instruments to Mauna Loa in Hawaii, where he began to measure carbon dioxide concentrations in the atmosphere using an infrared gas analyzer. Beginning in 1958, these data have provided the single most important piece of information on global warming. The data are referred to as the *Keeling curve* in honor of the scientist.

The Keeling curve shows that there has been more than a 15% increase in CO_2 concentration, which is a huge jump given the short time that the measurements have been taken. It is likely, if we extrapolate backward, that our present CO_2 levels are double what they were in pre–industrial revolution times, providing ample evidence that global warming is indeed occurring.

Keeling received an undergraduate degree from the University of Illinois and a Ph.D. from Northwestern University. He worked for many years at the Scripps Institute of Oceanography in La Jolla, California.

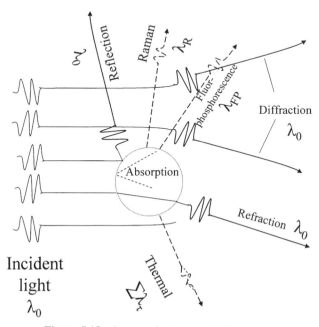

Figure 5.18   Patterns for heat and light energy.

and not allow it to pass through. Aerosols will scatter the light and provide "shade" for the Earth. It was this latter effect that caused many scientists in the 1970s to be concerned about the possibility of "global cooling."

The absorptive potential of several important gases is shown in Figure 5.19, along with the spectra for the incoming light (short-wavelength) radiation and the outgoing heat (long-wavelength) radiation. The incoming radiation is impeded by water vapor and oxygen and ozone, as discussed in the preceding section. Most of the light energy however, comes through unimpeded.

The heat energy, on the other hand, encounters several potential impediments. As it is trying to reach outer space, it finds that water vapor, CO_2, CH_4, O_3, and N_2O all have absorptive wavelengths in the middle of the heat spectrum. An increase in the concentration of any of these will greatly limit the amount of heat transmitted into space. These heat-trapping gases are appropriately called *greenhouse gases* because their presence will limit the heat escaping into space, much like the glass of a greenhouse or even the glass in your car limits the amount of heat that can escape, thus building up the temperature under the glass cover on a sunny day.

The effectiveness of a particular gas to promote global warming (or cooling, as is the case with aerosols) is known as *radiative forcing*. The atmospheric gases of most importance in forcing are listed in Table 5.1 and Appendix 2 (Radiative Forcing and Global Warming).

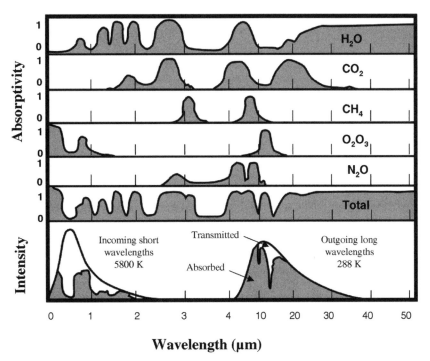

Wavelength (μm)

Figure 5.19 Absorptive potential of several important gases in the atmosphere. Also shown are the spectra for the incoming solar energy and the outgoing thermal energy from the Earth. Note that the wavelength scale changes at 4 μm. (From G. Masters, *Introduction to Environmental Engineering and Science,* Prentice Hall, Upper Saddle River, NJ, 1998.)

Table 5.1 Relative Radiative Forcing of Increased
Global Temperature

Gas	Percent of relative radiative forcing
Carbon dioxide, CO_2	64
Methane, CH_4	19
Halocarbons (mostly CFCs)	11
Nitrous oxide, N_2O	6

Carbon dioxide is the product of decomposition of organic material, whether biologically or through combustion. The effectiveness of CO_2 as a global warming gas has
been known for over 100 years, but the first useful measurements of atmospheric CO_2
were not taken until 1957. The data from Mauna Loa in Hawaii are exceptionally useful,
since they show that even in the 1950s the CO_2 concentration had increased from the
baseline 280 ppm to 315 ppm, and this has continued to climb over the last 50 years at
a constant rate of about 1.6 ppm per year. The most serious problem with CO_2 is that
the effects on global temperature due to its greenhouse effect are delayed. Even if we
stopped emitting any new CO_2 into the atmosphere, what we have already done will
increase CO_2 from our present 370 ppm to possibly higher than 600 ppm. Even if we
stopped producing excess CO_2 now, we have already contaminated the global atmosphere
to where we will have increased the CO_2 concentration to 450 ppm. The effect of this
is discussed below.

Methane is the product of anaerobic decomposition and human food production. One
of the largest producers of methane in the world is New Zealand, which has 80 million
sheep. Methane is also emitted during the combustion of fossil fuels and cutting and
clearing of forests. The concentration of CH_4 in the atmosphere has been steady at about
0.75 ppm for over 1000 years, but spiked to 0.85 ppm in 1900. Since then, in the space
of only 100 years, CH_4 has skyrocketed to 1.7 ppm. Methane is removed from the
atmosphere by reaction with the hydroxyl radical ($OH^{\cdot}$):

$$CH_4 + OH^{\cdot} + 9O_2 \rightarrow CO_2 + 0.5H_2 + 2H_2O + 5O_3 \qquad (5.31)$$

In so doing, it creates carbon dioxide, water vapor, and ozone, all of which are greenhouse
gases, so the effect of one molecule of methane is substantial to the production of the
greenhouse effect.

Halocarbons, the same chemical class involved in the destruction of atmospheric
ozone, are also at work in promoting global warming. The most effective global warming
gases are CFC-11 and CFC-12, both of which are no longer manufactured, and the
banning of these substances has shown a leveling off in the stratosphere.

Nitrous oxide (N_2O, known as laughing gas) is also in the atmosphere, primarily as
a result of human activities, especially the cutting and clearing of tropical forests. The
greatest problem with nitrous oxide is that there appear to be no natural removal processes
for this gas, so its residence time in the stratosphere is quite long.

The net effect of these global pollutants continues to be debated. Various atmospheric models used to predict temperature change over the next hundred years vary widely (about 1.5 to 5.5°C). They nevertheless agree that some increase will occur even if we do something drastic today (which does not seem likely). Even if we do not increase our production of greenhouse gases and if the Western nations were to agree to terms akin to those in the Kyoto Accord, which encourages the reduction in greenhouse gas production, the global temperature is likely to be between 0.5 and 1.5°C warmer by the year 2100. The effect of this seemingly small temperature increase on natural systems and ocean currents will be devastating, including shifts in biomass, altered habitats, and even changes in human disease transmission.

Emerging Health Endpoints of Concern

For most of the second half of the twentieth century, scientific research targeted cancer as its major chronic health endpoint. Environmental carcinogens received the most attention of the numerous etiologies addressed. As such, many of the toxic substances addressed by environmental professionals are carcinogenic. Recently, other health endpoints have also gained much attention. Since the 1970s a cascade of research has shown that lead (Pb), mercury (Hg), and other metals, as well as organophosphate and halogenated organic pesticides and other organics (e.g., PCBs, dioxins), are highly neurotoxic (i.e., they damage the brain and the peripheral and central nervous systems). This is particularly problematic for children, since the effects are more pronounced when tissue is proliferating and since neurotoxicity leads to learning disabilities and behavioral disorders.

More recently, numerous chemicals found in the environment have been linked to hormonal dysfunctions.[27] These chemicals, known as *endocrine disruptors* (see Table 5.2), act like natural hormones by binding to a cell's receptor, at which point they are known as *agonists.* Conversely, chemicals that inhibit the receptor are called *antagonists.* Environmental endocrine disruptors can be of either type. Another group of chemicals act as indirect disruptors, due to their effects on the immune or neural systems, such as the strong neurotoxin mercury, which does not bind to an estrogen or androgen site, but affects the neurological system, which in turn interferes with an organism's hormonal health. The neural, immunological, and endocrine systems are all chemical messaging systems that are so interconnected that a change in one can lead to changes to the others.

Environmental endocrine disruption was observed through much of the twentieth century. Abnormal mating patterns in bald eagles on the east coast of North America were observed in the 1940s. Rachel Carson in the classic 1962 book *Silent Spring* observed that predatory birds such as eagles accumulated chlorinated hydrocarbon pesticides. The eggshells of these birds were abnormally thin, and dichlorodiphenyltrichloroethane (DDT)-contaminated birds were less successful in hatching than those with lower concentrations of the pesticide. During the last two decades of the twentieth century, scientists increasingly found associations between exposure to chemical compounds and changes affecting endocrine systems in humans and animals. These compounds were first called *environmental estrogens,* but since other hormonal effects were found, the term *environmental hormone* gained usage. *Hormone mimicker* was sometimes used to

Table 5.2 Some Compounds Found in the Environment That Have Been Associated with Endocrine Disruption, Based on *In Vitro, In Vivo,* Cell Proliferation, or Receptor-Binding Studies

Compound[a]	Endocrine effect[b]	Potential source
2,2′,3,4′,5,5′-Hexachloro-4-biphenylol and other chlorinated biphenylols	Antiestrogenic	Degradation of PCBs released into the environment
4′,7-Dihydroxydaidzein and other isoflavones, flavones, and flavonals	Estrogenic	Natural flora
Aldrin*	Estrogenic	Insecticide
Alkylphenols	Estrogenic	Industrial uses, surfactants
Bisphenol A and phenolics	Estrogenic	Plastics manufacturing
DDE [1,1-dichoro-2,2-bis(*p*-chlorophenyl)ethylene]	Antiandrogenic	DDT metabolite
DDT and metabolites	Estrogenic	Insecticide
Dicofol	Estrogenic or antiandrogenic in top-predator wildlife	Insecticide
Dieldrin	Estrogenic	Insecticide
Diethylstilbestrol (DES)	Estrogenic	Pharmaceutical
Endosulfan	Estrogenic	Insecticide
Hydroxy-PCB congeners	Antiestrogenic (competitive binding at estrogen receptor)	Dielectric fluids
Kepone (Chlorodecone)	Estrogenic	Insecticide
Lindane (γ-hexachlorocyclohexane) and other HCH isomers	Estrogenic and thyroid agonistic	Miticide, insecticide
Lutolin, quercetin, and naringen	Antiestrogenic (e.g., uterine hyperplasia)	Natural dietary compounds
Malathion*	Thryroid antagonist	Insecticide
Methoxychlor	Estrogenic	Insecticide
Octachlorostyrene*	Thryroid agonist	Electrolyte production
Pentachloronitrobenzene*	Thyroid antagonist	Fungicide, herbicide
Pentachlorophenol	Antiestrogenic (competitive binding at estrogen receptor)	Preservative
Phthalates and their ester compounds	Estrogenic	Plasticizers, emulsifiers
Polychlorinated biphenyls (PCBs)	Estrogenic	Dielectric fluid
Polybrominated diphenyl ethers (PDBEs)*	Estrogenic	Fire retardants, including *in utero* exposures
Polycyclic aromatic hydrocarbons (PAHs)	Antiandrogenic (aryl hydrocarbon–receptor agonist)	Combustion by-products
Tetrachlorodibenzo-*para*-dioxin and other halogenated dioxins and furans*	Antiandrogenic (aryl hydrocarbon–receptor agonist)	Combustion and manufacturing (e.g., halogenation) by-product
Toxaphene	Estrogenic	Animal pesticide dip
Tributyl tin and tin organometallic compounds*	Sexual development of gastropods and other aquatic species	Paints and coatings

Table 5.2 (*Continued*)

Compound[a]	Endocrine effect[b]	Potential source
Vinclozolin and metabolites	Antiandrogenic	Fungicide
Zineb*	Thyroid antagonist	Fungicide, insecticide
Ziram*	Thyroid antagonist	Fungicide, insecticide

Source: D. A. Vallero, Environmental Endocrine Disruptors, in *McGraw-Hill Yearbook of Science and Technology,* McGraw-Hill, New York, 2004. For a full list, study references, study types, and cellular mechanisms of action, see Chapter 2 of National Research Council, *Hormonally Active Agents in the Environment,* National Academies Press, Washington, DC, 2000. The source for asterisked (*) compounds is T. Colburn et al., http://www.ourstolenfuture.org/Basics/chemlist.htm.

[a] Not every isomer or congener included in a listed chemical group (e.g., PAHs, PCBs, phenolics, phthlates, flavinoids) has been shown to have endocrine effects. However, since more than one compound has been associated with hormonal activity, the entire chemical group is listed here.

[b] Note that the antagonists' mechanisms result in an opposite net effect. In other words, an antiandrogen feminizes and an antiestrogen masculinizes an organism.

describe compounds that could elicit a response like that of a natural hormone. For example, certain pesticides bind very easily to estrogen receptors, resulting in increased feminization of the organism. For example, male fish exposed to estrogenic components will start producing vitellogenin, an egg-laying hormone. A hormonally active agent and an endocrine disruptor are more general classifications of any chemical that causes hormonal dysfunction.

Endocrine effects can vary dramatically within a population, with some groups being highly susceptible to endocrine disruptors, while other seemingly go completely unscathed. Sensitive subpopulations, especially adolescents and children, may be particularly threatened. Considerable scientific uncertainty exists, but numerous chemicals, including organochlorine pesticides such as DDT and persistent organic chemicals such as the polychlorinated biphenyls (PCBs) have been shown in laboratory studies at very low concentrations to block or mimic hormones during prenatal development, giving rise to concerns about birth defects and changes in a child's growth and development. Endocrine disruptors have also been associated with certain reproductive cancers. Like other environmental contaminants, endocrine disruptors vary in physical and chemical forms. These differences result in distinct ways that chemicals resist breaking down in the environment, known as *persistence.* Chemicals also vary in their ability to bioaccumulate as well as in their toxicity. Persistent, bioaccumulating toxics easily build up in the food chain and lead to toxic effects: in this instance, hormonal dysfunction. Thus, endocrine disruptors can be organic compounds [known as persistent organic pollutants (POPs)], inorganic compounds (certain metals and their salts), or organometallic compounds, such as the butylated or phenalated forms of the metal tin (Sn).

Endocrine disruptors enter the environment in numerous ways. Wastes from households and medical facilities may contain hormones that reach landfills and wastewater treatment plants, where they pass through untreated or incompletely treated and enter waterways. Fish downstream from treatment plants have shown symptoms of endocrine disruption. Engineers designing treatment facilities must consider the possibility that wastes will contain hormonally active chemicals and find ways to treat them. In addition, manufacturers of pharmaceuticals, plastics, and other sources of endocrine disruptors must find ways to eliminate them. One means of doing this is to change the chemical structure of compounds (known as *green chemistry*) so that they do not bind or block receptor sites on cells. The addition or deletion of a single atom or the arrangement of the same set of atoms (i.e., an isomer) can significantly reduce the likelihood of hormonal effects elicited by a compound.

Hundreds of thousands of chemicals in current or past use are present in the environment, meaning that human populations and ecosystems are at risk of being exposed to them. The large number and various forms of chemicals preclude regulators from evaluating every chemical with the most rigorous testing strategies. Instead, standard toxicity tests have been limited to a small number of chemicals, with the hope that the "worst" chemicals receive specific attention; or the chemicals that are tested may represent large classes of compounds, such as certain types of pesticides. The good news is that computational biology offers the possibility that with advances in its subdisciplines (e.g., genomics, proteomics, metabolomics, metabonomics), scientists may have the ability to develop a more detailed understanding of the risks posed by a much larger number of chemicals. Application of the tools of computational biology to assess the risk that chemicals pose to human health and the environment is termed *computational toxicology*. Computational toxicology applies mathematical and computer models to predict adverse effects and to better understand the mechanism(s) through which a given chemical induces harm (see Table 6.1).

Much of what is known about environmental problems and risks has been learned from laboratory or field studies, or from studies of exposures of populations. However, scientists are beginning to develop new "tools" to understand the processes that lead to environmental risks, including the blending of genomic technologies, sophisticated structure–activity analysis, and high-performance computer modeling of pharmacokinetic and pharmacodynamic pathways. Recent advances focus on breaking down the traditional dichotomy between approaches to evaluating cancer *versus* other disease endpoints, on addressing sensitive life stages, and on addressing aggregate and cumulative exposure to pollutants. The engineer and environmental professional will benefit from the new methods for predicting environmental problems and health threats.

Hazardous Waste Contamination

One of the biggest challenges to environmental professionals is what to do with contaminated sites. Over the past three millennia, civilization has generated wastes in exponentially increasing volumes. Engineering has its roots in these attempts at control. Knossos, Crete's burial program for solid wastes produced by the Minoan civilization (a precursor of the modern landfill where waste was buried in layers intermittently covered by soil), is one of the first recorded efforts at land disposal of wastes.[28] In the millennium before Christ, the city-state of Athens required that citizens be responsible for the refuse and garbage that they produced, and that they transport the wastes at least 1500 meters from the city walls for disposal. The ancient Greeks and Romans also addressed the need for potable water supplies.[29] Vitruvius, for example, recognized in the first century B.C. that water would become polluted in stationary ponds left to evaporate, a process we now refer to as *eutrophication*. He also noted the generation of "poisonous vapors," probably the generation of methane from the anaerobic, reduced conditions of eutrophic water bodies. Ironically, Vitruvius may well have avoided recommending that the neurotoxic lead be used for water supplies, not due to its toxicity (unknown until this century), but because bronze could better withstand the pressures on the closed pipe systems used to move water relatively long distances.[30]

Incineration applied to wastes was led by Europeans, especially Britain and Germany, in the nineteenth century, with the first municipal garbage incineration program established in Nottingham, England, in 1874, followed in a couple of decades by a Britain's first "waste-to-energy" incinerator in the 1890s.[31]

For centuries, wastes that would now be categorized as hazardous, were simply stored above ground, in pits, ponds, and lagoons, or buried under thin layers of soil. In the 1950s and 1960s, initiatives to eliminate open dumps called upon engineers to begin designing sanitary landfills. These engineered systems were a response to public health concerns, but the concern has moved beyond infectious disease agents to a myriad of toxic chemicals.

Hazardous waste can be generated from many types of processes (see Table 5.3) and generally has been considered a subset of solid waste and has been distinguished from municipal wastes and nonhazardous industrial wastes. This characterization is not based completely on science and engineering, but also on the regulatory history of hazardous wastes. For example, in the United States in the mid-1970s, statutes already existed to control air and water pollution, as well as the design of landfills and other facilities to address solid wastes, but the problem of hazardous wastes was not being addressed. As a result, the U.S. Congress enacted the Resource Conservation and Recovery Act (RCRA) in 1976. The law's primary goals are to protect human health and the environment from the potential hazards of waste disposal, to conserve energy and natural resources, to reduce the amount of waste generated, and to ensure that wastes are managed in an environmentally sound manner.[32] Of the 13 billion tons of industrial, agricultural, commercial, and household wastes generated annually in the United States, 2% (i.e., more than 279 million tons) are "hazardous" as defined by RCRA regulations.

One of the chief concerns from hazardous wastes is the potential to contaminate groundwater, which can be affected by any type of hazardous waste from numerous sources. Thus, hazardous wastes may be of any phase: solid, liquid, gas, or mixtures. Also, the hazardous of the waste can result from its inherent properties, such as its physicochemical properties, including its likelihood to ignite, explode, react, cause irritations, or elicit toxic effects. A waste's hazardous inherent properties can also be biological, such as that of infectious medical wastes.

Since wastes are transported by fluids (i.e., liquids and gases), especially in air and water, the likelihood of contamination depends strongly on physical properties, especially solubility, density, and vapor pressure. If the substances comprising the hazardous waste are quite soluble (i.e., easily dissolved in water under normal environmental conditions of temperature and pressure), they are known to be *hydrophilic*. If, conversely, a substance is not easily dissolved in water under these conditions, it is said to be *hydrophobic*. Since many contaminants are organic (i.e., consisting of molecules containing covalent carbon-to-carbon bonds and/or carbon-to-hydrogen bonds), the solubility can be further differentiated as to whether under normal environmental conditions of temperature and pressure, the substance is easily dissolved in organic solvents. Such substances are said

Table 5.3   Typical Hazardous Wastes Generated by Selected Industries

Waste generator	Types of wastes produced
Chemical manufacturers	Strong acids and bases
	Reactive wastes
	Ignitable wastes
	Discarded commercial chemical products
Vehicle maintenance shops	Paint wastes
	Ignitable wastes
	Spent solvents
	Acids and bases
Printing industry	Photography waste with heavy metals
	Heavy metal solutions
	Waste inks
	Spent solvents
Paper industry	Ignitable wastes
	Corrosive wastes
	Ink wastes, including solvents and metals
Construction industry	Ignitable wastes
	Paint wastes
	Spent solvents
	Strong acids and bases
Cleaning agents and cosmetic manufacturing	Heavy metal dusts and sludges
	Ignitable wastes
	Solvents
	Strong acids and bases
Furniture and wood manufacturing and refinishing	Ignitable wastes
	Spent solvents
	Paint wastes
Metal manufacturing	Paint wastes containing heavy metals
	Strong acids and bases
	Cyanide wastes
	Sludges containing heavy metals

Source: U.S. Environmental Protection Agency, *RCRA: Reducing Risks from Wastes,* EPA/530/K-97/004, U.S. EPA, Washington, DC, September 1997.

to be *lipophilic* (i.e., readily dissolved in lipids). If, conversely, a substance is not easily dissolved in organic solvents under these conditions, it is said to be *lipophobic*. When a fluid that is dense and miscible (i.e., able to be mixed in any concentration without separation of physical phases) seeps underground through the vadose zone, below the water table, and into the zone of saturation, the dense contaminants move downward (see Figure 5.20). When these contaminants reach the bottom of the aquifer, the shape continued movement is determined by the slope of the underlying bedrock or other relatively impervious layer. Solution and dispersion near the boundaries of the plume will generate a secondary plume that will generally follow the direction of groundwater flow (see Figure 5.21). Organics more dense than water [called dense nonaqueous-phase liquids,

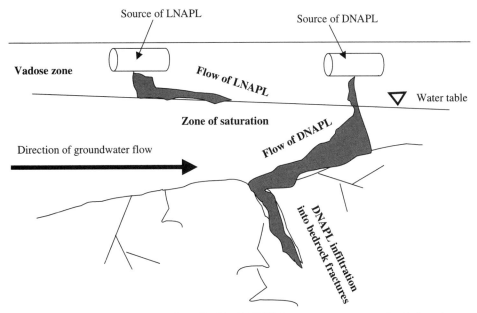

Figure 5.20 Dense nonaqueous-phase liquids (DNAPLs) can penetrate more deeply into the aquifer than do the light nonaqueous-phase liquids (LNAPLs). The density reference for whether a compound is a DNAPL or an LNAPL is whether it is denser or lighter than water, respectively. The DNAPL may even be against the general flow of the groundwater. (From D. A. Vallero, *Environmental Contaminants: Assessment and Control,* Elsevier Academic Press, Burlington, MA, 2004; and H. Hemond and E. Fechner-Levy, *Chemical Fate and Transport in the Environment,* Academic Press, San Diego, CA, 2000.)

(DNAPLs)] will penetrate more deeply, whereas the lighter organics [light nonaqueous-phase liquids (LNAPLs)] will float near the top of the zone of saturation.

The physics of this system determine the direction of the contaminant plume's movement. For example, the movement is greatly affected by the solubility and density of the contaminants, but other factors also influence the rate of transport (e.g., sorption, presence of other solvents besides water, and the amount of organic matter in the soil). Figure 5.2 indicates the importance of vapor pressure. The volatile contaminants (i.e., those with relatively high vapor pressure), can move upward from the plume, often reaching the atmosphere. Thus, hazardous wastes can contaminate the air, soil, aquifers, and surface waters.

A promising development in recent years has been improvements in ways to reclaim previously contaminated sites. *Brownfields* are properties that have been polluted by a hazardous substance. Cleaning up and reinvesting in these properties takes development pressures off undeveloped open land, and both improves and protects the environment. In Houston, Texas, for example, the inner-city and downtown areas that were neglected as the city experienced an economic boom in the 1970s and 1980s precipitated the move of businesses and residents away from the urban core to the expanding development of outlying suburbs.[33] The abandoned properties downtown were feared to have environ-

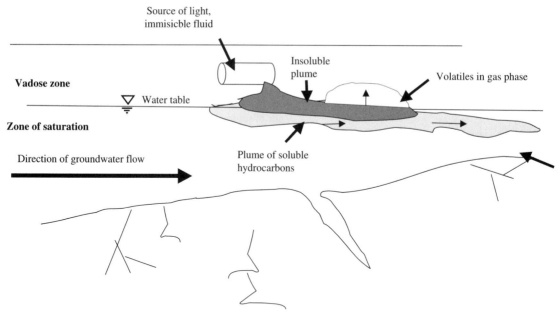

Figure 5.21   Hypothetical plume of hydrophobic fluid. [From D. A. Vallero, *Environmental Contaminants: Assessment and Control,* Elsevier Academic Press, Burlington, MA, 2004; and M. N. Sara, Groundwater Monitoring System Design, in *Practical Handbook of Ground-Water Monitoring,* D. M. Nielsen (Ed.), Lewis Publishers, Chelsea, MI, 1991.]

mental contamination, which prevented efforts to redevelop these properties for several years. Houston's Brownfields Redevelopment Program has since led to these inner-city brownfields being assessed, cleaned up, and redeveloped for such uses as a major league baseball park, a 450-acre golf course, a performing arts center, an aquarium and entertainment complex, and nearly 1000 new housing units. March 30, 2000 saw the opening of Astros Field, a 42,000-seat baseball stadium built on a 38-acre former brownfield. The redeveloped site also includes a railroad station, an industrial facility, and a number of corrugated-metal buildings. The site's owner funded environmental assessments at the site, entered the property into the Texas Voluntary Cleanup Program, and funded the site's cleanup. Located on the east side of downtown Houston, the ballpark offers spectators a spectacular view of the Houston skyline. The Astros Field property also includes the renovated Union Station, with restaurants, shops, and a theater. The $310 million redevelopment project has created 223 new jobs and is a leading force behind the revitalization of downtown Houston.

Before cleanup and redevelopment of a waste site begin, engineers must conduct site investigations to determine the degree and extent of contamination. These investigations are eligible for financial assistance from local, state, and federal agencies. A number of factors must be considered to determine if a contaminated site is a good candidate for redevelopment and whether to conduct a Brownfield investigation, including:

- The surrounding neighborhood's economic status
- The degree of uncertainty surrounding the site's contamination
- The likelihood that the site investigation will reduce such uncertainty
- The feasibility of proceeding with the assessment within the established time frame
- Plans for reuse
- Benefits to the community
- General geographic diversity and financial need

Diversity, Productivity, and Sustainability of Ecosystems

The health of the global ecosystem depends on diversity, productivity, and sustainability. If a productive and highly diverse system of interlocking species shows sustainability, the ecosystem is able to withstand stresses or perturbations. If, on the other hand, species die out and diversity is reduced or the productivity is stunted, the ecosystem is susceptible to failure and collapse.

Ecosystems can be stressed by numerous human activities that introduce stressors to the environment. As is the case for human health risk assessments, the stressors may be chemicals, but physical and biological stressors also exist. For example, the placement of a roadway or the changes brought about by bulldozers and earthmovers are considered to be physical stressors to habitats. The accidental or intentional introduction of invasive biota [e.g., grass carp (fauna) and kudzu (flora) in the southern United States] are examples of biological stressors.

The identification of possible adverse outcomes is crucial. These outcomes alter essential structures or functions of an ecosystem. The severity of outcomes is characterized as to their type, intensity, and scale of the effect and the likelihood of an ecosystem to recover from the damage imposed by a single or multiple stressors. The characterization of adverse ecological outcomes can range from qualitative, expert judgments to statistical probabilities. The emergent fields of eco-toxicology and eco-risk have several things in common with human toxicology and risk assessment, such as concern about ambient concentrations of contaminants and uptake in water, air, and soil. In some ways, however, ecological dose–response and exposure research differs from that in human systems. First, ecologists deal with many different species, some more sensitive than others, to the effects of contaminants. Second, the means of calculating exposure are different, especially if one is concerned about the exposure of an entire ecosystem.

Ecosystems are complex. Ecologists characterize them by evaluating their composition, structure, and functions. Ecosystem composition is a list, a taxonomy if you will, of every living and nonliving part of the ecosystem. As the term implies, *ecological structure* is how all of the parts of the system are linked to form physical patterns of life-forms from single forest stands to biological associations and plant communities. A single wetland or prairie, for example, is a much simpler structure than a multilayered forest, which consists of plant and microbial life in the detritus, herbs, saplings, newer trees, and canopy trees. Ecosystem functions include cycles of nitrogen, carbon, and phosphorus that lead to biotic processes such as production, consumption, and decomposition.

Indicators of an ecosystem's condition include:

- *Diversity. Biodiversity* has been defined as the ". . . composition, structure, and function [that] determine, and in fact constitute, the biodiversity of an area. Composition has to do with the identity and variety of elements in a collection, and includes species lists and measures of species diversity and genetic diversity. Structure is the physical organization or pattern of a system, from habitat complexity as measured within communities to the pattern of patches and other elements at a landscape scale. Function involves ecological and evolutionary processes, including gene flow, disturbances, and nutrient cycling."[34]

  The diversity of this stream community can be found using the *Shannon–Weiner index*:

$$D = -\sum_{i=1}^{m} P_i \log_2 P_i \qquad (5.32)$$

  or

$$D = -1.44 \sum_{i=1}^{m} \frac{n_i}{N} \ln \frac{n_i}{N} \qquad (5.33)$$

  where D is the index of community diversity; $P_i = n_i/N$; n_i is the number (i.e., density) of the ith genera or species; N is the total number (i.e., density) of all organisms in the sample; $i = 1, 2, \ldots, m$; and m is the number of genera or species. The higher the value of D, community diversity, the greater is the stability of the ecosystem.

- *Productivity.* This is an expression of how economical a system is with its energy. How much biomass is produced from abiotic (e.g., nutrients and minerals) and biotic resources (from microbial populations to canopy plant species to top predator fauna). One common measure is *net primary productivity*, which is the difference between two energy rates:

$$P_1 = k_p - k_e \qquad (5.34)$$

  where P_1 is the net primary productivity, k_p is the rate of chemical energy storage by primary producers, and k_e is the rate at which the producers use energy (via respiration).

- *Sustainability.* How likely is it that the diversity and productivity will hold up? Even though an ecosystem appears to be diverse and highly productive, is there something looming that threatens the continuation of these conditions? For example, is an essential nutrient being leached out of the soil, or are atmospheric conditions changing that may threaten a key species of animal, plant, or microbe? Sustainability is difficult to quantify precisely.

Example:

Assessing Ecological Risk[34]

In 2000, your environmental assessment of microbes in a small stream at your plant found a total of 1510 organisms, represented by seven different species. Your *actual number count of each microbial species in stream community was 10, 50, 75, 125, 200, 350, and 750 m L^{-1}.*

Construct a table to derive the values needed to find D, using equation (5.33):

$$D = -1.44 \sum_{i=1}^{m} \frac{n_i}{N} \ln \frac{n_i}{N}$$

i	n_i	n_i/N	$-1.44 \ln(n_i/N)$	$-1.44(n_i/N) \ln(n_i/N)$
1	10	0.006623	7.224883	0.047847
2	50	0.033113	4.907292	0.162493
3	75	0.049669	4.323423	0.21474
4	125	0.082781	3.587834	0.297006
5	200	0.13245	2.911028	0.385567
6	350	0.231788	2.105182	0.487956
7	700	0.463576	1.10705	0.513202
Σ	1510	1		2.10881

So the diversity index is 2.1. The index is most useful when comparing systems. So if your stream is 2.1 and the surrounding streams are all around 4, you may have a problem. Generally, D values range from about 1.5 to 4.5.

What will happen if all of the numbers of species quadrupled?

Construct a new table with four times each species:

i	n_i	n_i/N	$-1.44 \ln(n_i/N)$	$-1.44(n_i/N) \ln(n_i/N)$
1	40	0.006623	7.224883	0.047847
2	200	0.033113	4.907292	0.162493
3	300	0.049669	4.323423	0.21474
4	500	0.082781	3.587834	0.297006
5	800	0.13245	2.911028	0.385567
6	1400	0.231788	2.105182	0.487956
7	2800	0.463576	1.10705	0.513202
Σ	6040	1		2.10881

So nothing changed; the index remains 2.1. The index indicates, correctly, that the total species abundance does not affect diversity.

You conducted a follow-up study in 2005 which indicates that the density of these species had changed to 3000, 50, 35, 40, 30, 70, and 15 L^{-1} of the same species of microbes as the early study. How had the numbers and diversity changed in five years?

Again, calculate D by constructing a new table:

i	n_i	n_i/N	$-1.44 \ln(n_i/N)$	$-1.44(n_i/N) \ln(n_i/N)$
1	3000	0.92593	0.11082	0.10261
2	50	0.01543	6.00668	0.0927
3	35	0.0108	6.52029	0.07044
4	40	0.01235	6.32801	0.07812
5	30	0.00926	6.74227	0.06243
6	70	0.0216	5.52216	0.11931
7	15	0.00463	7.7404	0.03584
Σ	3240	1		0.56144

This shows that in five years, the actual number of microbes is more than doubling, but the diversity has been reduced drastically ($D = 0.6$ *versus* 2.1). This may indicate that conditions favorable to one species [e.g., the presence of a toxic chemical or a change in habitat (e.g., a road)] are detrimental to the other six species.

A follow-up question must be asked: Were they conducted in the same season (some microbes grow better in warmer conditions, whereas others may compete more effectively in cooler waters)? If the studies are comparable, further investigation is needed, but this certainly is an indication that things are amiss, since Shannon values usually range from about 1.5 to 4.5. Diversity, like productivity, is often an important indicator of change. If diversity as measured by the index shows a decline, it may well indicate than an ecosystem is under increasing stress.

Ecological risk assessments may be prospective or retrospective, but often are both. The Florida Everglades provides an example of an integrated risk approach. The population of panthers, a top terrestrial carnivore in southern Florida, was found to contain elevated concentrations of mercury (Hg) in the 1990s. This was observed through retrospective eco-epidemiological studies. The findings were also used as scientists recommended possible measures to reduce Hg concentrations in sediment and water in Florida. Prospective risk assessments can help to estimate expected changes in Hg concentrations in panthers and other organisms in the food chain from a mass balance perspective. That is, as the Hg mass entering the environment through the air, water, and soil is reduced, how has the risk to sensitive species been reduced concomitantly? Integrated retrospective/prospective risk assessments are employed where ecosystems have a history of previous impacts and the potential for future effects from a wide range of stressors. This may be the case for hazardous waste sites.

The ecological risk assessment process embodies two elements: characterizing the adverse outcomes and characterizing the exposures. From these elements, three steps are undertaken:

1. Problem formulation
2. Analysis
3. Risk characterization

In problem formulation, the need to conduct the assessment is fully described, the specific problem or problems are defined, and the plan for analysis and risk characterization is laid out. Problem formulation includes integrating available information about the potential sources, the description of all stressors, the effects, and the characterization of the ecosystem and the receptors. Two basic products result from this stage of eco-risk assessment: assessment endpoints and conceptual models.

The analysis phase consists of evaluating the available data to conduct an exposure assessment (i.e., exposure to stressors is likely to occur or to have occurred). From these exposure assessments, the next step is to determine the possible effects and how widespread and severe these outcomes will be. During analysis, the engineer should investigate the strengths and limitations of data on exposure, effects, and ecosystem and receptor characteristics. Using these data, the nature of potential or actual exposure and the ecological changes under the circumstances defined in the conceptual model can be determined. The analysis phase provides an exposure profile and stressor–response profile, which together form the basis for risk characterization.

Thus, the ecological risk assessment provides valuable information to the engineer by:

- Providing information to complement the human health information, thereby improving environmental decision making
- Expressing changes in ecological effects as a function of changes in exposure to stressors, which is particularly useful to the decision maker who must evaluate trade-offs, examine different options, and determine the extent to which stressors must be reduced to achieve a given outcome
- Characterizing uncertainty as a degree of confidence in the assessment, which aid the engineer's focus on those areas that will lead to the greatest reductions in uncertainty
- Providing a basis for comparing, ranking, and prioritizing risks, as well as information to conduct cost-benefit and cost-effectiveness analyses of various remedial options
- Considering management needs, goals, and objectives, in combination with engineering and scientific principles to develop assessment endpoints and conceptual models during problem formulation

Environmental problems are complex. As we learn more about them, engineers and environmental professionals must find innovative ways to address them. Engineers play a particularly important role in helping society manage these risks.

REFERENCES AND NOTES

1. Formerly an environmental impact statement reviewer for the U.S. Environmental Protection Agency and presently an environmental policymaker with the U.S. Fish and Wildlife Service.
2. T. S. Kuhn, *The Structure of Scientific Revolutions,* University of Chicago Press, Chicago, 1962.
3. Quoted in J. Schell, Our Fragile Earth, *Discover,* p. 47, October 1987.
4. C. P. Snow, *The Search,* Charles Scribner's Sons, New York, 1959.

5. The principal source for this discussion is B. Cooper, J. Hayes, and S. LeRoy, Science Fiction or Science Fact? The Grizzly Biology Behind Parks Canada Management Models, *Frasier Institute Critical Issues Bulletin,* Frasier Institute, Vancouver, BC, Canada, 2002.

6. Articles included: S. A. Karl and B. W. Bowen, Evolutionary Significant Units *versus* Geopolitical Taxonomy: Molecular Systematics of an Endangered Sea Turtle (genus *Chelonia*), pp. 990–999; P. C. H. Pritchard, Comments on Evolutionary Significant Units *versus* Geopolitical Taxonomy, pp. 1000–1003; J. M. Grady and J. M. Quattro, Using Character Concordance to Define Taxonomic and Conservation Units, pp. 1004–1007; K. Shrader-Frechette and E. D. McCoy, Molecular Systematics, Ethics, and Biological Decision Making Under Uncertainty, pp. 1008–1010; and B. W. Bowen and S. A. Karl, In War, Truth Is the First Casualty, pp. 1013–1016; *Conservation Biology* 13(5), 1999.

7. Bowen and Karl, note 6, p. 1015.

8. Shrader-Frechette and McCoy, note 6, p. 1012.

9. Personal communication with John Ahearne, former Commissioner, Nuclear Regulatory Commission, April 24, 2006.

10. The National Environmental Policy Act of 1969, 42 U.S.C. §4321 et seq.

11. L. A. Canter, *Environmental Impact Assessment.* McGraw-Hill, New York, 1996.

12. See *Code of Federal Regulations,* Title 40, *Protection of Environment;* Chapter I, Environmental Protection Agency, Subchapter A—General, Part 6: Procedures for Implementing the Requirements of the Council on Environmental Quality of the National Environmental Policy Act. This guidance can be accessed at http://www.epa.gov/compliance/nepa/epacompliance/index.html.

13. For example, the Leopold index is an interaction matrix developed by the U.S. Geological Survey for environmental assessment applications.

14. For an interesting application of the NEPA process in an area important to environmental justice communities (i.e., nuclear power facility siting), consider the Nuclear Regulatory Commission (NRC) environmental justice policy. The policy articulates the NRC's commitment to complying the requirements of the National Environmental Policy Act (NEPA) in all of its regulatory and licensing actions, especially as these actions may incur differential impacts greater than those of the general population, due to a community's distinct cultural characteristics or practices. See http://www.epa.gov/fedrgstr/EPA-IMPACT/2003/November/Day-05/i27805.htm.

15. Actually, many acid–base systems in the environment are an equilibrium condition among acids, bases, and ionic and other chemical species. For example, in natural waters, carbonic acid, which is responsible for most natural acidity in surface waters, is indicated as $H_2CO_3^*$. The asterisk indicates that the actual acid is only part of the equilibrium, which also consists of CO_2 dissolved in the water, CO_2 dissolved in the air, and ionic forms, especially carbonates (CO_3^{2-}) and bicarbonates (HCO_3^-).

16. Environmental Protection Agency, U.S. EPA, Washington, DC.

17. B. C. Levin, A Summary of the NBS Literature Reviews on the Chemical Nature and Toxicity of Pyrolysis and Combustion Products from Seven Plastics: Acrylonitrile–Butadiene–Styrenes; Nylons; Polyesters; Polyethylenes; Polystyrenes; Poly(vinyl chlorides) and Rigid Polyurethane Foams, *Fire Materials,* 11:143–157, 1987.

18. The major source of information about Rio Grande brickmaking is *Environmental Health Perspectives,* 104(5), May 1996.

19. The Coalition for Justice in the Maquiladoras, a cross-border group that organizes maquiladora workers, traces the term *maquiladora* to *maquilar,* a popular form of the verb *maquinar,* which means roughly "to submit something to the action of a machine, as when rural Mexicans speak of *maquilar* with regard to the grain that is transported to a mill for processing. The farmer owns the grain; yet someone else, who owns the mill, keeps a portion of the value of the grain

for milling. So the origin of maquiladora can be found in this division of labor. The term has more recently been applied to the small factories opened by U.S. companies that provide labor-intensive jobs on the Mexican side of the border. Thus, *maquilar* has changed to include this process of labor, especially assembling parts from various sources, and the maquiladoras are those small assembling operations along the border. While the maquiladoras have fueled the growth of entrepreneurs along the Mexico–U.S. border, they have also provided an opportunity for the workers and their families to be exploited in the interests of profit and economic gain.

20. S. Mukerjee, Communication Strategy of Transboundary Air Pollution Findings in a US–Mexico Border XXI Program Project, *Environmental Management,* 29(1):34–56, 2002.

21. Often, these barrels were the 55-gallon drum variety, so the first burning probably volatilized some very toxic compounds, depending on the residues remaining in the drum. These contents could have been solvents (including halogenated compounds such as chlorinated aliphatics and aromatics), plastic residues (e.g., phthalates), and petroleum distillates. They may even have contained substances with elevated concentrations of heavy metals such as mercury, lead, cadmium, and chromium. The barrels (drums) themselves were often perforated to allow for higher rates of oxidation (combustion) and to take advantage of the smokestack effect (i.e., driving the flame upward and pushing the products of incomplete combustion out of the barrel and into the plume). Vallero recalls neighbors not being happy about burning trash while their wash was drying on the clothesline. They would complain of ash (aerosols) blackening their clothes and the odor from the incomplete combustion products on their newly washed laundry. Both of these complaints are evidence that the plume leaving the barrel contained harmful contaminants that were transported in the atmosphere.

22. The principal source for this section is National Toxicology Program, Eleventh Report on Carcinogens, Coke Oven Emissions, Substance Profile, http://ntp.niehs.nih.gov/ntp/roc/eleventh/profiles/s049coke.pdf, accessed May 11, 2005.

23. S. S. Schiffman and C. M. Williams, Science of Odor as a Potential Health Issue, *Journal of Environmental Quality,* 34:129–138, 2005.

24. For a survey of the state of the science in electronic-odor-sensing technologies, see H. T. Nagle, S. S. Schiffman, and R. Gutierrez-Osuna, The How and Why of Electronic Noses, *IEEE Spectrum,* 35(9):22–34, 1998.

25. G. Hardin, Tragedy of the Commons, *Science,* 162. 1243–1248, Dec. 13, 1968.

26. A common reason for halogenating compounds is to reduce their flammability, making them safer to use than their nonhalogenated counterparts. Unfortunately, halogenation often increases chronic toxicity, such as carcinogenicity, and makes the compounds more persistent in the environment (i.e., more resistant to biodegradation).

27. The endocrine disruptor discussion is based on D. A. Vallero, Environmental Endocrine Disruptors, *McGraw-Hill Yearbook of Science and Technology,* McGraw-Hill, New York, 2004.

28. D. G. Wilson, History of Solid Waste Management, in *Handbook of Solid Waste Management,* Van Nostrand Reinhold, New York, 1997, pp. 1–9.

29. J. G. Landels, *Engineering in the Ancient World,* Barnes & Noble Books, New York, 1978. This book contains an excellent discussion of water supplies as recorded by the Roman architect/engineer Vitruvius in his eighth book of *De Architectura* in the first century B.C.

30. Landels, 1978. The two major challenges to delivering water in sprawling, ancient Rome were pressure and sediment. The large head needed to transport water over distances would split the lead and earthenware pipes, so the harder bronze was recommended. The siltation in the aqueducts was addressed using cisterns similar to sedimentation basins in modern wastewater treatment plants.

31. H. Tammenagi, *The Waste Crisis: Landfills, Incinerators, and the Search for a Sustainable Future,* Oxford Univeristy Press, Oxford, 1999, pp. 22–24.

32. The RCRA regulations are found in the *Code of Federal Regulations* at Title 40, Parts 260 through 280. In fact, RCRA is an amendment to the Solid Waste Disposal Act of 1956. The 1984 amendments to RCRA are known as the Hazardous and Solid Waste Amendments (HSWA). Subtitle C (hazardous waste) and Subtitle D (solid, primarily nonhazardous waste) provide the structure for comprehensive waste management programs. In addition, RCRA also regulates underground storage tanks under Subtitle I and medical waste under Subtitle J. Many other countries have similar programs to control hazardous wastes.

33. The source for this discussion is the U.S. Environmental Protection Agency's brownfield Web site, http://www.epa.gov/swerosps/bf/success/houston.pdf, accessed June 22, 2005.

34. R. Noss, Indicators for Monitoring Biodiversity: A Hierarchical Approach, *Conservation Biology*, 4(4):355–364, 1990.

35. This example is based on information from D. A. Vallero, *Environmental Contaminants: Assessment and Control*, Elsevier Academic Press, Burlington, MA, 2004.

BIBLIOGRAPHY

American Society of Civil Engineers, Environmental Impact Analysis Research Council of the Technical Council on Research, A Proposed Eighth Fundamental Canon for the ASCE Code of Ethics, *Journal of Professional Issues in Engineering*, 110:3, 1984.

Golding, M., Obligations to Future Generations, *Monist*, 56:85–99, 1972.

MacLean, D. (Ed.), *Values at Risk*, Rowman & Littlefield, Totowa, NJ, 1986.

Martin, M. W., and R. Schinzinger, *Ethics in Engineering*, McGraw-Hill, New York, 1989.

Petulla, J. M., *American Environmentalism*, Texas A&M University Press, College Station, TX, 1980.

Schwartz, S., Acid Deposition: Unraveling a Regional Phenomenon, *Science*, 243, February 1989.

Vallero, D. A., *Environmental Contaminants: Assessment and Control*, Elsevier Academic Press, Burlington, MA, 2004.

Vallero, D. A., *Paradigms Lost: Learning from Environmental Mistakes, Mishaps, and Misdeeds*, Butterworth-Heinemann, Burlington, MA, 2006.

Vesilind, P. A., and A. S. Gunn, *Engineering, Ethics, and the Environment*, Cambridge University Press, New York, 1998.

Vesilind, P. A., and A. S. Gunn, Sustainable Development and the ASCE Code of Ethics, *Journal of Professional Issues in Engineering Education and Practice*, 124:72–78, 1999.

6

Green Engineering

For in the true nature of things, if we rightly consider, every green tree is far more glorious than if it were made of gold and silver.

Martin Luther (1483–1546)

What do we as individuals and as a society truly value? In recent years the color green has been a metaphor for environmental consciousness. Why is that? Perhaps it is because when ecosystems are thriving, such as forests and prairies, they exhibit green growth. But green is also a color of pollution, such as the blooms of green algae in eutrophic lakes and ponds, and the growth of green mold on bread. All of these instances, however, do reflect the presence of chlorophyll, the pigment that is part of the energy transformation process, *photosynthesis.* This is the principal means on Earth of storing and converting solar energy into food, which is the energy source for all living creatures. So it makes sense that green represents sustenance from the Earth, hence sustainability.

The two important discoveries that elucidated the photosynthetic pathway were made by Joseph Priestley and Julius Mayer (see their biographical sketchs).

Whatever the reason, green has become recognized as a code for sustainable programs. So a *green engineer* is no longer a term for a neophyte to the profession (opposite of a "gray beard"); it is now more likely to mean an environmentally oriented engineer. One of the principles of "green engineering" is a recognition of the importance of *sustainability.*

SUSTAINABILITY

Their recognition of an impending and assured global disaster led the World Commission on Environment and Development, sponsored by the United Nations, to conduct a study of the world's resources. Also known as the Brundtland Commission, their 1987 report, *Our Common Future,* introduced the term *sustainable development* and defined it as "development that meets the needs of the present without compromising the ability of future generations to meet their own needs."[1] The United Nations Conference on Environment and Development (UNCED), the Earth Summit held in Rio de Janeiro in 1992, communicated the idea that sustainable development is both a scientific concept and a philosophical ideal. The document *Agenda 21* was endorsed by 178 governments (not including the United States) and hailed as a blueprint for sustainable development. In 2002, the World Summit on Sustainable Development (WSSD) identified five major areas that are considered fundamental for moving sustainable development plans forward.

Biographical Sketch: Joseph Priestley

 Joseph Priestley (1733–1804) was born in Yorkshire and studied languages, history, and philosophy as a young man. He became a nonconformist in religion, but decided nevertheless to try to earn a living being a preacher. He was singularly unsuccessful, due in part to his stuttering and in part to his unorthodox religious views. He decided to teach at a liberal school, and started to publish books on history, science, and educational theory.

His marriage to Mary Wilkinson forced him to relocate to Leeds, where he once again became a preacher. His house was next to a brewery, which provided him the occasion to conduct experiments on the gases given off during the fermentation of beer. He discovered that the gas (which we now know as CO_2) extinguished lighted wood chips and was heavier than air. If dissolved in water, this gas gave the water a pleasant taste, and Priestley started to serve the drink to his friends, producing the first carbonated beverage.

In 1774, Priestley, continuing his experiments with gases, discovered oxygen, unaware that Carl Wilhelm Scheele had made the same discovery a year earlier. In fact, Scheele's experiments referred to "aerial acid" (CO_2) and "fire air" (O_2). Priestley published his work before Scheele, however, and thus received credit for the discovery.

Priestley also gave us a rudimentary understanding of chlorophyll. Conducting experiments on mint plants, he wrote in 1780 that a plant is able to "restore air which has been injured by the burning of candles." In what would today be animal testing, Priestley placed a mouse beneath a glass vessel underwater, so that the only oxygen (Priestley's forte, you will recall) in the glass was available to the mouse. After a few days, he observed that "the air would neither extinguish a candle, nor was it all inconvenient to a mouse which I put into it." Thus, Priestley concluded, the plant was the source of free oxygen.

Priestley was a dissident in many ways, including his opposition to the war in America. After moving to Birmingham, he continued his intemperate preaching, often lauding the goals of the French Revolution and the religious freedom of the United States. This got him driven out of town, and he moved to London. His three sons, meanwhile, had emigrated to the United States, and in 1793, Joseph Priestley followed, moving to central Pennsylvania and establishing a Unitarian fellowship in Northumberland.

The underlying purpose of sustainable development is to help developing nations manage their resources, such as rain forests, without depleting these resources and making them unusable for future generations. In short, the objective is to prevent the collapse of the global ecosystems. The Brundtland report presumes that we have a core ethic of intergenerational equity, and that future generations should have an equal opportunity to achieve a high quality of life. The report is silent, however, on just why we should embrace the ideal of intergenerational equity, or why one should be concerned about the

Biographical Sketch: Julius Robert Mayer

The process of photosynthesis was documented by a German surgeon, Julius Robert Mayer (1814–1878), who wrote: "Nature has put itself the problem of how to catch in flight light streaming to the Earth and to store the most elusive of all powers in rigid form. The plants take in one form of power, light; and produce another power, chemical difference."

Like many biological and environmental processes, photosynthesis is deceptively simple; the chlorophyll molecule absorbs sunlight and uses its energy to synthesize carbohydrates from carbon dioxide (CO_2) and water:

$$6CO_2 + 6H_2O \xrightarrow[\text{chlorophyll}]{\text{sunlight}} C_6H_{12}O_6 + 6O_2 \qquad (6.1)$$

Also, note the 6 moles of molecular oxygen produced from this reaction, in addition to the carbohydrate (glucose shown).

Mayer was a classic case of good science, yet poor communication. His inability to describe what he did and his apparent unsociability estranged him from the scientific establishment and even caused some to ridicule his findings.

survival of the human species. The goal is a sustainable global ecologic and economic system, achieved in part by the wise use of available resources.

Although this goal has been applied principally at developing nations, sustainable development applies to all human developments. We are creatures that have different needs. Maslow[2] articulated this as a hierarchy of needs, consisting of two classes: basic and growth (see Figure 6.1). The *basic needs* must first be satisfied before a person can progress toward higher-level *growth needs*. Within the basic needs classification, Maslow separated the most basic physiological needs, such as water, food, and oxygen, from the need for safety. Therefore, one must first avoid starvation and thirst, satisfying minimum caloric and water intake, before being concerned about the quality of the air, food, and water. The latter is the province of environmental protection. The most basic of needs must first be satisfied before we can strive for more advanced needs. Thus, we need to ensure adequate quantities and certain ranges of quality of air, water, and food. Providing food requires ranges of soil and water quality for agriculture. Thus, any person and any culture that is unable to satisfy these most basic needs cannot be expected to "advance" toward higher-order values, such as free markets and peaceful societies. In fact, the inability to provide basic needs militates against peace. This means that when basic needs go unmet, societies are frustrated even if they strive toward freedom and peace. And even those societies that begin to advance may enter into vicious cycles wherein any progress is undone by episodes of scarcity. We generally think of peace and justice as the province of religion and theology, but engineers will increasingly be called upon to "build a better world." And, one aspect of "better" is "sustainable."

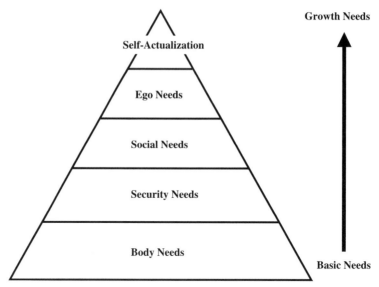

Figure 6.1 Maslow's hierarchy of needs. The lower part of the hierarchy (i.e., basic needs) must first be satisfied before a person can advance to the next growth levels.

Even mechanical engineers, whom we may at first blush think of as being concerned primarily about nonliving things, are embracing sustainable design in a large way. In fact, in many ways the mechanical engineering profession is out in front on sustainable design. For example, the American Society of Mechanical Engineers (ASME) Web site draws a systematic example from ecology: "To an engineer, a sustainable system is one that is in equilibrium or changing at a tolerably slow rate. In the food chain, for example, plants are fed by sunlight, moisture and nutrients, and then become food themselves for insects and herbivores, which in turn act as food for larger animals. The waste from these animals replenishes the soil, which nourishes plants, and the cycle begins again,"[3] Sustainability is, therefore, a systematic phenomenon, so it is not surprising that engineers have embraced the concept of sustainable design. At the largest scale, manufacturing, transportation, commerce, and other human activities that promote high consumption and wastefulness of finite resources cannot be sustained. On the individual designer scale, the products and processes that engineers design must be considered for their entire lifetimes and beyond.

Hardin's parable demonstrates that even though the person sees the utility of preservation (no new cows) in a collective sense, the ethical egoistic view may well push the decision toward the immediate gratification of the individual at the expense of the collective good. Arguably, this is why we pollute.

FROM GREEN ENGINEERING TO SUSTAINABILITY[5]

To attain sustainability, people need to adopt new and better means of using materials and energy. The operationalizing of the quest for sustainability is defined as *green en-*

Biographical Sketch: Garrett Hardin

Garrett Hardin (1915–2003) received a B.S. in zoology from the University of Chicago and then went to Stanford University where he received his Ph.D. in microbiology. He is best remembered as a curmudgeon—a person who was not afraid to speak what he considered to be the truth, however unpopular the truth might be. In 1968 he wrote a hugely influential article entitled "The Tragedy of the Commons,"[4] which has become a must-read item in every ecology course. In this article, Hardin imagines an English village with a common area where everyone's cow may graze. The common is able to sustain the cows, and village life is stable until one of the villagers figures out that if he gets two cows instead of one, the cost of the extra cow will be shared by everyone, whereas the profit will be his alone. So he procures two cows and prospers, but others see this and similarly each wants two cows. If two, why not three—and so on—until the village common is no longer able to support the large number of cows, and everyone suffers. In other words, systems are elastic up to a point after which they begin to crash. Common goods are particularly vulnerable.

A similar argument can be made for the use of nonrenewable resources. If we treat diminishing resources such as oil and minerals as capital gains, we will soon find ourselves in the "common" pickle of resource of expenditure outstripping availabe resources.

A thread running all through Hardin's books is that ethics has to be based on rational argument and not on emotion. He argues that for ethics to be useful, people have to be literate, they must use words correctly, and they must appreciate the power of numbers. His most interesting book is *Stalking the Wild Taboo,* in which he takes on any number of what he considers to be social misconceptions that demand rational reasoning. Engineers should be careful in applying reason, since the assumptions can drive the expected results. For example, some scientists had argued throughout recent history that the world cannot sustain even half of our present population. Scientific advances have undone some of these assumptions (e.g., improved food supply). On the other hand, engineers should be aware that prudence is essential since resources are finite.

gineering, a term that recognizes that engineers are central to the practical application of the principles sustainability to everyday life. The relationship between sustainable development, sustainability, and green engineering is as follows:

$$\text{sustainable development} \;\rightarrow\; \text{green engineering} \;\rightarrow\; \text{sustainability}$$

Sustainable development is an ideal that can lead to sustainability, but this can only be done through green engineering.

Green engineering[6] treats environmental quality as an end in itself. The U.S. EPA has defined *green engineering* as ". . . the design, commercialization, and use of processes and products, which are feasible and economical while minimizing (1) generation of pollution at the source and (2) risk to human health and the environment. The discipline embraces the concept that decisions to protect human health and the environment can have the greatest impact and cost effectiveness when applied early to the design and development phase of a process or product."[7]

Green engineering approaches are being linked to improved computational abilities (see Table 6.1) and other tools that were not available at the outset of the environmental movement. Increasingly, companies have come to recognize that improved efficiencies

Table 6.1 Principles of Green Programs

Principle	Description	Example	Role of computational toxicology
Waste prevention	Design chemical syntheses and select processes to prevent waste, leaving no waste to treat or clean up.	Use a water-based process instead of an organic solvent-based process.	Informatics and data mining can provide candidate syntheses and processes.
Safe design	Design products to be fully effective, yet have little or no toxicity.	Using microstructures, instead of toxic pigments, to give color to products. Microstructures bend, reflect, and absorb light in ways that allow for a full range of colors.	Systems biology and "omics" technologies can support predictions of cumulative risk from products used in various scenarios.
Low-hazard chemical synthesis	Design syntheses to use and generate substances with little or no toxicity to humans and the environment.	Select chemical synthesis with toxicity of the reagents in mind up front. If a reagent ordinarily required in the synthesis is acutely or chronically toxic, find another reagent or new reaction with less toxic reagents.	Computational chemistry can help predict unintended product formation and reaction rates of optional reactions.
Renewable material use	Use raw materials and feedstocks that are renewable rather than those that deplete nonrenewable natural resources. Renewable feedstocks are often made from agricultural products or are the wastes of other processes; depleting feedstocks are made from fossil fuels (petroleum, natural gas, or coal) or that must be extracted by mining.	Construction materials can be from renewable and depleting sources. Linoleum flooring, for example, is highly durable, can be maintained with nontoxic cleaning products, and is manufactured from renewable resources amenable to being recycled. Upon demolition or reflooring, the linoleum can be composted.	Systems biology, informatics, and "omics" technologies can provide insights into the possible chemical reactions and toxicity of the compounds produced when switching from depleting to renewable materials.

Table 6.1   (*Continued*)

Principle	Description	Example	Role of computational toxicology
Catalysis	Minimize waste by using catalytic reactions. Catalysts are used in small amounts and can carry out a single reaction many times. They are preferable to stoichiometric reagents, which are used in excess and work only once.	The Brookhaven National Laboratory recently reported that it has found a "green catalyst" that works by removing one stage of the reaction, eliminating the need to use solvents in the process by which many organic compounds are synthesized. The catalyst dissolves into the reactants. Also, the catalyst has the unique ability of being easily removed and recycled, because at the end of the reaction, the catalyst precipitates out of products as a solid material, allowing it to be separated from the products without using additional chemical solvents.[a]	Computation chemistry can help to compare rates of chemical reactions using various catalysts.
Avoiding chemical derivatives	Avoid using blocking or protecting groups or any temporary modifications, if possible. Derivatives use additional reagents and generate waste.	Derivativization is a common analytical method in environmental chemistry (i.e., forming new compounds that can be detected by chromatography). However, chemists must be aware of possible toxic compounds formed, including leftover reagents that are inherently dangerous.	Computational methods and natural products chemistry can help scientists start with a better synthetic framework.
Atom economy	Design syntheses so that the final product contains the maximum proportion of the starting materials. There should be few, if any, wasted atoms.	Single atomic- and molecular-scale logic used to develop electronic devices that incorporate design for disassembly, design for recycling, and design for safe and environmentally optimized use.	The same amount of value (e.g., information storage and application) is available on a much smaller scale. Thus, devices are smarter and smaller, and more economical in the long term. Computational toxicology enhances the ability to make product decisions with better predictions of possible adverse effects, based on the logic.

Table 6.1   (*Continued*)

Principle	Description	Example	Role of computational toxicology
Nano-materials	Tailormade materials and processes for specific designs and intent at the nanometer scale ($\leq$100 nm).	Emissions, effluent, and other environmental controls; design for extremely long life cycles. Limits and provides better control of production and avoids overproduction (i.e., "throwaway economy").	Improved, systematic catalysis in emission reductions (e.g., large sources like power plants and small sources like automobile exhaust systems). Zeolite and other sorbing materials used in hazardous waste and emergency response situations can be better designed by taking advantage of surface effects; this decreases the volume of material used.
Selection of safer solvents and reaction conditions	Avoid using solvents, separation agents, or other auxiliary chemicals. If these chemicals are necessary, use innocuous chemicals.	Supercritical chemistry and physics, especially that of carbon dioxide and other safer alternatives to halogenated solvents, are finding their way into the more mainstream processes, most notably dry cleaning.	To date, most of the progress as been the result of wet chemistry and bench research. Computational methods will streamline the process, including quicker "scale-up."
Improved energy efficiencies	Run chemical reactions and other processes at ambient temperature and pressure whenever possible.	To date, chemical engineering and other reactor-based systems have relied on "cheap" fuels and thus have optimized on the basis of thermodynamics. Other factors (e.g., pressure, catalysis, photovoltaics, and fusion) should also be emphasized in reactor optimization protocols.	Heat will always be important in reactions, but computational methods can help with relative economies of scale. Computational models can test the feasibility of new energy-efficient systems, including intrinsic and extrinsic hazards (e.g., to test certain scale-ups of hydrogen and other economies). Energy behaviors are scale-dependent. For example, recent measurements of H_2SO_4 bubbles when reacting with water have temperatures in the range of those found the surface of the sun.[b]

Table 6.1 (*Continued*)

Principle	Description	Example	Role of computational toxicology
Design for degradation	Design chemical products to break down to innocuous substances after use so that they do not accumulate in the environment.	Biopolymers (e.g., starch-based polymers) can replace styrene and other halogen-based polymers in many uses. Geopolymers (e.g., silane-based polymers) can provide inorganic alternatives to organic polymers in pigments, paints, etc. These substances, when returned to the environment, become their original parent form.	Computation approaches can simulate the degradation of substances as they enter various components of the environment. Computational science can be used to calculate the interplanar spaces within the polymer framework. This will help to predict persistence and to build environmentally friendly products (e.g., those where space is adequate for microbes to fit and biodegrade the substances).
Real-time analysis to prevent pollution and concurrent engineering	Include in-process real-time monitoring and control during syntheses to minimize or eliminate the formation of by-products.	Remote sensing and satellite techniques can be linked to real-time data repositories to determine problems. The application to terrorism using nanoscale sensors is promising.	Real-time environmental mass spectrometry can be used to analyze whole products, obviating the need for any further sample preparation and analytical steps. Transgenic species, although controversial, can also serve as biological sentries (e.g., fish that change colors in the presence of toxic substances).
Accident prevention	Design processes using chemicals and their forms (solid, liquid, or gas) to minimize the potential for chemical accidents, including explosions, fires, and releases to the environment.	Scenarios that increase probability of accidents can be tested.	Rather than waiting for an accident to occur and conducting failure analyses, computational methods can be applied in prospective and predictive mode; that is, the conditions conducive to an accident can be characterized computationally.

Source: D. A. Vallero, *Paradigms Lost: Learning from Environmental Mistakes, Mishaps, and Misdeeds,* Butterworth Heinemann, Burlington, MA, 2005. First two columns, except "nano-materials," adapted from U.S. Environmental Protection Agency, Green Chemistry, http://www.epa.gov/greenchemistry/principles.html, accessed April 12, 2005. Other information from discussions with Michael Hays, U.S. EPA, National Risk Management Research Laboratory, April 28, 2005.

[a]U.S. Department of Energy, Research News, http://www.eurekalert.org/features/doe/2004-05/dnl-brc050604.php, accessed March 22, 2005.

[b]D. J. Flannigan and K. S. Suslick, Plasma Formation and Temperature Measurement During Single-Bubble Cavitation, *Nature* 434: 52–55, 2005.

save time, money, and other resources in the long run. Hence, companies are thinking systematically about the entire product stream in numerous ways:

- Applying sustainable development concepts, including the framework and foundations of "green" design and engineering models
- Applying the design process within the context of a sustainable framework, including considerations of commercial and institutional influences
- Considering practical problems and solutions from a comprehensive standpoint to achieve sustainable products and processes
- Characterizing waste streams resulting from designs
- Understanding how first principles of science, including thermodynamics, must be integral to sustainable designs in terms of mass and energy relationships, including reactors, heat exchangers, and separation processes
- Applying creativity and originality in group product and building design projects

There are numerous industrial, commercial, and governmental green initiatives, including Design for the Environment (DFE), Design for Disassembly (DFD), and Design for Recycling (DFR).[8] These are replacing or at least changing pollution control paradigms. For example, the concept of a "cap and trade" has been tested and works well for some pollutants. This is a system where companies are allowed to place a "bubble" over an entire manufacturing complex or to trade pollution credits with other companies in their industry instead of a "stack-by-stack" and "pipe-by-pipe" approach: the *command and control perspective.* Such policy and regulatory innovations call for some improved technology-based approaches as well as better quality-based approaches, such as leveling out the pollutant loadings and using less expensive technologies to remove the first large bulk of pollutants, followed by higher operation and maintenance technologies for the more difficult-to-treat stacks and pipes. But the net effect can be a greater reduction of pollutant emissions and effluents than when treating each stack or pipe as an independent entity. This is a foundation for most sustainable design approaches: conducting a life-cycle analysis, prioritizing the most important problems, and matching the technologies and operations to address them. The problems will vary by size (e.g., pollutant loading), difficulty in treating, and feasibility. The easiest ones are the big ones that are easy to treat (so-called "low-hanging fruit"). You can do these first with immediate gratification. However, the most intractable problems are often those that are small but very expensive and difficult to treat (i.e., less feasible). Of course, as with all paradigm shifts, expectations must be managed from both a technical and an operational perspective. Not least, the expectations of the client, the government, and those of the individual engineer must be realistic in how rapidly the new approaches can be incorporated.

Historically, environmental considerations have been approached by engineers as constraints on their designs. For example, hazardous substances generated by a manufacturing process were dealt with as a waste stream that must be contained and treated. The hazardous waste production had to be constrained by selecting certain manufacturing types, by increasing waste-handling facilities, and if these did not do the job entirely, by limiting rates of production. Green engineering emphasizes that these processes are often inefficient economically and environmentally, calling for a comprehensive, systematic life-cycle approach. Green engineering attempts to achieve four goals:

1. Waste reduction
2. Materials management
3. Pollution prevention
4. Product enhancement

Waste reduction involves finding efficient material uses. It is compatible with other engineering efficiency improvement programs, such as total quality management and real-time or just-in-time manufacturing. The overall rationale for waste reduction is that if materials and processes are chosen intelligently at the beginning, less waste will result. In fact, a relatively new approach to engineering is to design and manufacture a product simultaneously rather than sequentially, known as *concurrent engineering*. Combined with DFE and life-cycle analysis, concurrent engineering approaches may allow environmental improvements under real-life manufacturing conditions. However, changes made in any step must consider possible effects on the remainder of the design and implementation.

CASE STUDY: SIDS, A CONCURRENT ENGINEERING FAILURE

One of the most perplexing and tragic medical mysteries of the past 50 years has been sudden infant death syndrome (SIDS). The syndrome was first identified in the early 1950s. Numerous etiologies have been proposed for SIDS, including a number of environmental causes. A recent study, for example, found a statistically significant link between exposure of newborn infants to fine aerosols and SIDS.[9] The study found that approximately 500 of the 3800 SIDS cases in 1994 were associated with elevated concentrations of particle matter with aerodynamic diameters less than 10 micrometers (PM_{10}) in the United States. This estimate is based only on metropolitan areas in counties with standard PM_{10} monitors. Based on the metropolitan area with the lowest particle concentrations, there appears to be a threshold, that is, particulate-related infant deaths occurred when PM_{10} levels were below 11.9 $\mu g\ m^{-3}$.

Extrapolations from these data show that almost 20% of all SIDS cases each year in the top 12 most polluted metro areas in the United States are associated with PM_{10} pollution. The number of annual SIDS cases associated with PM_{10} in Los Angeles, New York, Chicago, Philadelphia, and Detroit metropolitan areas range from 20 to 44. The study found that 10 states accounted for more than 60% of the particle-related SIDS cases, with 93 in California, 37 in Texas, and 32 in Illinois.

Since particle matter has been linked to SIDS cases, a logical extension would be to suspect the role of environmental tobacco smoke (i.e., "sidestream" exposure) in some cases, since this smoke contains both particulate and gas-phase contaminants that are released into the infant's breathing zone. Also, in utero exposures to toxic substances when a pregnant woman smokes (e.g., nicotine and other organic and inorganic toxins) may make the baby more vulnerable.

Another suspected etiology for SIDS is the exposure to pollutants via consumer products. For example, polyvinyl chloride (PVC) products have been indirectly linked to SIDS. The most interesting link is not the PVC itself, but the result of an engineering "solution."

Plastics came into their own in the 1950s, replacing many other substances, because of their light weight and durability. However, being a polymer, physical and chemical conditions affect the ability of PVC to stay "hooked together." This can be a big problem for plastics used for protection, such as waterproofing. One such use was as a tent material.

Serendipity often plays a role in linking harmful effects to possible causes. In 1988, Barry Richardson was in the process of renting a tent for his daughter's wedding. While renting a tent from proprietor Peter Mitchell, Richardson, an expert in material science and deterioration, inquired about its durability and found that PVC tents tend to break down. Richardson surmised that the rapid degradation was microbial and in fact due to fungi. The tent manufacturers decided to correct the PVC durability by changing the manufacturing process, that is, by concurrent engineering. In this case, they decided to increase the amount of fungicide, 10,10'-oxybis(phenoxarsine) (OBPA).

10,10'-oxybis(phenoxarsine)

A quick glance at the OBPA structure shows that when it breaks down, it is likely to release arsenic compounds. In this case, it is arsine (AsH_3), a toxic gas (vapor pressure = 11 mmHg at 20°C). It is rapidly absorbed when inhaled, and easily crosses the alveolocapillary membrane and enters red blood cells. Arsine depletes the reduced glutathione content of red blood cells, leading to the oxidation of sulfhydryl groups in hemoglobin and, possibly, red cell membranes. These effects produce membrane instability with rapid and massive intravascular hemolysis. It also binds to hemoglobin, forming a metalloid–hemaglobin complex.[10] These can lead to acute cardiovascular, neurotoxic, and respiratory effects.

Increasing the OBPA to address the problem of PVC disintegration is an example of the problem of ignoring the life-cycle and systematic aspects of most engineering problems. In this case, production and marketing would greatly benefit from a type of PVC that does not break down readily under ambient conditions. In fact, if that problem cannot be solved, the entire camping market might be lost, since fungi are ubiquitous in the places where these products are used.

Had the engineers and planners considered the chemical structure and the possible uses, however, they at least might have restricted the PVC treated with high concentrations of OBPA to certain uses, such as only on tent materials, and not in materials that come in contact with or near humans (bedding materials, toys, etc.). To the contrary, the PVC manufacturers blatantly disregarded the science. Richardson, the expert, from the outset had warned that increasing the amount of fungicide would not only increase the hazard and risk but would make the product less efficacious (even more vulnerable to fungal attack). He stated: "The biocide won't kill this fungus—instead, the fungus will consume the biocide as well as the plasticizer. Since the biocide contains arsenic, the fungus will generate a very poisonous gas which would be harmful to your staff working with the marquees." Plasticizers are semivolatile organic compounds (e.g., phthalates) that can serve as a food source for microbes once the microbes become acclimated. The engineers should have known this, since it is one of the biological principles upon which much wastewater treatment is based. But the manufacturers wanted to approach the situation as a linear problem with a simple solution, that is, increase fungicide and decrease fungus. As a kicker, the PVC manufacturer argued that the fungicide was even approved for use in baby mattresses.

The extent to which arsine gas released by the degradation of OBPA was a causative agent in SIDS cases is a matter of debate. But the fact that a toxic gas *could* be released, leading to exposures of a highly susceptible population (babies), is not debatable. Pollution and consumer products are only some of the possible causes of SIDS. Others include breathing position (probably increased carbon dioxide inhalation), poor nutrition, and physiological stress (e.g., overheating).[11]

The overall lesson is that there are many advantages to concurrent engineering, such as real-time feedback between design and build stages, adaptive approaches, and continuous improvement. However, concurrent engineering works best when the entire life cycle is considered. The designer must ask how even a small change to improve one element in the process can affect other steps and systems within the design and build process.

LIFE-CYCLE ANALYSIS

One means of understanding questions of material and product use and waste production is to conduct what has become known as a *life-cycle assessment*. Such an assessment is a holistic approach to pollution prevention by analyzing the entire life of a product, process, or activity, encompassing raw materials, manufacturing, transportation, distribution, use, maintenance, recycling, and final disposal. In other words, assessing its *life cycle* should yield a complete picture of the environmental impact of a product.

The first step in a life-cycle assessment is to gather data on the flow of a material through an identifiable society. Once the quantities of various components of such a flow are known, the environmental effect of each step in the production, manufacture, use, and recovery/disposal is estimated.

Life-cycle analyses are performed for several reasons, including the comparison of products for purchasing and a comparison of products by industry. In the former case, the total environmental effect of glass-returnable bottles, for example, could be compared to the environmental effect of nonrecyclable plastic bottles. If all of the factors going into the manufacture, distribution, and disposal of both types of bottles are considered, one container might be shown to be clearly superior. In the case of comparing the products of an industry, we might determine if the use of phosphate builders in detergents is more detrimental than the use of substitutes that have their own problems in treatment and disposal.

One problem with such studies is that they are often conducted by industry groups or individual corporations, and (predictably) the results often favor their own product. For example, Proctor & Gamble, the manufacturer of a popular brand of disposable baby diapers, found in a study conducted for them that the cloth diapers consume three times more energy than the disposable kind. But a study by the National Association of Diaper Services found that disposable diapers consume 70% more energy than cloth diapers. The difference was in the accounting procedure. If one uses the energy contained in the disposable diaper as recoverable in a waste-to-energy facility, the disposable diaper is more energy efficient.[12] A lesson here is to use consistent standards and metrics to compare systems.

Life-cycle analyses also suffer from a dearth of data. Some of the information critical to the calculations is virtually impossible to obtain. For example, something as simple as the tonnage of solid waste collected in the United States is not readily calculable or measurable. Even if the data *were* there, the procedure suffers from the unavailability of a single accounting system. Is there an optimal level of pollution, or must all pollutants be removed completely (a virtual impossibility)? If there is both air and water pollution, how must they be compared? This lack of data is particulary problematic when comparing a conventional system that has been in use for decades or centuries to an emerging technology, where only experimental data are available. For example, there is a large disparity of data available for light emitting diode (LED) televisions *versus* the information known about nanotechnologies (e.g., nanotubes) used for the same products.

A recent study supported by the U.S. EPA developed complex models using principles of life-cycle analysis to estimate the cost of materials recycling. The models were able to calculate the dollar cost as well as the cost in environmental damage caused at various levels of recycling. Contrary to intuition and the stated public policy of the U.S. EPA, it seems that there is a breakpoint at about 25% diversion. That is, as shown in Figure 6.2, the cost in dollars and adverse environmental impact start to increase at an exponential rate at about 25% diversion. Should we therefore even strive for greater diversion rates, if this results in unreasonable cost in dollars and actually does harm to the environment?

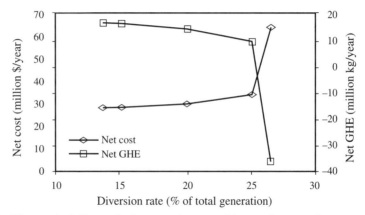

Figure 6.2   The cost in dollars and adverse environmental impact (i.e., greenhouse gas emissions) increases dramatically when the fraction of solid waste recycled exceeds 25%. GHE is greenhouse equivalents, that is, the weighted sum of CO_2 and CH_4 emissions. (From E. Solano, R. D. Dumas, K. W. Harrison, S. Ranjithan, M. A. Barlaz, and E. D. Brill, Integrated Solid Waste Management Using a Life-Cycle Methodology for Considering Cost, Energy, and Environmental Emissions, 2: Illustrative Applications, Department of Civil Engineering, North Carolina State University, Raleigh NC, 2002.)

Discussion: The Coffee Cup Debate

A simple example of the difficulties in life-cycle analysis would be in finding a solution to the great coffee cup debate: whether to use paper or polystyrene coffee cups. The answer most people would give is not to use either, but instead, to rely on the permanent mug. But there nevertheless are times when disposable cups are necessary (e.g., in hospitals), and a decision must be made as to which type to choose.[13] So let's use life-cycle analysis to make a decision.

The paper cup comes from trees, but the act of cutting trees results in environmental degradation. The foam cup comes from hydrocarbon sources such as oil and gas, and this also results in adverse environmental impact, including the use of nonrenewable resources. The production of the paper cup results in significant water pollution, whereas the production of the foam cup contributes significantly less to water pollution. The production of the paper cup results in the emission of chlorine, chlorine dioxide, reduced sulfides, and particulate, whereas the production of the foam cup results in none of these. The paper cup does not require chlorofluorocarbons (CFCs), but neither do the newer foam cups ever since the CFCs in polystyrene were phased out. The foam cup results in the emission of pentane, however, whereas the paper cup contributes none. From a materials separation perspective, the recyclability of the foam cup is much higher than the paper cup since the latter is made from several materials, including the plastic coating on the paper. They both burn well, although the foam cup produces 17,200 Btu/lb (40,000 kJ kg^{-1}), whereas the paper cup produces only 8600 Btu/lb (20,000 kJ kg^{-1}). In the landfill, the paper cup degrades into carbon dioxide (CO_2) and methane (CH_4), both green-house gases, whereas the foam cup is relatively inert. Since it is non-reactive, the foam cup will remain in the landfill for a very long time, whereas the paper cup will eventually (but very slowly) decompose. If the landfill is considered a waste storage receptacle, the foam cup is superior, since it does not participate in the reaction, whereas the paper cup produces gases and probably leachate. If, on the other hand, the landfill is thought of as a treatment facility, the foam cup is highly detrimental since it does not biodegrade.

So which cup is better for the environment? If you wanted to do the right thing, which cup should you use? This question, like so many others in this book, is not an easy one to answer. Private individuals can, of course, practice pollution prevention by such a simple expedient as not using either plastic or paper disposable coffee cups but by using a refillable mug instead (although a thorough life cycle analysis would be needed for this option as well). The argument as to which kind of cup, plastic or paper, is better is then moot. It is better not to produce the waste in the first place. In addition, the coffee tastes better from a mug! We win by doing the right thing.

Once the life cycle of a material or product has been analyzed, the next (engineering) step is to manage the life cycle. If the objective is to use the least energy and cause the least detrimental effect on the environment, it is clear that much of the onus is on the manufacturers of these products. The users of the products can have the best intentions for reducing adverse environmental effects, but if the products are manufactured in such a way as to make this impossible, the fault is with the manufacturers. On the other hand, if the manufactured materials are easy to separate and recycle, energy is probably saved and the environment is protected. This process has become known as *pollution prevention* in industry, and there are numerous examples of how industrial firms have reduced emissions or the production of other wastes, or have made it easy to recover waste products and in the process saved money. Some automobile manufacturers, for example, are modularizing the engines so that junked parts can be easily reconditioned and reused, a process known as *design for disassembly* (DFD). Printer cartridge manufacturers have found that refilling cartridges is far cheaper than remanufacturing them, and now offer trade-ins. All of the efforts by industry to reduce waste (and save money in the process) will influence the solid waste stream in the future.

CASE STUDY: THE TRABI

Bad examples of industrial neglect of environmental concerns for the sake of short-term economic gain abound. The use and manufacture of nonre-cyclable beverage containers, for example, is perhaps the most ubiquitous example. It might be instructive to establish some baseline of absolutely unconscionable industrial behavior to measure how far we have come in the pollution prevention process. The authors of this book recommend that there be an award established, called the *Trabi Award,* which could commemorate the worst, most environmentally unfriendly product ever manufactured.

Figure 6.3 The Trabant, an East German car designed without any concern for environmental impact or ease of final disposal.

The Trabant, affectionately known as the Trabi, was manufactured in East Germany during the 1970s and 1980s. This homely-looking car (Figure 6.3) was designed to be the East German version of the Volkswagen (the People's Car), and its design objectives were to make it as cheaply as possible. So, to power the Trabi, the engineers used a two-stroke engine that was notoriously polluting. All of the components were designed at the least cost, and few survived normal use. Worst of all, the body was made of a fiberglass composite that was impossible to fix (except with duct tape, the engineer's cure-all!), and as far as the solid waste management was concerned, had absolutely no recycling value since it could not be melted down or reprocessed in any other way, nor could it be burned in incinerators.

After the reunification of Germany, thousands of Trabis were abandoned on the streets and had to be disposed of in landfills. The Trabi is the best example of engineering design when the sole objective is production cost and when environmental concerns are nonexistent, and the Trabi deserves to be immortalized as the best example ever of environmentally destructive design.

POLLUTION PREVENTION

The present methods of disposing of hazardous wastes are woefully inadequate. All we are doing is simply storing them until a better idea (or more funds, or stricter laws) comes along. Would it be better not to create waste in the first place? That is, why not *prevent* pollution?

The EPA defines *pollution prevention* as the following: "The use of materials, processes, or practices that reduce or eliminate the creation of pollutants or wastes at the source. It includes practices that reduce the use of hazardous materials, energy, water, or other resources and practices that protect natural resources through conservation or more efficient use."[14] In the widest sense, pollution prevention is the idea of eliminating waste, regardless of how this might be done.

Originally, pollution prevention was applied to industrial operations with the idea of reducing either the amount of the wastes being produced or to change their characteristics to make them more readily disposable. Many industries changed to water-soluble paints, for example, thereby eliminating organic solvents, cleanup time, and other expenses and often ended up saving considerable money. In fact, the concept was introduced as "pollution prevention pays," emphasizing that many of the changes would actually save money for the companies. In addition, the elimination or reduction of hazardous and otherwise difficult wastes has a long-term effect—it reduces the liability the company carries as a consequence of its disposal operations.

With the passage of the Pollution Prevention Act of 1990, the U.S. EPA was directed to encourage pollution prevention by setting appropriate standards for pollution prevention activities, assisting federal agencies in reducing wastes generated, working with industry and promoting the elimination of wastes by creating waste exchanges and other programs, seeking out and eliminating barriers to the efficient transfer of potential wastes, and doing all this with the cooperation of the individual states.

In general, the procedure for the implementation of pollution prevention activities is to:

1. Recognize a need
2. Assess the problem
3. Evaluate the alteratives
4. Implement the solutions

These are steps common to any good design. Contrary to most pollution control activities, industries generally have welcomed this governmental action, recognizing that pollution prevention can and often does result in the reduction of costs to the industry. Thus, recognition of the need quite often is internal and the company seeks to initiate the pollution prevention procedure.

During the assessment phase, a common procedure is to perform a *waste audit,* which is merely a black box mass balance, using the company as the black box.

Example:

Waste Audit

A manufacturing company is concerned about the air emissions of volatile organic carbons (VOCs). These chemicals can volatilize during the manufacturing process, but there is no way of estimating just how much, or which chemicals. The company conducts an audit of three of their most widely used volatile organic chemicals, with the following results:

Purchasing Department Records

Material	Purchase quantity (barrels)
Carbon tetrachloride[a] (CCl_4)	48
Methyl chloride[b] (CH_2Cl_3)	228
Trichloroethylene (C_2HCl_3)	505

[a] The correct name is tetrachloromethane, but the compound was in such common use throughout the century, referred to as carbon tetrachloride, that the name is still frequently used in the engineering and environmental professions.

[b] Also known as chloromethane.

Wastewater Treatment Plant Influent

Material	Average concentration (mg L^{-1})
Carbon tetrachloride	0.343
Methylene chloride	4.04
Trichloroethylene	3.23

The average influent flow rate to the treatment plant is 0.076 m^3 s^{-1}.

Hazardous waste manifests (what leaves the company by truck, headed to a hazardous waste treatment facility):

Material	Barrels	Concentration (%)
Carbon tetrachloride	48	80
Methyl chloride	228	25
Trichloroethylene	505	80

Unused barrels at the end of the year:

Material	Barrels
Carbon tetrachloride	1
Methyl chloride	8
Trichloroethylene	13

How much VOC is escaping?

Conduct a black box mass balance:

$$A_{acc} = A_{in} - A_{out} + A_{prod} - A_{cons}$$

where A_{acc} is the mass of A per unit time accumulated, A_{in} is the mass of A per unit time in, A_{out} is the mass of A per unit time out, A_{prod} is the mass of A per unit time produced, A_{cons} is the mass of A per unit time consumed. The materials A are, in this example the three VOCs.

We also need to know the conversion from barrels to cubic meters and the density of each chemical. Each barrel is 0.12 m², and the density of the three chemicals is 1548, 1326, and 1476 kg m⁻³. The mass per year of carbon tetrachloride accumulated is

$$A_{acc} = 1 \text{ barrel/year} \times 0.12 \text{ m}^3/\text{barrel} \times 1548 \text{ kg m}^{-3} = 186 \text{ kg yr}^{-1}$$

Similarly,

$$A_{in} = 48 \times 0.12 \times 1548 = 8916 \text{ kg yr}^{-1}$$

The mass out is in three parts: the mass discharge to the wastewater treatment plant, the mass leaving on the trucks to the hazardous waste disposal facility, and the mass volatilizing. So

$$A_{out} = (0.343 \text{ g m}^{-3} \times 0.076 \text{ m}^3 \text{ s}^{-1} \times 86{,}400 \text{ s day}^{-1} \times 365 \text{ days yr}^{-1}$$

$$\times 10^{-3} \text{ kg g}^{-1}) + (48 \times 0.12 \times 1548 \times 0.80) + A_{air}$$

$$= 822.1 + 7133 + A_{air}$$

where A_{air} is the mass per unit time emitted to the air. Since no carbon tetrachloride is consumed or produced,

$$186 = 8916 - (822.1 + 7133 + A_{air}) + 0 - 0$$

and $A_{air} = 775 \text{ kg yr}^{-1}$.

If a similar balance is conducted for the other chemicals, it appears that the loss to air of methyl chloride is about 16,000 kg yr⁻¹ and of the trichloroethylene is about 7800 kg yr⁻¹. If the intent is to cut total VOC emissions, it is clear that the first target should be the methyl chloride, at least in terms of the mass released.

However, as we discuss in Chapters 4 and 5, another approach to preventing pollution is relative risk. Although methyl chloride is two orders of magnitude more volatile, all three compounds are likely to be found in the atmosphere, so inhalation is a probable exposure pathway.

Since risk is the product of exposure times hazard ($R = E \times H$), we can compare the risks by applying a hazard value (e.g., cancer potency). We can use the air emissions calculated above as a reasonable approximation of exposure via the inhalation pathway[15] and the inhalation cancer slope factors to represent the hazard. These slope factors for the three compounds are:

$$\text{carbon tetrachloride} = 0.053$$

$$\text{methyl chloride} = 0.0035$$

$$\text{trichloroethylene} = 0.0063$$

Thus, the relative risk for the three compounds can be estimated by removing the units (i.e., we are not actually calculating the risk, only comparing the three compounds against each other, so we do not need units). If we were calculating risks, the units for exposure would be mass of contaminant per body mass per time (e.g., mg kg⁻¹ day⁻¹), whereas the slope factor unit is the inverse of this: kg · day mg⁻¹, so risk itself includes unitless probability (exposure) (see Chapters 4 and 5). The units of risk are often the number of adverse consequences in a population (e.g., one additional cancer per million or 10^{-6}).

$$\text{carbon tetrachloride} = 0.053 \times 775 = 41$$

$$\text{methyl chloride} = 0.0035 \times 16,000 = 56$$

$$\text{trichloroethylene} = 0.0063 \times 7800 = 49$$

Thus, in terms of relative risk, methyl chloride is again the most important target chemical, but the other two are much closer. In fact, given the uncertainties and assumptions, from a relative cancer risk perspective, the importance of removing the three compounds is nearly identical, owing to the much higher cancer potency of CCl_4.

The same approach can be used for non-cancer risk, but rather than using slope factors as the hazard, reference doses (RfDs) and concentrations (RfCs), and average daily doses could represent the potential exposures. Either approach provides a screening or prospective view to aid in pollution prevention.

Once we know what and where the problems are, the next step is to determine useful options. These options fall generally into three categories:

1. Operational changes
2. Materials changes
3. Process modifications

Operational changes might consist simply of better housekeeping, plugging leaks, eliminating spills, and so on. A better schedule for cleaning, and segregating the water might directly yield a large return on a minor investment.

Materials changes often involve the substitution of one chemical for another which is less toxic or requires less hazardous materials for cleanup. The use of trivalent chromium (Cr^{3+}) for chrome plating instead of the much more toxic hexavalent chrome has found favor, as has the use of water-soluble dyes and paints. Note that this is the same element, Cr, but its oxidative state differs, illustrating the importance of details in sustainable decision-making. In some instances, ultraviolet radiation has been substituted for biocides in cooling water, resulting in better-quality water and no waste cooling-water disposal problems. In one North Carolina textile plant, biocides were used in air washes to control algal growth. Periodic "blowdown" and cleaning fluids were discharged to the stream, but this discharge proved toxic to the stream and the state of North Carolina revoked the plant's discharge permit. The town would not accept the waste into its sewers, rightly arguing that this might have serious adverse effects on its biological wastewater treatment operations. The industry was about to shut down when it decided to try ultraviolet radiation as a disinfectant in its air wash system. Fortunately, they found that the ultraviolet radiation effectively disinfected the cooling water and that the biocide was no longer needed. This not only eliminated the discharge, but it eliminated the use of biocides altogether, thus saving the company money. The payback was 1.77 years.[16]

Process modifications usually involve the greatest initial monetary investments and can result in the most rewards. For example, a countercurrent wash water use instead of a once-through batch operation can significantly reduce the amount of wash water needing treatment, but such a change requires pipes, valves, and a new process protocol. In industries where materials are dipped into solutions, such as in metal plating, the use of dragout recovery tanks, an intermediate step, has resulted in the savings of the plating solution and reduction in waste generated.

In any case, the most marvelous thing about pollution prevention is that most of the time a company not only eliminates or greatly reduces the discharge of hazardous materials but also saves money. Such savings are in several forms, including of course direct savings in processing costs, as with the ultraviolet disinfection example above. But there are other savings, including the savings in not having to spend time on submitting compliance permits and suffering potential fines for noncompliance. Future liabilities weigh heavily where hazardous wastes have to be buried or injected below ground (a type of "environmental time bomb"). Additionally, there are the intangible benefits of employee relations and safety. Finally, of course, there is the benefit that comes from doing the right thing, something not to be sneezed at.

MOTIVATIONS FOR PRACTICING GREEN ENGINEERING[17]

To understand the reasons why humans behave as they do, one must identify the driving forces that lead to particular activities. The concept of the driving force can also be used to explain engineering processes. For example, in gas transfer the driving force is the difference in concentrations of a particular gas on either side of an interface. We express the rate of this transfer mathematically as $dM/dt = k(\Delta C)$ where M is mass, t is time, k is a proportionality constant, and ΔC is the difference in concentrations on either side of the interface. The rate at which the gas moves across the interface is thus directly proportional to the difference in concentrations. If ΔC approaches zero, the rate drops until no net transfer occurs. The driving force is therefore ΔC, the difference in concentrations.

Similarly in engineering, driving forces spur the adoption of new technologies or practices. The objective here is to understand what the motivational forces are for adopting green engineering practices. We propose that the three diving forces supporting green engineering are legal considerations, financial considerations, and ethical considerations.

Legal Considerations

At the simplest and most basic level, green engineering is practiced to comply with the law. For example, a supermarket recycles corrugated cardboard because it is the law—either a state law such as in North Carolina, or a local ordinance as in Bucks County, Pennsylvania. This behavior is, at best, "morality lite." Engineers and managers comply with the law because of the threat of punishment for noncompliance. The decision to comply with the law is thus largely a nonmoral decision. Complying with the law is not morally good or morally bad, although not complying may be considered morally bad. So in this situation, managers and engineers choose to do the "right thing," not because it is the right thing to do—but simply because they feel it is their only choice.

We should point out that the vast majority of firms will comply with the law regardless of the financial consequences. Most will not even bother to conduct a cost–benefit analysis because it assumes that breaking the law is not worth the cost.

Occasionally, however, firms may prioritize financial concerns over legal concerns and the managers may determine that by adopting an illegal practice (or failing to adopt a practice codified in law) they can enhance profitability. In such cases they argue that either the chances of getting caught are low or that the potential for profit is large enough to override the penalty if they do get caught (e.g., paying a $10,000 fine each month is preferred to making a $10 million upgrade to meet an environmental standard).

For example, in November 1999, the U.S. EPA sued seven electric utility companies: American Electric Power, Cinergy, FirstEnergy, Illinois Power, Southern Indiana Gas & Electric Company, Southern Company, and Tampa Electric Company for violating "the Clean Air Act by making major modifications to many of their coal burning plants without installing the equipment required to control smog, acid rain and soot."[18] On August 7, 2003, "Judge Edmund Sargus of the U.S. District Court for the Southern District of Ohio found that Ohio Edison, an affiliate of FirstEnergy Corp., violated Clean Air Act's New Source Review (NSR) provisions by undertaking 11 construction projects at one of its coal-fired plants from 1984 to 1998 without obtaining necessary air pollution permits and installing modern pollution controls on the facility."[19] Given the number of violations, it seems obvious that the companies had calculated that breaking the law and possibly getting caught was the least-cost solution and thus behaving illegally was the "right answer." A recent attempt to change such behaviors is that environmental penalty decisions now can include financial advantages gained in noncompliance. The fine will be assessed and additional penalties will be added to make environmental compliance *fair*. Thus, a $10,000 fine may be increased to $100,00 if the regulatory agency believes a company gained $90,000 advantage over their competitors of not complying for the past five years.

In some cases private firms can take advantage of loopholes in tax law that inadvertently allow companies to pretend to be environmentally green while in reality doing nothing but gouging the taxpayer. An example of this is the great synfuel scam.[20] In the 1970s the U.S. Congress decided to promote the use of cleaner fuels in order to take advantage of both the huge coal reserves in the United States and the environmental benefits derived from burning a clean gaseous fossil fuel made from coal. Producing such synfuel from coal had been implemented successfully in Canada, and the U.S. government wanted to encourage our power companies to enter the synfuel business. To promote this industry, Congress wrote into law substantial tax credits for companies that would produce synfuel and defined a *synfuel* as chemically altered coal, anticipating that the conversion would be to a combustible gas that could be used much as natural gas is used today.

Unfortunately, the synfuel industry in the United States did not develop as expected because cheaper natural gas supplies became available. The synfuel tax credit idea remained dormant until the 1990s, when a number of corporations (including some giants like the Marriott hotel chain) found the tax break and went into the synfuel business. Since the only requirement was to change the chemical nature of the fuel, it became evident that even spraying the coal with diesel oil or pine tar would alter the fuel chemistry and that this fuel would then be legally classified as a synfuel. The product of these synfuel plants was still coal, and more expensive coal than raw coal at that, but the tax credits were quite large. Some companies formed specifically to take advantage of the tax break, often with environmentally attractive names such as Earthco, and made huge profits by selling their tax credits to other corporations that needed them. The synfuels industry presently is receiving over $1 billion annually in tax credits, while doing nothing illegal, but also while doing little to benefit the environment.

Financial Considerations

Decisions about the adoption of green practices are also driven by financial concerns. This level of involvement with "greening" is at the level promoted by the economist Milton Friedman, who stated famously: "The one and only social responsibility of business [is] to use its resources and engage in activities designed to increase its profits so long as it . . . engages in open and free competition, without deception or fraud."[21] In line with this stance, the firm calculates the financial costs and benefits of adopting a particular practice and makes its decision based on whether the benefits outweigh the costs, or vice versa. In fact, this is the most common metric in Western nations for determining whether an activity is acceptable or unacceptable.

Many companies seek out green engineering opportunities solely on the basis of their providing a means of lowering expenses, thereby increasing profitability. Here are some examples[22]:

- In one of its facilities at Deepwater, New Jersey, DuPont uses phosgene, an extremely hazardous gas, and used to ship the gas to the plant. In an effort to reduce the chance of accidents, DuPont redesigned the plant to produce phosgene on site and to use almost all of it in the manufacturing process, avoiding costs associated with hazardous gas transport and disposal.

- Polaroid did a study of all of the materials it used in manufacturing and grouped them into five categories based on risk and toxicity. Managers are encouraged to alter product lines to reduce the amount of material in the most toxic groups. In the first five years, the program resulted in a reduction of 37% of the most toxic chemicals and saved over $19 million in money not spent on waste disposal.

- Dow Chemical challenged its subsidiaries in Louisiana to reduce energy use and sought ideas on how this should be done. Following up on the best ideas, Dow invested $1.7 million and received a 173% return on its investment.

Other firms may believe that adopting a particular green engineering technology will provide them with public relations opportunities: Green engineering is a useful tool for enhancing a company's reputation and community standing. If the result is likely to be an increase in sales for the business, and if sales are projected to rise *more* than expenses, so that profits rise, the firm is likely to adopt such a technology. The same is true if the public relations opportunities can be exploited to provide the firm with expense reductions, such as decreased enforcement penalties or tax liabilities. Similarly, green technologies that not only yield increased sales but decrease expenses at the same time are the perfect recipes for the adoption of green practices by a company whose primary driving forces are financial concerns. For instance:

- DuPont's well-publicized decision to discontinue its $750 million a year business producing chlorofluorocarbons (CFCs) was a public relations bonanza. Not only did DuPont make it politically possible for the United States to become a signatory to the Montreal Protocol on ozone depletion, but it already had alternative refrigerants in the production stage and were able to transition smoothly to these. In 1990, the U.S. EPA gave DuPont the Stratospheric Protection Award in recognition of their decision to get out of CFC manufacturing.[23] The fact that the decision also proved to be highly profitable for DuPont apparently did not matter to the judges.

- The seven electric companies sued by the EPA in November 1999 for Clean Air Act violations (mentioned earlier) heavily publicized their efforts to reduce greenhouse gas emissions. For example, American Electric Power (AEP) issued news releases on May 8, June 11, and November 21, 2002 regarding emissions reduction efforts at various plants. Not coincidentally, the U.S. government, which had sued earlier, was in the process of revising the portions of the Clean Air Act that the company had violated previously. Presumably, regulators were favorably impressed with the company's hard work; in August 2003, the EPA announced that it was dropping the suit and revamping that portion of the act.

These examples clearly demonstrate bottom-line thinking: Cases in which managers were simply trying to practice "good business," seeking ways to increase the difference between revenues and expenses so that profits would rise. As far as we can tell, these decisions were not influenced by the desire to "do the right thing" for the environment. It certainly did not seem to be the *primary* factor. Here we again have examples of nonmoral decisionmaking. Businesses are organized around the idea that they will either make money or cease to survive; in the "financial concerns" illustrations provided so far, green practices were adopted as a means of making more money.

On occasion, though, managers are *forced* into considering the adoption of greener practices by the threat that not doing so will cause expenses to rise and/or revenues to fall. For example, in October 1998, ELF (Earth Liberation Front) targeted Vail Ski Resort, burning a $12 million expansion project to the ground.[24] In the wake of this damage, the National Ski Areas Association (NSAA) began developing its Environmental Charter in 1999 with "input from stakeholders, including . . . environmental groups"[25] and officially adopted the charter in June 2000.[26] In accordance with the charter, NSAA has produced its Sustainable Slopes Annual Report each year since 2001.[27] Apparently, the driving force behind the decision to adopt the Environmental Charter was largely a response to financial concerns rather than by the desire to treat the environment responsibly—it was a non-moral decision. That is, NSAA was spurred to create the Environmental Charter by concerns about member companies' bottom lines: Further "ecoterrorist" activity could occur, thereby causing expenses to rise; and the ELF action may have sufficiently highlighted the environmental consequences of resort development to the point that environmentally minded skiers might pause before deciding to patronize resorts where development was occurring, thereby causing revenues to fall. Incidentally, the act of ELF is considered by many ethicists to be *immoral;* that is, unethical means to achieve an end that the group sees as a higher value.

Similarly, for firms trying to do business in Europe, adopting ISO 14000 is close to a required management practice. The ISO network has penetrated so deeply into business practices that firms are nearly locked out if they do not gain ISO 14000 certification.

There is ample evidence that one of the reasons businesses participate in the quest for sustainability is because it is good for business. The leaders of eight leading firms that adopted an environmentally proactive stance on sustainability were asked in one study to justify the firms' adoption of such a strategy.[28] All companies reported that they were motivated first by regulations such as the control of air emissions, pretreatment of wastewater, and the disposal of hazardous materials. One engineer in the study admitted that "the [waste disposal] requirements became so onerous that many firms recognized that benefits of altering their production processes to generate less waste."

The second motivator identified in this study was competitive advantage. Lawrence and Morell quote one director of a microprocessor company, who noted that "by reducing pollution, we can cut costs and improve our operating efficiencies." The company recognized the advantage of cutting costs by reducing its hazardous waste stream.[29]

Another study, conducted by PriceWaterhouse Coopers, confirmed these findings.[30] When companies were asked to self-report on their stance on sustainable principles, the top two reasons for adopting sustainable development were found to be (1) enhanced reputation (90%), and (2) competitive advantage (cost savings) (75%). It is not clear if the respondents were given the option of responding that they practiced sustainable operations because this was mandated by law. If it had, there is no doubt that all companies would have publicly stated that they are, indeed, law abiding.

So it seems likely that the two primary driving forces behind the adoption of green business and engineering practices are (1) legal concerns and (2) financial concerns. Can we argue that such behavior is morally admirable simply on the basis that the outcomes (e.g., cleaner air and water) are morally preferable? We say no. In accord with Sethi,[31] we argue instead that actions undertaken in response to legal and financial concerns are actually *obligatory,* in that society essentially demands that businesses make their decisions within legal and financial constraints. For an action to be morally admirable, however, the motivating force driving the decisions has to be far different in character.

Ethical Considerations

The first indication that some engineers and business leaders are making decisions where the driving force may not be due to legal or financial concerns comes from several cases in American business. Although most business or engineering decisions are made on the basis of legal or financial concerns, some companies believe that behaving more environmentally responsibly is simply the right thing to do. They believe that saving resources, and perhaps even the planet, for the generations that will follow is an important part of their job. When making decisions, they are guided by the "triple bottom line." Their goal is to balance the financial, social, and environmental impacts of each decision.

A prime example of this sort of thinking is the case of Interface Carpet Company.[32] Founded in 1973, its founder and CEO until 2001 was Ray Anderson, now chairman of the board. By the mid-1990s, Interface had grown to nearly $1.3 billion in sales, employed some 6600 people, manufactured on four continents, and sold its products in over 100 countries worldwide. In 1994, several members of Interface's research group asked Anderson to give a kick-off speech for a task force meeting on sustainability; they wanted him to provide Interface's environmental vision. Despite his reluctance to do so—

Anderson had no "environmental vision" for the company except to comply with the law—he agreed. Fortuitously, as Anderson struggled to determine what to say, someone sent him a copy of Paul Hawken's *The Ecology of Commerce*[33]; Anderson read it, and it completely changed not only his view of the natural environment, not only his vision for Interface Carpet Company, but his entire conception of business. In the coming years, he held meetings with employees throughout the Interface organization explaining to them his desire to see the company spearhead a sustainability revolution. No longer would they be content to keep pollutant emissions at or below regulatory levels. Instead, they were going to strive to be a company that created zero waste and did not emit any pollutants *at all*. The company began to employ "The Natural Step"[34] and notions of "Natural Capitalism"[35] as part of its efforts to become truly sustainable. The program continues today, and although the company has saved many millions of dollars as a result of adopting green engineering technologies and practices, the reason for adopting these principles was not to earn more money, but rather, to do the right thing.

Yet another example is that of Herman Miller, an office furniture manufacturing company located in western Michigan. Its pledge in 1993 to stop sending any materials to landfills by 2003 has resulted in the company's adoption of numerous progressive but sometimes expensive practices. For example, the company ceased taking scrap fabric to the landfill and began shredding it and trucking it to a firm in North Carolina that processes it into automobile insulation. This environmentally friendly process costs Herman Miller $50,000 each year, but the company leaders agree that a decision that is right for the environment is the right decision. Similarly, the company's new waste-to-energy plant has increased costs, but again company leaders feel that it is worth the cost, as employees and managers are proud of the company's leadership in preserving the natural environment in their state.[36]

Our point here is this: The decisions made by the leadership of Interface Carpet Company and Herman Miller were not morally admirable simply because they enabled these companies to reduce toxic emissions (among many other positive outcomes for the environment); they were morally admirable because the *driving force* behind those decisions was the desire to stop harming the Earth, to protect it so that future generations would be able to enjoy it as much as, or even more than, we do today. Conversely, in the cases of DuPont, Polaroid, and Dow Chemical cited earlier, the *driving force* behind their decisions to adopt green technologies was a desire to save the company money; the benefits to the Earth were simply an ancillary by-product of those decisions.

THE MORAL CHALLENGE OF GREEN ENGINEERING[37]

We identify three primary driving forces behind corporate decisions to adopt green engineering practices: legal, financial, and ethical considerations. Most firms do not even consider disobeying the law. Legal concerns are their top priority. Financial concerns are nearly as high on the priority scale: Managers consider it their duty to shareholders or owners to assure that the company makes an adequate profit, so they base decisions about green practices on a cost–benefit analysis of the probable consequences. These firms are not concerned with "doing the right thing" except inasmuch as the "right thing" means obeying the law and making money. In other words, these firms may decide to adopt green engineering and business practices strictly on the basis of legal and financial factors, without being significantly influenced by the desire to protect the natural environment. Only when the driving forces involve the desire to do good for all people do such decisions become moral in character.

This observation suggests that it might be possible to develop a normative model of green engineering. Such a normative view would ask the question: What *ought* to be the driving forces for adopting green engineering practices? Our proposed normative model is rooted in the work of developmental constructivist thinkers such as Kohlberg,[38] Piaget,[39] Rest,[40] and others who noted that moral action is a complex process entailing four components: moral awareness (or sensitivity), moral judgment, moral motivation, and moral character. The actor must first be aware that the situation is moral in nature: that is, at the least, that the actions considered would have consequences for others. Second, the actor must have the ability to judge which of the potential actions would yield the best outcome, giving consideration to those likely to be affected. Third, the actor must be motivated to prioritize moral values above other sorts of values, such as wealth or power. And fourth, the actor must have the strength of character to follow through on a decision to act morally.

Piaget, Kohlberg, and others (e.g., Duska[41]) have noted that the two most important factors in determining a person's likelihood of behaving morally—that is, of being morally aware, making moral judgments, prioritizing moral values, and following through on moral decisions—are age and education. These seem to be particularly critical regarding moral judgment: A person's ability to make moral judgments tends to grow with maturity as they pursue further education, generally reaching its final and highest stage of development in early adulthood. This theory of moral development is illustrated by Kohlberg's stages of moral development:

Pre-conventional level

    1. Punishment–obedience orientation

    2. Personal reward orientation

Conventional level

    3. "Good boy"–"nice girl" orientation

    4. Law and order orientation

Post-conventional level

    5. Social contract orientation

    6. Universal ethical principle orientation

Kohlberg insisted that these steps are progressive. He noted that in the two earliest stages of moral development, which he combined under the heading *pre-conventional level,* a person is motivated primarily by the desire to seek pleasure and avoid pain. The *conventional level* consists of stages 3 and 4: In stage 3, the consequences that actions have for peers and their feelings about these actions; in stage 4, considering how the wider community will view the actions and be affected by them. Few people reach the *post-conventional stage,* wherein they have an even broader perspective. Their moral decision making is guided by universal moral principles[42]: that is, by principles that reasonable people would agree should bind the actions of all people who find themselves in similar situations.

We propose that the normative model of green engineering can be developed along the same lines. The moral need to consider the impact that one's actions will have on others forms the basis for the normative model we are proposing. Pursuing an activity with the goal of obeying the law has as its driving force the avoidance of punishment, and pursuing an activity with the goal of improving profitability is a goal clearly in line with stockholders' desires; presumably customers', suppliers', and employees' desires must also be met at some level. And finally, pursuing an activity with the goal of "doing the right thing," behaving in a way that is morally right and just, can be the highest level of green engineering behavior. This normative model of green engineering can be illustrated as shown in Figure 6.4.

There is a striking similarity between Kohlberg's model of moral development and the model of moral green engineering. Avoiding punishment in the moral development model is similar to a corporation staying out of trouble by obeying the law. The preconventional level and our legal concern level have similar driving forces.

At the second level in the moral development model is a concern with peers and community, while in our model the corporation undertakes green business practices in order to make more money for the stockholders and to provide a service or product for their customers that will in turn make the corporation more profitable. At this level, as in the previous one, self-centeredness and personal well-being govern decisions.

Finally, at the highest level of moral development, a concern with universal moral principles begins to govern actions, while for the corporate model, fundamental moral principles having to do with environmental issues control corporate decisions. In both of these cases the driving force or motivation is trying to do the right thing on a moral (not legal or financial) basis.

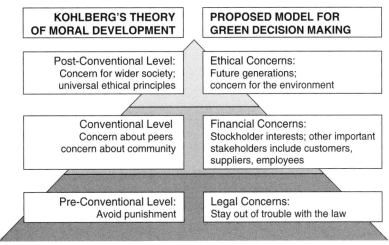

Figure 6.4 Model for understanding the motivations for green decision making. (From P. A. Vesilind, L. Heine, and S. Hamill, Kermit's Lament: It's Not Easy Being Green, *ASCE Journal of Professional Issues in Education and Engineering*, Forthcoming.)

We suggest that moral green engineering occurs only when engineers and managers base their decisions about the adoption of green business and engineering principles on ethical considerations. That is, they recognize the broad impact that their decisions may have, and they act in such a way that their actions will be in the best interest not only of themselves, their companies, and their companies' direct stakeholders, but also the broader society and even future generations.

Green engineering will eventually lead the world to sustainability, but green engineering today occurs most often when doing the right thing also results in adherence to laws and regulations and in achieving greater profitability for the organization. This kind of green engineering, although often beneficial to society, is more business acumen than morally admirable.

The true heroes of green engineering are those leaders who believe deeply in the principles of green engineering and who try to work within these principles while still helping their corporations to be profitable. They enjoy working to promote sustainability, and do so not for show or profit but because it gives them pleasure to do the right thing.

FUTURE PEOPLE

One of the unique characteristics of humans is that we have self-awareness. We can see ourselves in the world today, and we know that humans existed in days gone by, and our species will (we hope) exist tomorrow. We thus are able to plan for the future and accept delayed gratification.

But there will come a time in the future where we individually are long dead and we can no longer personally benefit from any actions that we might have taken on our own behalf. For that matter, there will come a time, after our death, when we are not longer burdened by the ill-considered actions that might have led to unhappiness. Why, then, worry about the future?

We can, based on empirical evidence, assume that there will be a future, of some sort, and we have some confidence that this future will be inhabited by human beings. It is this future—the future without you and me—that we now address.

Although the "client" for engineers is almost always an existing person or organization, the work in which engineers engage can have far-reaching consequences for persons who are not yet born: future people. It is easy to argue that engineers have a moral responsibility to existing people by virtue of their position in society, but does this extend to these future persons, those as yet unborn, who may or may not even exist?

We believe that there are two reasons why the engineer has moral responsibilities to future people:

- Many engineering works, be they small gadgets or huge buildings, will certainly last for more than one generation and will be used by people who were not yet born when the product or facility was constructed.
- Engineers can and do appreciably alter the environment, and the health, safety, and welfare of future people will depend on maintaining a sustainable environment.

Engineers conceive, design, and construct products and facilities that last for generations. Indeed, many engineering decisions have no effects until decades later. For example, suppose that engineers choose to dispose of some hazardous waste in steel containers buried underground. It may take generations for waste containers to corrode, for their contents to leach, for the leachate to migrate and pollute groundwater, and for toxic effects to occur in people coming in contact with the water. Such a problem is not of concern for present people since it will be decades before the effects are felt. The only persons to be adversely affected by such an engineering decision are future people, and they are the only ones who have no say in the decision.

Some would argue that we owe no moral obligations to future generations because they do not exist and the alleged obligation has no basis because we do not form a moral community with them. This is a fallacious argument, however. Even if future generations do not yet exist (by definition), we can still have obligations to them. If we agree that we have moral obligations to distant peoples whom we do not know, it would be reasonable to argue that we have similar moral obligations to people who are yet unborn.

Vesilind and Gunn[43] use an analogy to illustrate this point. Consider a terrorist who plants a bomb in a primary school. Plainly the act is wrong, and in breach of a general obligation not to cause (or recklessly risk) harm to fellow citizens. Even though the terrorist may not know the identities of the children, we would all agree that this is an evil act. And the same would be true if the terrorist bomb had a very long fuse, say 20 years. This would be equally heinous, even though the children, at the time the bomb was placed, had not yet been born. Some engineering works, such as the hazardous waste disposal alluded to above, have very long fuses, and there is no doubt that future people can be harmed by irresponsible engineering activities. The act of burying wastes in the ground where they will not find their way to drinking water supplies for some decades is no different from the act pouring the wastes down a well, except in terms of time.

The second way that engineers have responsibility to future generations is by consciously working to maintain a sustaining environment. Global warming is one instance where the damage done to date is so severe that the effects will not be felt until many years from now. Most models predict that by building up greenhouse gases at the present time, the temperature of the Earth will be slowly getting warmer even if and when we begin to reduce the emission of such gases. This is analogous to heating a pot of water on an electric stove. The burner is turned on and the water begins to heat. When the burner is turned off the temperature of the water does not drop immediately to room temperature. The burner is still warm, heat continues to be transmitted to the pot, and the temperature of the water continues to rise even after the burner is turned off. This effect will also occur with global warming (although rather than heat being buffered, the decrease in the concentrations of greenhouse gases will resist change even after the sources are removed). We therefore may have already exceeded the level of sustainability with regard to the Earth's temperature but we will not know about it until decades from now.[44]

Some argue that we have no obligation to maintain a good-quality environment for future generations because we cannot know what kind of an environment they will want. Our sole responsibility to future generations is therefore not to plan for them.[45]

But this is a hollow argument. Just like the Vestal Virgin who was buried alive with only a little bread and water, we know very well that future generations will *not* want contaminated air or water, dramatically reduced number of species, or global warming. Certainly, there will be changes in style and fashion, and future generations will no doubt have different views on many of our present moral issues, but we also know that they will want a sustainable environment for themselves and their children. Irreparable global warming, or large-scale radiation, or the destruction of the ozone layer are not, under any circumstances, what our progeny would want. Parents do not know what careers their children will choose when they grow up, whom they will marry, or what their life-style will be like, but the parents *do* know that their children will want to be healthy, and thus the parents are morally obligated to provide heath care for their children. Also, future generations probably will not want to suffer genetic damage or to produce babies with severe birth defects, and thus our obligation to them is to control chemical pollution. The argument that because we don't know the desires of future people, our only obligation is to not plan for them is therefore wrong.

The engineers' responsibilities to society are the control and prevention of pollution, and they are therefore entrusted to help maintain a healthy environment. Because this responsibility extends into the future, the "public" in the first canon in the codes of ethics should refer to all people, present and future.

The future, therefore, is the future beyond the careers of present engineers. But unlike some laborers or tradespeople, the effect of their work will last long after they are no longer around. Is it important to you, today, to know that what you do will have a positive effect on future people? Sustainability is, afterall, a temporal concept since we do now will have lasting effects.

The profession and practice of engineering is changing, but we will always be re-quired to have strong analytical skills. The engineer of the future will increasingly need "practical ingenuity" as well as the ability to find new ways of doing things (i.e., crea-tivity) built on a framework of high ethical standards, professionalism, and lifelong learn-ing.[46] These are the qualities of a *good* engineer. New tools are becoming available to assit us, such as computational methods (e.g., toxicology and fluid dynamics), quantitative structural activity relationships (QSARS), and ever-improving preductive models.

Although the Viking society of northern Europe was in many ways cruel and crude, they had a very simple code of honor. Their goal was to live their life so that when they died, others would say "He was a good man." The definition of what they meant by a "good man" might be quite different by contemporary standards, but the principle is important. Conversely, the Talmudic precautions regarding the "sins of the fathers" sug-gests the chaotic (e.g., "Butterfly Effect") of today's decisions. If we live our professional engineering lives so as to uphold the exemplary values of engineering, the greatest pro-fessional honor we could receive would be to be remembered as a *good* engineer.

REFERENCES AND NOTES

1. World Commission on Environment and Development, United Nations, *Our Common Future,* Oxford Paperbacks, Oxford, 1987.
2. A. Maslow, *Motivation and Personality,* 2nd ed., Harper & Row, New York, 1970.
3. American Society of Mechanical Engineers, Professional Practice Curriculum: Sustainabil-ity, http://www.professionalpractice.asme.org/communications/sustainability/index.htm, ac-cessed November 2, 2004.
4. G. Hardin, Tragedy of the Commons, *Science,* 162:1243–1248, Dec. 13, 1968.
5. This section is based on a paper originally authored by P. A. Vesilind, L. Heine, J. R. Hendry, and S. A. Hamill.
6. The source for this discussion is S. B. Billatos and N. A. Basaly, *Green Technology and Design for the Environment,* Taylor & Francis, Bristol, PA, 1997.

7. U.S. Environmental Protection Agency, What Is Green Engineering? http://www.epa.gov/oppt/greenengineering/whats_ge.html, accessed November 2, 2004.

8. See S. B. Billatos and N. A. Basaly, *Green Technology and Design for the Environment,* Taylor & Francis, Washington, DC, 1997; and V. Allada, Preparing Engineering Students to Meet the Ecological Challenges Through Sustainable Product Design, *Proceedings of the 2000 International Conference on Engineering Education,* Taipei, Taiwan, 2000.

9. T. J. Woodruff, J. Grillo, and K. C. Schoendorf, 1997, The Relationship Between Selected Causes of Postneonatal Infant Mortality and Particulate Air Pollution in the United States, *Environmental Health Perspectives,* 105(6), June 1997.

10. R. E. Gosselin, R. P. Smith, and H. C. Hodge, *Clinical Toxicology of Commercial Products,* 5th ed., Williams & Wilkins, Baltimore, 1984.

11. Since we brought it up, the SIDS Alliance recommends a number of risk reduction measures that should be taken to protect infants from SIDS:
    - *Place your baby on his or her back to sleep.* The American Academy of Pediatrics recommends that healthy infants sleep on their backs or sides to reduce the risk for sudden infant death syndrome (SIDS). This is considered to be most important during the first six months of age, when baby's risk of SIDS is greatest.
    - *Stop smoking around the baby.* SIDs has long been associated with women who smoke during pregnancy. A new study at Duke University warns against use of nicotine patches during pregnancy as well. Findings from the National Center for Health Statistics now demonstrate that women who quit smoking during pregnancy, but resume after delivery, put their babies at risk for SIDS, too.
    - *Use firm bedding materials.* The U.S. Consumer Product Safety Commission has issued a series of advisories for parents regarding hazards posed to infants sleeping on top of beanbag cushions, sheepskins, sofa cushions, adult pillows, and fluffy comforters. Waterbeds have also been identified as unsafe sleep surfaces for infants. Parents are advised to use a firm, flat mattress in a safety-approved crib for their baby's sleep.
    - *Avoid overheating, especially when your baby is ill.* SIDS is associated with the presence of colds and infections, although colds are not more common among babies who die of SIDS than babies in general. Now research findings indicate that overheating too much clothing, too heavy bedding, and too warm a room may greatly increase the risk of SIDS for a baby who is ill.
    - *If possible, breastfeed.* Studies by the National Institutes of Health show that babies who died of SIDS were less likely to be breastfed. In fact, a more recent study at the University of California–San Diego found breast milk to be protective against SIDS among nonsmokers but not among smokers. Parents should be advised to provide nicotine-free breast milk if breastfeeding, and to stop smoking around your baby, particularly while breastfeeding.
    - *Mother and baby need care.* Maintaining good prenatal care and constant communication with your health care professional about changes in your baby's behavior and health are of the utmost importance.

12. Dy-Dee Diaper Service, http://www.dy-dee.com/, accessed April 22, 2005.

13. M. S. Pritchard, *On Being Responsible,* University Press of Kansas, Lawrence KS, 1991.

14. U.S. Environmental Protection Agency Pollution Prevention Directive, May 13, 1990, quoted in H. Freeman et al., Industrial Pollution Prevention: A Critical Review, presented at the Air and Waste Management Association Meeting, Kansas City, MO 1992.

15. Even without calculating the releases, is probably reasonable to assume that the exposures will be similar since the three compounds have high vapor pressures (more likely to be inhaled): carbon tetrachloride, 115 mmHg; methyl chloride, 4300 mmHg; trichloroethylene, 69 mmHg.

16. S. Richardson, Pollution Prevention in Textile Wet Processing: An Approach and Case Studies, *Proceedings: Environmental Challenges of the 1990's,* EPA/66/9-90/039, U.S. EPA, Washington, DC, September 1990.

17. Much of this discussion appeared in an earlier paper, "Ethics of Green Engineering" by P. A. Vesilind, L. Heine, J. R. Herndry, and S. A. Hamill.

18. C. Lazaroff, U.S. Government Sues Power Plants to Clear Dirty Air, *Environment News Service,* http://ens.lycos.com/ens/nov99/1999L-11-03-06.html, 1999.

19. D. Fowler, 2003, Bush Administration, Environmentalists Battle over "New Source Review" Air Rules, *Group Against Smog and Pollution Hotline,* Fall 2003, http://www.gasp-pgh.org/hotline/fall03_4.html, accessed January 3, 2004.

20. D. L. Barlett and J. B. Steele, The Great Energy Scam, *Time,* 162(15):60–70, October 13, 2003.

21. M. Friedman, *Capitalism and Freedom,* University of Chicago Press, Chicago, 1962, p. 133.

22. K. Gibney, Sustainable Development: A New Way of Doing Business, *Prism* (American Society of Engineering Education), January 2003.

23. Billatos and Basaly, note 6.

24. J. Faust, 2004, Earth Liberation Who? ABCNEWS.com Web site, http://more.abcnews.go.com/sections/us/DailyNews/elf981022.html, accessed January 3, 2004.

25. National Ski Areas Association, Environmental Charter, http://www.nsaa.org/nsaa2002/_environmental_charter.asp, 2002.

26. J. Jesitus, Charter Promotes Environmental Responsibility in Ski Areas, *Hotel and Motel Management,* 215(19):64, 2000.

27. National Ski Areas Association, Sustainable Slopes Annual Report, http://www.nsaa.org/nsaa2002/_environmental_charter.asp?mode=s, 2002.

28. A. T. Lawrence and D. Morell, Leading-Edge Environmental Management: Motivation, Opportunity, Resources and Processes, in *Research in Corporate Social Performance and Policy,* J. E. Post, D. Collins, and M. Starik (Eds.), Supplement 1, JAI Press, Greenwich, CT, 1995.

29. Lawrence and Morell, note 28.

30. PriceWaterhouse Cooper, *Sustainability Survey Report,* PWC, New York, August 2002.

31. S. P. Sethi, Dimensions of Corporate Social Performance: An Analytical Framework, *California Management Review,* 17(3):58–64, Spring 1975.

32. R. C. Anderson, *Mid-Course Correction—Toward a Sustainable Enterprise: The Interface Model,* Peregrinzilla Press, Atlanta, GA, 1998.

33. P. Hawken, *The Ecology of Commerce: A Declaration of Sustainability,* Harper Business, New York, 1994.

34. See K.-H. Robért, 1991, Educating a Nation: The Natural Step, *In Context,* 28, Spring 1991. According to their website (http://www.naturalstep.org/): "Since 1988, The Natural Step has worked to accelerate global sustainability by guiding companies, communities, and governments onto an ecologically, socially, and economically sustainable path. More than 70 people in 12 countries work with an international network of sustainability experts, scientists, universities, and businesses to create solutions, innovative models, and tools that will lead the transition to a sustainable future."

35. A. B. Lovins, L. H. Lovins, and P. Hawken, A Road Map for Natural Capitalism, *Harvard Business Review,* 7(3):145–158, 1999.

36. L. T. Hosmer, Herman Miller and the Protection of the Environment, in *The Ethics of Management.* McGraw-Hill Irwin, Boston, 2003.

37. An earlier version of this discussion appeared as "Ethical Motivations for Green Business and Engineering" by P. A. Vesilind and J. R. Hendry. *Clean Technology and Environmental Policy.*

38. L. Kohlberg, *The Philosophy of Moral Development,* Vol. 1, Harper & Row, San Francisco, CA, 1981.

39. J. Piaget, *The Moral Judgment of the Child,* Free Press, New York, 1965.

40. J. R. Rest, *Moral Development: Advances in Research and Theory,* Praeger, New York, 1986; and J. D. Rest, D. Narvaez, M. J. Bebeau, and S. J. Thoma, *Postconventional Moral Thinking: A Neo-Kohlbergian Approach,* Lawrence Erlbaum Associates, Mahwah, NJ, 1999.

41. R. Duska and M. Whelan, *Moral Development: A Guide to Piaget and Kohlberg,* Paulist Press, New York, 1975.

42. J. A. Rawls, *A Theory of Justice.* Harvard University Press, Cambridge, MA, 1785; and I. Kant, *Foundations of the Metaphysics of Morals,* translated by L. W. Beck (1951), Bobbs-Merrill, Indianapolis, IN, 1959.

43. P. A. Vesilind and A. S. Gunn, *Engineering, Ethics, and the Environment,* Cambridge University Press, New York, 1998 p. 39.

44. This is one of the hottest topics of debate in environmental circles. It is not our purpose here to take one side or another, but *if* the predictions are correct, the most dramatic and harmful effects would be in the coming decades. Our point is that even with the gaps in knowledge, these effects would only be avoided by prudent decisions now. After the warming has reached severity, several decades of recovery may well be needed to reach a new chemical and energy atmospheric equilibrium. This dilemma is the thrust behind the *precautionary principle,* that is, if the potential harm is sufficiently severe the prudent course of action is to avoid the action. However, in some decisions, the precautionary principle does not work well, such as the current debate regarding potential environmental impacts from emerging technologies (e.g., nanotechnology). Simply not allowing advancing technologies is unsatisfying since society would lose the benefits (i.e., *opportunity risks*), including environmental and public health benefits (e.g., improved sensors, sentinel systems, treatment of cancer, and improved waste site clean-up).

45. M. Golding, Obligations to Future Generations, *Monist,* 56:85–99, 1972.

46. National Academy of Engineering, *The Engineer of 2020: Visions of Engineering in the New Century,* National Academies Press, Washington, DC, 2004.

7

Environmental
Justice Reconsidered

You don't need a weather man to know which way the wind blows.
Subterranean Homesick Blues, Bob Dylan, 1965

"Okay, I get it!" we can hear you saying at this point. So what can engineering and design professionals do to ensure that our designs and projects are fair to the entire community and do not cause disproportionate burdens on certain members of society? The first clue is to consider the subtitle of this book. We must manage risks *and* be socially responsible. In fact, to be socially responsible *requires* that we manage environmental risks properly and that we do so in a fair manner.

We must take our professional responsibilities seriously, but as Dylan's lyrics seem to say, we must avoid condescension and underestimation of the "common sense" and problem-solving abilities of those outside the profession. And as Dylan also seems to imply, we may miss some very obvious truths since we are so sharply focused on the specific project at hand. Indeed, we live and die by the manner and extent that we attend to the details, but we must not lose sight of the different trade-offs and value judgments we make as we select from the engineering options available to us.

One of the first things for us to keep in mind is that many of historically disadvantaged communities lack the "voice" of other groups who have long made use of the system. For example, one can argue that much of the environmental movement that grew out of the 1960s was in many ways elitist. Environmental considerations have often been thrown into the mix of protesting a plan, design, or project, but the principle underlying protest is something else, like property values and maintaining the current demographics and *status quo*.

Discussion: Professional versus Personal Values

Here's a test. Drive through three neighborhoods in your town, other than your own, that vary from one another in terms of socioeconomics, property values, and land uses (e.g., predominantly residential, industrial, agricultural, rural buffer, mixed). Assume that the community, without question, needs a new facility (e.g., a very dangerous waste needs to be stored, but once it is stored the potential risks are reduced by five orders of magnitude from the present situation). Based on your first inclinations, similar to responding to the images in a Rorschach test (called *free association* by psychologists), where are you most likely from a professional standpoint to locate a hazardous waste site? Where do you consider, among these three diverse neighborhoods, the best location for a rock quarry or a power plant? What steps would you undertake to convince the people now residing there that they need to get with the program and help the overall community reduce its overall risk (or provide jobs or electricity)?

Now drive around your own neighborhood and do the same. Pick the best site for the waste site, quarry, or power plant. Would you prefer it be in one of the other neighborhoods? If you *must* have it in your neighborhood (i.e., as a professional, your client *demands* it), how close would you locate it to your own home? Can you *justify* this decision from an environmental perspective? How about from the design criteria? You probably could.

Think about your answers to this test and what they mean in terms of professional responsibility. If we go into a project with preconceived ideas of the worth and worthlessness of certain communities, are we not more likely to suggest actions that we would not allow in other areas that we deem to be more *valuable?* And to exacerbate the situation, are we not ready to use the tools of our trade (e.g., the environmental assessment report, the plans and specifications, and the land-use plan) to propagate injustice? For example, once an area has allowed one noxious activity, is it not easier to permit another? Whereas will not a pristine area or one with high property values continued to be buffered from even a first entrée or encroachment by noxious land uses?

Can you imagine your client's reaction if you were to suggest a zoning amendment or exclusion to allow a hazardous waste site in your town's richest neighborhood? But such exclusions are requested in poorer neighborhoods, sometimes because they abut industrial or institutional zones, but sometimes because the powers that be perceive them to be of lesser value.

An empathic view can help. If you were dead set against a project in your neighborhood, how would you use your gifts (talents, education, experience, and contacts) to stop the project? Applying such a model is a good way to gauge the amount of fairness in the professional advice being dispensed. If you would protest such a project in your own neighborhood, should you be advocating and putting the force of your profession behind a similar project in another?

JUSTICE BY DESIGN

After conceiving the project, the first thing a project designer or engineer should do is to learn what people, all people, think of the proposed project. Each step in the design process allows for community involvement and participation.

Project Goals and Objectives

Of course, the practicing professional must pay close attention to the expectations of the client, but must remember that the first canon of the engineering profession gives primacy to our larger client, the public. Thus, even in the earliest stages in identifying project goals and objectives, certain constraints and accommodations must be built in, such as ensuring that the project does not induce disproportionate costs and risks to certain groups.

We advise engineers to begin with the end in mind, but to be adaptive. Ownership is good, but defensiveness is not. Rigorous science is required, but engineering "rigor mortis" is not. We must be able to listen, really listen, even if it means changing our plans to ensure fairness.

Needs Assessment/Data Collection

Once the goals and objectives of a project are clear, the difference between the existing conditions, such as the current housing stock, land-use patterns, and product clientele, must be evaluated to determine just what needs to be done for the project to be a success.

When considering these gaps and needs, the professional should be mindful of the effect that addressing these will have on various groups. For example, if the project calls for the relocation of people currently living in the area, the rerouting of roadways, or changes in land use, such as rezoning and exemptions from existing zoning requirements, the needs assessment must clearly identify how the project will accommodate, in a fair and just manner, those people affected. In addition, since we are talking about "environmental justice" we should also consider the effects on nonhuman species, such as changes in ecosystem function and structure, changes in habitat (including hydrological changes and creation of barriers that affect predator–prey dynamics, species richness, and biodiversity), and modification of natural cycles (e.g., nitrogen, phosphorus, and micronutrients).

Preliminary Review/Feasibility Study/Economic Analysis

It is best to involve the community as early as possible in an engineering project. This can prevent misconceptions or mistrust that can result, especially if rumors or information are leaked. The project engineer may find the need to play "catch up" if this occurs. Also, communities can be a great resource. Not everything is written and available through publications. For example, we have found that in the southeastern United States, much (most?) important information is passed down by oral tradition. This might help explain, in addition to the blatant ignorance of many northerners about things "southern," why southern culture is so misunderstood or even reviled. The North seems to have invested more in written tradition, but the South relies more heavily on storytelling and other oral methodologies, peppered with a bit of written tradition (e.g., entries in the family Bible).

This also applies to the implementation of any project. For example, it is common for a safety manual to be written by a competent engineer, only to be ignored or not well understood by those needing protection. If the manual sits on the shelf, it does no good. Even worse, it may be an excuse for management not to do more. Many people learn by hearing and doing (i.e., interactive learners), so simply having a written plan is insufficient.

The engineer must use the entire community's resources to gather such important information such as previous failures, old and abandoned facilities, past land uses, water quality (see the discussion of arsenic and Bangladesh in Chapter 3), and the key influential people (e.g., African American clergy, Native American elders, and Mexican American "soccer club" members).

Planning, Problem Formulation, and Design Specifications

As mentioned in our discussion of the needs assessment step, a diverse mix of perspectives must be part of project planning. The problem being addressed (e.g., hazardous

Biographical Sketch: Ian McHarg

Ian McHarg founded the University of Pennsylvania's Department of Landscape Architecture and Regional Planning 46 years ago and ran it for three decades. He died March 5, 2001 at the age of 80. The Penn program attracted graduate students from around the world who wished to emulate McHarg's environmental approach to design, which he conveyed with a memorable mix of polished urbanity and missionary zeal and using any medium available, including books and television, imparted his essential message: that any human action, be it building a highway, city, housing development, subdivision, or park, must account for the suitability of the site, as represented by the slope, contours, vegetative cover, surface water and groundwater, fauna, and other natural features.

McHarg was unhappy with the job that humans were doing in protecting the environment. His most famous quote was: "Man is a blind, witless, low brow, anthropocentric clod who inflicts lesions upon the Earth." He was a strong advocate of holistic planning, arguing that no construction or land use should occur without a study of the suitability to the topography, vegetation, wildlife, and beneficial use of land for other purposes.

McHarg used visual relationships to press his point, establishing the groundwork for spatial analytical tools such as geographic information systems and even computer-assisted drafting, which are so important to environmental planning, engineering, and design today. More than that, perhaps, his work reminds us of the importance of spatial synergies and vulnerabilities that are not readily apparent without employing the right spatial tools. Thus, it should come as no surprise that tools such as geographic information systems can implement environmental justice, especially as they are increasingly used to point out potential problems *before* any decisions on siting are made.

waste disposal, energy production, land-use change) must be properly defined and characterized to match it to the appropriate solution (the project). If key people are not included in this phase, there is a likelihood that the project will be of little value or will have ancillary harm.

Problem formulation must build in ways to measure success and the standards of performance for the project. These include the development of design specifications, which are of numerous types:

1. *Physical.* This includes size, volume, mass, and other dimensions, such as the amount of waste processed daily and the maximum size of a recycling facility. (Size can be good or bad; for example, larger size means more capacity but increased displacement of people and problems with aesthetics.)

2. *Functional.* This includes reliability, duration, durability, production rates, processing rates, transportation efficiency, pollutant removal rates, and energy production. (Function should be sustainable and should be viewed as a life cycle, so if a project does not meet all of its functional specifications, it has failed.)

3. *Safety, Risk, and Environmental.* While meeting the physical and functional specs, the project must not cause undue environmental and health problems. Justice demands that no one be put at undue risk, including those building and operating the project as well as those affected by the project operation (e.g., the adjacent landowners).

4. *Economic.* All projects have a budget and time frame for completion. The engineer is remiss if the important physical, functional, safety, and environmental specifications are not completed because human and capital resources needed are not available. This is particularly problematic at a time when many engineering consultants operate on a "design-only" basis (i.e., the consultant agrees only to design the project, with no responsibility for its construction or operation, not even overseeing the project to ensure that the design is followed properly). We contend that the ethical engineer never truly engages in a design-only project, but is responsible to some extent to ensure that the design is properly built and not misused. Often, only the design engineer knows the design's finer points to ensure that the construction and maintenance steps are adequate.

All specifications can be affected by input from those using and those affected by a project, so the best rule is transparency (i.e., let everyone know what the specs are early and often throughout the project). The specifications are in fact measures of success. If they are not met, the project is at least in part a failure. Thus, if the project is to be truly successful, some specifications will need to measure the social acceptability of the project.

Design Abstraction and Synthesis

The design must not fail for lack of imagination. Just because it has been done this way 10 or 100 times before does not mean that this approach will work in a specific situation or be acceptable to a particular community. Every design is unique in some ways. When

Biographical Sketch: Danie Krige

In the late 1940s, South African engineer Danie Krige (pronounced "krig") was working as an inspector of mining leases for the South African Mining Engineer's department. This experience provided him with access and perspective into sampling and production statistics required to be submitted regularly by the gold mining interests. Others saw the same data, but Krige was interested in spatial interpretation and ways to model large amounts of information to find statistical trends and patterns. Such insights allowed Krige to conduct extensive basic statistical research into the valuation problems of the mines. From this work, Krige discovered ways to improve the quality of ore block valuations and to make logical estimates of the block grade distributions. Previously, only borehole values were available to site new mines. As a result of this work, an entirely new discipline of statistics, *geostatistics,* was born, especially the method bearing his name, *Kriging.*

Kriging is a technique presently used to interpolate of spatial information to describe the increasing difference or decreasing *correlation* between sample values as separation between them increases, to determine the value of a point in a heterogeneous *grid* from known values nearby (see Figure 7.1). It is linear since the estimated values are weighted linear combinations of available data. It is unbiased because the mean of the error is zero. It is best since it aims at minimizing the variance of the errors. *Kriging* has become synonymous with optimal prediction and is used to represent uncertainty, applying a *variogram,* a two-point statistical function that describes the increasing difference or decreasing correlation, or continuity, between sample values as separation between them increases. The technique is now highly automated and is part of many computerized geographic information systems (GISs).

Krige continued his work in the geostatistical field for some 30 years in the Anglovaal Head, before retiring for 10 years as professor of mineral economics at Wits. He remains quite active as a consultant.

putting different things together from the physical and social sciences, the result may lead to unexpected and unintended consequences. Testing a design prototype only under the best conditions is an invitation to costly and dangerous consequences down the road. Examples of such surprises are seen in the news almost daily (e.g., a drug that has a side effect that was not found during clinical trials, a part on a truck that failed due to unforeseen but real-life driving conditions, a sport utility vehicle (SUV) that overturns because tires are underinflated with air to give a smoother ride, or malicious tampering with an over-the-counter pain reliever). Synthesis can result in something other than additive effects. The effects may be muted (antagonism) or increased (synergism). The designer should be careful to consider such effects.

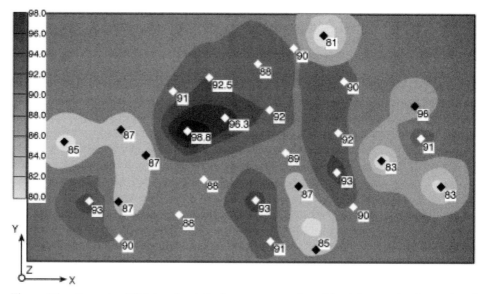

Figure 7.1. Example of Kriging: isopleths for percentage of amphipod that survive for a reduced number of sampling stations in a hypothetical harbor. This spatial interpolation allows the use of fewer sampling sites, reducing costs and giving wider spatial coverage. (From C. J. Leadon, *Kriging in Eco-Risk Assessments,* Issue Paper, Argonne National Laboratory, Argonne, IL, June 21, 2000.)

Implementation/Construction

The design engineer is often called upon during construction to help to interpret specifics of a design or even to make adjustments to address real-world concerns. Certainly, one type of implementation concern is the acceptability of the project by the local or larger community where it is being built. Witness the fact that hazardous waste site and nuclear waste site designs are fairly straightforward in terms of their physical and functional specifications (e.g., thickness of barriers, soil type, hydrological conditions, resistance to terrorist attack, separation of wastes and other materials). However, much of the concern is in trying to address the profound concerns of those fearing the effects of leaks. The Yucca Mountain nuclear waste site in Nevada has been straddling the design/construction phases for some time now, with no clear consensus on how to proceed.

Communication/Feedback

The other steps all include open and effective risk communication, but after the project is ready to go online, a formal feedback system is needed. This will help managers and decision makers to ascertain the project's effectiveness and adherence to design specifications and even whether new success measures are needed. The feedback from this communication can be used to make important adjustments and improvements and to ensure that the affected community is protected from possible problems (e.g., one of the major problems in the Bhopal toxic cloud disaster was that the poor people living near the plant were ignored in their complaints prior to the all-out disaster).[1]

Operation and Maintenance

As long as the project is in place, the professional must monitor its performance. For example, the Superfund and Resource Conservation and Recovery Acts require closure and postclosure monitoring even after a hazardous waste site has received a "clean bill of health" (i.e., it has met its cleanup standards for soil, water, air, or sediment). Operation and maintenance requires the assurance that the project not only performs as designed, but also looks for any previously unforeseen consequences. In fact, many of the cases in this book are not failures of design so much as they are failures of anticipation. Hindsight is more likely to be better than foresight, so even the most conscientious professional is likely to be surprised by the interplay of events. This is not a place for blame and defensiveness, but calls for vigilance. As long as the reasonable safeguards and best judgments are part of the design and implementation of a project, most review boards and oversight authorities are unlikely to reprimand the professional. However, once a problem is identified, even after the project is "completed," it behooves the professional to do all that is possible to address it. In fact, the exemplars are those who look after their previous designs, such as that exhibited by William LeMessurier after some recal-

Biographical Sketch: William LeMessurier

William LeMessurier (born 1927) had a distinguished career as a structural engineer and in 1978 was asked to design a new skyscraper that Citicorp wanted to use as its New York headquarters. With a degree from the Harvard Graduate School of Design and a masters' degree from MIT, LeMessurier had built his firm to be one of the most respected engineering consultants and designers for such projects.

The design called for an attractive, functional, and imaginative 59-story building. Because of space and light restrictions, the architects designed a building that seemed to float on four columns nine stories high, providing light and space below, and enhancing the visual appearance of a new church building on the corner of the lot. To achieve this, the architects suggested that the four columns be placed in the middle of each side instead of at the corners. LeMessurier decided to use a unique form of construction, with the forces being transferred to the four columns by means of V-shaped beams (see Figure 7.2).

Engineering design is a trial-and-error procedure. A structure is postulated, and the loads on that structure are then estimated. Using mathematical principles and well-tested equations, the effect of these loads on the structure are calculated. In the case of the Citicorp building, LeMessurier's engineers calculated, in addition to other live loads, the effect of wind and decided that with a damper mechanism in the attic of the building, the building would be able to withstand high winds.

The Citicorp building was constructed and occupied, and the client was very pleased with the result. Then out of the blue, LeMessurier got a telephone call from a student who told LeMessurier about a homework assignment he had done. The student had calculated the ability of the building to withstand wind loads. As long as the winds were from the side of the building, the structure seemed to be secure, but if the winds hit the building at its corners, called *quartering winds*, it would be possible to topple the building at moderate winds.

LeMessurier told the student that in effect, he didn't know what he was talking about. But the call put him to thinking, and he redid some calculations. To his surprise, it seemed that the effects of 45-degree winds were much greater than he had originally estimated. But this would not have been a problem had the method of structural construction not been changed during the erection procedure. Instead of welding the joints, the construction engineers had substituted a newer standard using bolted joints. If the effect of these joints was now included in the analysis, it became painfully clear that some of the beams in the building, should it be hit by quartering winds, would not be able to withstand the live load, and the building would topple. Weather records showed that such winds might occur once in every 16 years.

That design reliability was unacceptably low and the risk far too high. Should the building fall, thousands of people would die, and only LeMessurier knew the full story. He contemplated his options, and in his words: "Thank you Lord for making this problem so sharply defined that there's no choice to make."[2]

With the consent of the owners, he went with his news to disaster engineers, who planned for evacuation should a storm be imminent. He instrumented the entire building with strain gauges and set a 24/7 watch on the damper mechanism in the top floor to make sure that it functioned perfectly. Then LeMessurier started to strengthen each of the V-joints where bolts had been used by welding in supporting plates. Since the structural members were all inside the building, he could do all the construction from the inside and none of it would be visible from the outside, thus avoiding embarrassing questions and possible panic. Within months, all of the joints had been strengthened and the building became one of the safest in New York, able to withstand the highest winds that could reasonably be expected to occur.

After discovering the problem, LeMessurier could have done nothing, believing that by revealing this information he would have lost stature and respect in the engineering community and hoping that the series of events that cumulatively would have led to a catastrophe would not occur within his lifetime. Instead, he chose the honorable alternative, perhaps remembering the engineering code of ethics: "The engineer shall hold paramount the health, safety, and welfare of the public." The risk was so great, in terms of both its probability and its magnitude, that there was little choice for him or for the owners of the building.

As it was, by conducting himself in such an honorable manner, he actually gained considerable stature in the profession and in the public's eye. Since the "fifty-nine-story crisis," he has made himself widely available as a lecturer in colleges and universities, always speaking with candor about what could have been the greatest disaster in his otherwise illustrious career, but which turned out to be his greatest engineering triumph.

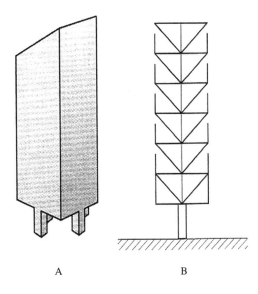

Figure 7.2. Building design of Citicorp Building in New York City. The building appears to float on four columns (A) using V-shaped beams to transfer forces (B).

A   B

culations showed significant design flaws in New York's Citicorp Building in the event of high winds.

CASE STUDY: WILLIAM LeMESSURIER AND THE FIFTY-NINE-STORY CRISIS

Visit the Online Ethics Web site: http://onlineethics.org/moral/lemessurier/index.html. Read all eight parts (they are short but very interesting) and then view the brief videoclip of LeMessurier. Consider what happened and what could have happened. Answer the following questions about the Citicorp Towers and LeMessurier's professional behavior.

1. Why do engineers "push the envelope" so frequently? Does that in some way tell you something about yourself and why you want to become an engineer?

2. Where do you think LeMessurier learned about his responsibilities and obligations as an engineer? What is the difference between professional obligation and the moral responsibilities of other members of society?

3. To whom are engineers "accountable"? To whom was LeMessurier accountable? Prioritize these responsibilities. Do any patterns jump out at you?

4. If you were faced with a $2 million personal loss (although eventually covered by insurance), would you have done the same thing that LeMessurier did, or would you have weighed the risks and hoped that the confluence of factors (e.g., winds and resonance) would not occur? Can you think of a way that "no action" (no retrofit) could have been rationalized or even argued for legally?

5. Do engineers' responsibilities go beyond legal requirements? Why?

6. Some say that there are two types of wrongdoing: voluntary wrong-doing and negligence. Another way to put it is that professionals can commit either sins of commission or sins of omission. Which was more likely to happen in the Citicorp case? How were these wrong-doings avoided?

7. Do you believe that LeMessurier's behavior is rare among people in general? Is it rare among engineers? Is it rare among engineers who are given large responsibilities, such as the design of a major commercial facility? What does this tell you about the possible rewards of the profession you are pursuing?

Extra credit: Leslie Robertson was one of the structural engineers called in to do the damage assessment. What other recent famous engineering case involves Robertson?

While the Citicorp Building example is dramatic, there are countless other projects with unexpected and undesired outcomes. Many of these have nothing to do with how well the engineer has applied the physical sciences. The unforeseen problems may not even be directly the result of the project. For example, the engineer may have assumed, for good or bad reasons, that the land use and population distributions would remain stable or at least follow a particular type of projection. However, if a roadway or water line creates access to the area, or even worse, if the engineer's project in some way induces growth, environmental exposures and public health risks may increase. Such contingencies should have been foreseen. So in a way, an engineer's project is never completely done.

TRUST: OUR MOST IMPORTANT PROFESSIONAL COMMODITY

Once the engineer has a good idea of what is perceived to be good or bad about a project, a word of caution is in order. Most people in the neighborhoods that are potentially affected by the project are not required to be technically literate in their daily lives. The engineer is put in the position of trust.

Note the difference in professional ethics *versus* individual ethics. The vendors' credo, *caveat emptor,* places the onus on the buyer (client). The professional credo is very different. Ours is *credat emptor,* roughly translated from Latin as "Let the client trust." Arguably, the engineer's principal client is the public. And the clients need know little about the practice of engineering because, owing to the expertise, they have delegated the authority to the engineer. Just as society allows a patient to undergo brain surgery even if that person has no understanding of the fundamentals of brain surgery, so our society cedes authority to engineers for design decisions. With that authority comes a commensurate amount of responsibility, and when things go wrong, culpability. The first canon of most engineering ethical codes requires that we "hold paramount" the health and welfare of the public. The public is an aggregate, not an "average." So leaving out any segment violates this credo. Thus, even though most people favor a particular approach, it is still up to the engineer to ensure that any approach recommended is scientifically sound. In other words, no design, even a very popular one, is to be recommended unless it meets acceptable, scientifically sound standards. Conversely, the most scientifically sound approach may have unacceptable social side effects. Once again, the engineer is put in the position of balancing trade-offs.[3]

In every engineering office or department there is a designated "engineer in responsible charge" whose job it is to make sure that every project is completed successfully and within budget. This responsibility is often indicated by the fact that the engineer in responsible charge places his or her professional engineering seal on the design drawings or the final reports. By this action the engineer is telling the world that the drawings or plans or programs or whatever are correct, accurate, and that they will work. (In some countries not too many years ago, the engineer in responsible charge of building a bridge was actually required to stand under the bridge while it was being tested for bearing capacity!) In sealing drawings or otherwise accepting responsibility, the engineer in charge places his or her professional integrity and professional honor on the line. There is nowhere the engineer in charge can hide if something goes wrong. If something *does* go wrong, "One of my younger engineers screwed up," is not a reasonable defense, because it is the engineer in charge who is supposed to have overseen the calculations or the design.

For very large projects where the responsible engineer may not even know all the engineers working on the project, much less be able to oversee their calculations, this is clearly impossible. In a typical engineering office the responsible engineer depends on a team of senior engineers who oversee other engineers, who oversee others, and so on down the line. How can the responsible engineer at the top of the pyramid be confident that the product of collective engineering skills meets the client's requirements?

Fortunately, the rules governing this activity are fairly simple. Central to the rules is the concept of truthfulness in engineering communication. Such technical communication up and down the organization requires an uncompromising commitment to tell the truth no matter what the next level of engineering wants to hear.

It is theoretically possible for an engineer in the lower ranks to develop spurious data, lie about test results, or generally manipulate the basic design components. Such information might not be readily detected by supervisory engineers if the bogus information is beneficial to the completion of the project. If the information is not beneficial, on the other hand, everyone along the chain of engineering responsibility will give it a hard, critical look. Therefore, the inaccurate information, if it is the desired information, can steadily move up the engineering ladder because at every level the tendency is not to question good news. The superiors at the next level also want good news, and want to know that everything is going well with the project. They do not want to know that things may have gone wrong somewhere at the basic level. Knowing this, and fearing being shot as the messenger, engineers tend to accept good news and question bad news. In short, the axiom that "good news travels up the organization—bad new travels down" holds for engineering as well. And bad news will travel down very quickly (for example: "You're fired!").

The only correcting mechanism in engineering exists at the very end of the project if failure occurs: The software crashes, the bridge falls down, the project is grossly overbid, or the refinery explodes. And then the search begins for what went wrong. Eventually, the truth emerges, and often the problems can be traced to the initial level of engineering design, the development of data and the interpretation of test results.

It is for this reason that engineers, especially young engineers, must be extremely careful of their work. It is one thing to make a mistake (we all do), but it is another thing totally to use misinformation in the design. Fabricated or spurious test results can lead to catastrophic failures because there is an absence of a failure detection mechanism in engineering until the project is completed. Without trust and truthfulness in engineering, the system will fail. To paraphrase the eminent scientist Joseph Bronowski:[4]

> All engineering projects are communal; there would be no computers, there would be no airplanes, there would not even be civilization, if engineering were a solitary activity. What follows? It follows that we must be able to rely on other engineers; we must be able to trust their work. That is, it follows that there is a principle which binds engineering together, because without it the individual engineer would be helpless. This principle is truthfulness.

BEYOND EXISTENTIAL PLEASURES

In his highly readable and important work *The Existential Pleasure of Engineering,*[5] civil engineer Samuel Florman argues that the application of the physical sciences is a sufficient dedication to the greater good of society. This line of reasoning requires that the engineer merely do her or his job by ensuring that the math and science, and the design itself, are exemplary. The engineer need not worry about the larger social issues, since having chosen such a helping profession in the first place, there is no further societal obligation as long as the engineer's practice is ethical and competent.

We very much appreciate and use Florman's text in our classrooms, but it should come as no surprise that we strongly disagree with the premise. We believe that the value added by the engineer requires more than passive or coincidental attention to social responsibility. In fact, the major premise of *this* book is that we must be active in ensuring socially responsible designs and projects. To wit, Florman's postulates provide examples of why engineers must go beyond being competent in their technical specialities.

We follow two lines of reasoning, the first a reconsideration of what is meant by a professional calling, and the second, a brief consideration of where the engineering profession is today in dealing with "microethical" and "macroethical" issues. This reasoning can be extended to other environmental professionals and scientists.

The ancient Greeks' virtue ethics is a good place to start. Socrates *et al.* differentiated two contemporary professional virtues: technical competence and character. In fact, they devised the concept of "skill of character" (roughly translated from *ethike aretai*). A professional must at the same time be skillful (an engineer who is good at engineering) and enlightened (an engineer who does good while doing engineering). Florman may rightfully argue that this is embodied in the existential aspects of the profession. We agree to some extent, but it is still possible for a person to be skillful and very good at engineering and even follow the precepts of the profession, but for the overall outcome of the work to be unjust.

As discussed in Chapter 3, engineering failures can be of five kinds: (1) mistakes and miscalculations, (2) extraordinary natural forces, (3) unpredictable occurrences, (4) ignorance or carelessness, and (5) intentional accidents (terrorism). An unanticipated side effect in a type 3 failure often results from social, economic, or other nonengineering externalities. The Pruitt–Igoe housing project in St. Louis is an example of a type 3 engineering failure. Purely from the perspective of applying the physical sciences and mathematics, the project *should* not have failed. But the fact that the project was razed only a decade or so after construction belies this conclusion. It *was indeed a failure.* But the failure was the result of an incomplete and even inaccurate understanding of the social needs of the people the project was designed to help.

We need to extend the reasoning a bit further. Recently, the National Academy of Engineering bifurcated the professional responsibilities into two categories, those of the individual professional and those of the profession at large.[6] The first, the *microethics,* demand ethical behavior on the part of the practitioner. These are the "thou shalts" and "thou shalt nots" usually delineated in our codes of ethics. The professional responsibilities, the *macroethics,* address how the profession operates.

This thinking led the academy to ask: "Even if every engineer in our society abided by every canon and principle in our codes, would the profession adequately address the emerging societal issues?" The answer is "no." The "bottom-up" approach only gives us high quality and sound designs and projects; it does not completely prepare society for the bigger problems on the horizon or even those that are already here. Thus, a "top-down" strategy must complement the specific codes of practice for the individual professional.

Consider the engineer Kurt Prüfer. By many measures, Prüfer was a highly skilled (*aretai?*) engineer. In fact, had he made an enlightened choice (*ethike?*) at a key point in his career, this book may have included his biography as an exemplar of the profession. To the contrary, Prüfer was the engineer who designed the ovens at Auschwitz, where hundreds of thousands of innocent people were exterminated. In fact, he achieved his design specifications and "measures of success" by giving the Nazis the ability to gas and to incinerate 200 human beings daily.[7] Prüfer was loyal and abided by the wishes of his superiors, his clients, and the greater German society at the time. He let others worry about the application of his designs, trusting "others to assess those consequences . . . beyond the engineering task."[8]

Engineering has a special sensitivity to societal well-being. Our works, as Prüfer's tragically demonstrated, can be used and misused. In the Prüfer case, the design was used as designed, albeit nefariously. However, most of the cases in this book are examples of misuse of what under many circumstances would have been good designs. The microethical approach is limited in being able to see and to stop such misuse. Thus, the larger, macroethical view is needed to help to foresee scenarios of potential harm.

Engineers must not be naive and must be aware of hidden agendas and groupthink. In our professional ethics courses at Bucknell and Duke, we show films (DVDs actually, so that you don't think we are complete anachronisms!) depicting engineers going about their business and doing their best (microethically) to meet the client's and the boss's needs, only to find themselves before a board of review or government authority explaining their role in some disaster (plant explosion, release of a toxic substance, etc.). The failure analysis eventually shows that somewhere along the critical path someone had "used" the engineer's expertise to achieve some unjust goal. The engineer is confronted with questions about why he or she did not know about this. Those asking are not impressed by the engineer's answer: "It wasn't my job; my job was only to. . . ."[9]

SUSTAINABILTY: THE KEY TO JUSTICE

The Prüfer case is truly a "downer" but illustrates the importance of expanding our view. A more positive view is to see all of our designs and projects through the prism of sustainability. Ideally, all environmental engineering projects should be sustainable. The added benefit of enhanced sustainability is the increased attention to details that will support environmentally fair and just projects. We have talked a great deal about ensuring that projects be underpinned with sound and reliable science. This, if done without bias, is a big step in ensuring justice. Unfortunately, some of the biases are so subtle that we miss them.

A case in point is a recent political radio ad for an elected position in a wealthy town in the Research Triangle area of North Carolina. To paraphrase the ad, one candidate is warning the town's electorate of what would happen if his opponent were elected. Among the accusations, the candidate reminds the citizens that the opponent in a previous contest called for a greater amount of "public housing" in the wealthy suburb, and added that the opponent did not properly account for the "tens of thousands of dollars in lost property values." What makes such statements particularly interesting is that in this growing area of North Carolina, there is a great need for "affordable" housing, so most people would freely admit that the private construction sector has not provided such livable and affordable conditions. So then, what is wrong with the opponent's call for increased public housing? After all, is not the government's responsibility to be a "provider of last resort" when the private sector and free market do not provide for common welfare of its citizens? The ad could support an elitist view held by at least some of the citizens and elected officials of this town, which has an inordinate number of gated neighborhoods.

An even more dramatic, yet possibly more subtle example of how perpetuating the *status quo* can easily permeate decision making is also currently being debated in another wealthy town in the Triangle area. Ironically, the town is home to a very progressive university with a very strong urban planning program. A local chapter of Habitat for Humanity, which has a stellar reputation for helping people achieve home ownership and in creating neighborhoods that have proven track records of upkeep and sustainability, has proposed a development of 17 acres with 50 "affordable" homes near wealthy neighborhoods. The proposal has met with criticism and resistance from local neighbors.[10] However, they have been singularly unsuccessful in winning any support outside the town's northern tier. A rhetorical misstep by one of their leaders shows why. A critic compared Habitat's plan to "the projects of the '60s," a phrase that reminds people of the public housing that sprang up during that era in cities throughout the country. The target of the critics appears not so much to have been Habitat as the Orange Community Housing and Land Trust, a local group that Habitat wants to bring in to help it develop attached housing on the site. The critic claimed that the creation of attached homes threatens to throw "a bunch of low-income people into a specific area," creating a pocket of poverty as harmful to the development's residents as to its neighbors. Environmental professionals are likely to be brought in as "hired guns" on either side of the debate. The sustainable view is one means of ensuring that their advice is not misinterpreted or misused.

It is useful to consider the costs of ignoring sustainability by comparing an empathic view to that of the unencumbered ethical egoist. In "The Tragedy of the Commons" example described in Chapter 6, the pasture could be seen as "ripe for the picking" as an exhaustible resource.[11] So if one is to have a cow, one had better win the race to claim a piece of the pasture immediately. But the cost to all, even the herders who are first in line to exploit, is a persistent problem (i.e., a barren land where a pasture once existed). With time, neglecting or exploiting the "pasture" makes us all losers; consistent with Reverend Martin Luther King's observation mentioned in Chapter 1: "Injustice anywhere is a threat to justice everywhere."[12] Disparate risk to one group threatens justice for the entire "pasture" (i.e., we are all vulnerable to inequity and injustice). Thus, since environmental problems persist, so must their solutions be sustained.

Even if a person is not a conservationist or an environmentalist (or at least would never label oneself as one), the persistence of the problem may well motivate a sustainable approach. Sustainability may be viewed as a form of ensuring system reliability: in this instance, *ecosystem reliability*. For engineers, we know that the best way to ensure the reliability of systems is the life-cycle analysis. In it, we direct ourselves toward a goal but consider as many of the possible contingencies, decision points, blind alleys, and cul-de-sacs that may be encountered while reaching that goal. So for our engineering and scientific colleagues who see sustainability as a "soft" or nontechnical issue, they may prefer to consider that what is really being engineered is a system (ecosystem), and what is being sustained (i.e., the environmental resource), is the productivity and efficiency of that system. The metrics for the system reliability are quite quantitative and may be measured directly, including net primary productivity, biological diversity (e.g., as indicated by the Shannon index), and chemical and physical integrity.

Much of the undergraduate engineering curriculum in the United States addresses the actual and possible moral dilemmas, problems, and questions of the individual engineering student, and represents the possible successes and failures the students may encounter in their individual careers (microethics). There is good reason for engineering curricula to prepare students for such issues, and we are only beginning to incorporate ethical content into coursework and projects. However, ethical issues of societal import and global scale are also important to engineering. Such issues as nanotechnologies (or angstrom-scale technologies), neurotechnologies, and sustainability will probably influence most people's lives in the decades to come.

Actually, engineers will be at the nexus of the two scales. In fact, in many technologies, we are already there. Of course, we will continue to be concerned about the individual careers of engineers. We will also need to speak to present and coming ethical issues at large scales. For example, are we doing enough to help keep ethical issues and respect for persons as we push the envelope in biotechnological engineering and genetic research? Are we fully considering the cultural ramifications of blastocyst research, including the implantation of stem cells in people's brains to treat neurological diseases? Are we in some way changing the meaning of personhood with pharmacological and neurological advances? Are we fully and adequately assuming our roles as stewards (i.e., Latin *credat emptor*, "let the client trust") as we alter ecosystems and anthropogenic modifications of our food supplies by introducing genetically modified organisms?

Engineers do things![13] So many of the students currently enrolled in engineering schools will be doing things and making decisions about whether to do things during their careers whose nature we can only surmise. It is fairly certain, however, that engineers of the future will work in increasingly complex and technological work environments. As such, they may move away from and be tempted to forget "first scientific principles." For example, how often do high school students really think about the algorithms of long division when they press the "/" key on the calculator or keyboard?

The future engineer will be tempted to design and implement plans without sufficient regard for whether it is the right thing to do. Thus, we must continuously remind ourselves that ethics and justice are affirmative enterprises. To quote Socrates: "The unexamined life is not worth living." Aristotle put it more grimly: "The happy life is thought to be virtuous; now a virtuous life requires exertion, and does not consist of amusement." Finally, if we are reductionist in our view, we may say that good macroethical decisions will be the outgrowth of the collective soundness of individual engineers. We do not disagree completely. However, we will also need somehow to ensure that the engineering profession give wise advice and work with policy and governmental decisions on macroethical issues. The "hold paramount" canon extends to both the individual professional and to the profession itself.

Reinhold Niebuhr's quote in Chapter 1 is worth repeating here: "Man's capacity for justice makes democracy possible, but man's inclination to injustice makes democracy necessary. Democracy is finding proximate solutions to insoluble problems."[14] Special attention must be paid to the sectors of society that without affirmative approaches would be underrepresented. That is, certain people are excluded from the benefits of the marketplace either intentionally or tacitly. In a word, they are disenfranchised, calling to mind John Rawls' concept of the veil of ignorance.[15] To ensure fairness among society's stakeholders, no one knows the status of another. Thus, moral choices under the veil of ignorance would be made from the position of the most disadvantaged. Without such a constraint, engineers would design and build only for those who are sufficiently well off to pay. Niebuhr and Rawls thus make the case for affirmative justice in engineering. This also points to the need for *pro bono* work, since the principal guardian of democracy (i.e., the government) cannot provide for all engineering-related needs. In fact, in many nations, the government militates against those covered by Rawls' veil.[16]

Advocating for justice is not an invitation to "junk" science and unsound design. To the contrary, it requires the best application of physical *and* social sciences. If we as a profession are able to achieve success in this tandem, we will be agents for positive change. After all, what everyone expects from us is *just engineering*.

REFERENCES AND NOTES

1. T. Kletz, *Lessons from Disaster: How Organizations Have No Money and Accidents Recur,* Gulf Publishing Company, Houston, TX, pp. 114–115, 1993.

2. More on William LeMessurier can be found in J. Morganstern, The Fifty-Nine Story Crisis, *The New Yorker,* May 29, 1995.

3. Texts, manuals, and handbooks are valuable to the engineer, but only when experience and good listening skills are added to the mix can wise (and just) decisions be made. Not to be overly didactic, but this is the sage advice offered by the great thinkers and philosophers for the past three millennia. As mentioned on page 131, the counsel is akin to that of St. Peter (Acts 24:25 and II Peter 1:6), who linked maturity with greater "self-control" or "temperance" (Greek *kratos* for "strength"). Interestingly, St. Peter considered knowledge as a prerequisite

for temperance. Thus, from a professional point of view, he seemed to be arguing that one can really only understand and appropriately apply scientific theory and principles after one practices them (I realize that he was talking about spirituality, but anyone who even casually studied Peter's life would see that he fully integrated the physical and spiritual). Engineers who intend to practice must first submit to a rigorous curriculum (approved and accredited by the Accreditation Board for Engineering and Technology), then must sit for the Future Engineers (FE) examination. After some years in the profession (assuming tutelage by and intellectual osmosis with more seasoned professionals), the engineer has demonstrated the *kratos* (strength) to sit for the Professional Engineers (PE) exam. Only after passing the PE exam does the National Society for Professional Engineering certify that the engineer is a "professional engineer" and eligible to use the initials PE after one's name. The engineer is, supposedly, now schooled beyond textbook knowledge and knows more about why in many problems the correct answer is: "It depends." In fact, the mentored engineer even has some idea of what the answer depends on (i.e., beyond the first step of professional wisdom, "knowing that one does not know," as Socrates would say).

4. Quoted by I. Jackson, *Honor in Science,* Sigma Xi, Research Triangle Park, NC, 1956, p. 7, from J. Bronowski, *Science and Human Values,* Messner, New York, 1894, p. 73.

5. S. Florman, *The Existential Pleasure of Engineering*, St. Martin's Press, New York, 1972.

6. National Academy of Engineering, *Emerging Technologies and Ethical Issues in Engineering,* National Academies Press, Washington, DC, 2004.

7. P. A. Vesilind, Closure to "Why Do Engineers Wear Black Hats?" *Journal of Professional Issues in Engineering Education and Practice,* 119(1):331–332, 1994.

8. K. deRobertis, Discussion of "Why Do Engineers Wear Black Hats?" *Journal of Professional Issues in Engineering Education and Practice,* 119(1):330–331, 1994.

9. This discussion is actually a composite of numerous sources, especially the videos *Ethics in Engineering,* Center for Applied Ethics, Duke University, Durham, NC, and *Incident at Morales,* National Institute for Engineering Ethics, Alexandria, VA.

10. The sources for this discussion are A. Blythe, Developer Has Plans for Large Homes, *News and Observer,* Raleigh, NC, June 29, 2004; and editorials in the *Chapel Hill Herald,* Chapel Hill, NC, in June 2004.

11. G. Hardin, Tragedy of the Commons, *Science*, 162, December 13, 1968.

12. Martin Luther King, Letter from Birmingham Jail, *Why We Can't Wait*, HarperCollins, New York, 1963.

13. P. A. Vesilind, *Engineers Working for Peace*, Bucknell University, Lewisburg, PA, November 15, 2003.

14. R. Niebuhr, *The Children of Light and the Children of Darkness* (1844), Prentice-Hall, Englewood Cliffs, NJ, 1974.

15. J. Rawls, *A Theory of Justice* (1971), Belknap Press Reprint, Cambridge, MA, 1999.

16. The *pro bono* movement within engineering is gaining currency with such organizations as Engineers Without Frontiers (http://www.ewf-usa.org/) and Engineers Without Borders (http://www.ewb-usa.org/). It is worth noting that engineering has been a helping profession from its beginnings, but like the medical profession, we can benefit from organized and systematized efforts at bringing our talents to those in need, especially those that are not readily seen in our daily lives, such as people in need of clean water in the far reaches of the world or in parts of our own cities that we do not frequently visit.

Appendix 1

Physicochemical Properties Important in Hazard Identification of Chemical Compounds

Property of substance or environment	Chemical importance	Physical importance
Molecular weight (MW)	Contaminants with MW > 600 may not be bioavailable because they are too large to pass through membranes (known as *steric hindrance*). Larger molecules tend initially to be attacked and degraded at more vulnerable functional groups (e.g., microbial degradation often first removes certain functional groups).	The heavier the molecule, the lower the vapor pressures. For example, the more carbon atoms in an organic compound, the less likely that it will exist in the gas phase under common environmental conditions. Heavier molecules are more likely to remain sorbed to soil and sediment particles.
Chemical bonding	Chemical bonds determine the resistance to degradation. Ring structures are generally more stable than chains. Double and triple bonds add persistence to molecules compared to single-bonded molecules.	Large aromatic compounds have affinity for lipids in soil and sediment. Solubility in water is enhanced by the presence of polar groups in structure. Sorption is affected by the presence of functional groups and ionization potential.
Stereochemistry	Stereochemistry is the spatial configuration or shape of a molecule. Neutral molecules with cross-sectional dimensions > 9.5 Å have been considered to be sterically hindered in their ability to penetrate the polar surfaces of the cell membranes. A number of persistence, bioaccumulation, and toxicity properties of chemicals are determined, at least in part, by a molecule's stereochemistry.	Lipophilicity (i.e., solubility in fats) of neutral molecules generally increases with molecular mass, volume, or surface area. Solubility and transport across biological membranes are affected by a molecule's size and shape. Molecules that are planar, such as polycyclic aromatic hydrocarbons, dioxins, or certain forms of polychlorinated biphenyls, are generally more lipophilic than are globular molecules of similar molecular weight. However, the restricted rate of bioaccumulation of octachlorodibenzo-*p*-dioxin (9.8 Å) and decabromobiphenyl (9.6 Å) has been associated with these compounds' steric hindrance.

Property of substance or environment	Chemical importance	Physical importance
Solubility	Lipophilic compounds may be very difficult to remove from particles and may require highly destructive (e.g., combustion) remediation techniques. Insoluble forms (e.g., valence states) may precipitate out of the water column or be sorbed to particles.	Hydrophilic compounds are more likely to exist in surface water and in solution in the interstices of pore water of soil, vadose zone, and aquifers underground. Lipophilic compounds are more likely to exist in organic matter of soil and sediment.
Co-solvation	If a compound is hydrophobic and nonpolar but is easily dissolved in acetone or methanol, it can still be found in water because these organic solvents are highly miscible in water. The organic solvent and water mix easily, and a hydrophobic compound will remain in the water column because it is dissolved in the organic solvent, which in turn has mixed with the water.	Important mechanism for getting a highly lipophilic and hydrophobic compound into water, where the compound can then move by advection, dispersion, and diffusion. Like PCBs and dioxins, PBTs may be transported as co-solutes in water by this means.
Vapor pressure or volatility	Volatile organic compounds (VOCs) exist almost entirely in the gas phase since their vapor pressures in the environment are usually greater than 10^{-2} kilopascal, whereas semivolatile organic compounds (SVOCs) have vapor pressures between 10^{-2} and 10^{-5} kPa, and nonvolatile organic compounds (NVOCs) have vapor pressures $<10^{-5}$ kPa.	Volatility is a major factor in where a compound is likely to be found in the environment. Higher vapor pressures mean larger fluxes from the soil and water to the atmosphere. Lower vapor pressures, conversely, cause chemicals to have a greater affinity for the aerosol phase.
Fugacity	Often expressed as Henry's law constant (K_H; i.e., the vapor pressure of the chemical divided by its solubility of water). Thus, high-fugacity compounds are likely candidates for remediation using the air (e.g., pump and treat, air stripping).	Compounds with high fugacity have a greater affinity for the gas phase and are more likely to be transported in the atmosphere than are those with low fugacity. Care must be taken not to allow these compounds to escape prior to treatment.
Octanol–water coefficient (K_{ow})	Substances with high K_{ow} values are more likely to be found in the organic phase of soil and sediment complexes than in the aqueous phase. They may also be more likely to accumulate in organic tissue.	Transport of substances with higher K_{ow} values is more likely to be on particles (aerosols in the atmosphere and sorbed to fugitive soil and sediment particles in water) rather than in water solutions.

Property of substance or environment	Chemical importance	Physical importance
Sorption	Adsorption (onto surfaces) dominates in soils and sediments low in organic carbon (solutes precipitate onto soil surface). Absorption (three-dimensional sorption) is important in soils and sediments high in organic carbon (partitioning into organic phase/aqueous phase matrix surrounding mineral particles), so the organic partitioning coefficient (K_{oc}) is often a good indicator of the sorption potential of a PBT.	Partitioning determines which environmental media will dominate. Strong sorption constants indicate that soil and sediment may need to be treated in place. Phase distributions favoring the gas phase indicate that contaminants may be offgassed and treated in their vapor phase. This is particularly important for "semivolatile" PBTs, which under typical environmental conditions exist in both the gas and solid phases.
Substitution, addition, and elimination	These processes are important for treatment and remediation of PBT contamination. For example, dehalogenation (e.g., removal of chlorine atoms) of organic compounds by anaerobic treatment processes often renders them much less toxic. Adding or substituting a functional group can make the compound more or less toxic. Hydrolysis is an important substitution mechanism where a water molecule or hydroxide ion substitutes for an atom or group on molecule. Phase 1 metabolism by organisms also uses hydrolysis and redox reactions (discussed below) to breakdown complex molecules at the cellular level.	These processes can change the physical phase of a compound (e.g., dechlorination can change an organic compound from a liquid to a gas) and can change their affinity to or from one medium (e.g., air, soil, and water) to another. That is, properties such as fugacity, solubility, and sorption will change and may allow for more efficient treatment and disposal. New species produced by hydrolysis are more polar and thus more hydrophilic than their parent compounds, so they are more likely to be found in the water column.
Dissociation	Molecules break down by a number of types of dissociation, including hydrolysis, acid–base reactions, photolysis, dissociation of complexes, and nucleophilic substitution [i.e., a nucleophile ("nucleus lover") is attracted to a positive charge in a chemical reaction and donates electrons to the other compound (i.e., an electrophile) to form a chemical bond].	Hydrolysis involves the dissociation of compounds via acid–base equilibria among hydroxyl ions and protons and weak and strong acids and bases. Dissociation may also occur by photolysis "directly" by the molecules absorbing light energy, and "indirectly" by energy or electrons transferred from another molecule that has been broken down photolytically.

Property of substance or environment	Chemical importance	Physical importance
Reduction–oxidation (redox)	*Reduction* is the chemical process where at least one electron is transferred to another compound. *Oxidation* is the companion reaction where an electron is transferred from a molecule. These reactions are important in hazardous waste remediation. Often, toxic organic compounds can be broken down ultimately to CO_2 and H_2O by oxidation processes, including the reagents ozone, hydrogen peroxide, and molecular oxygen (i.e., aeration). Reduction is also used in treatment processes. For example, hexavalent chromium is reduced to the less toxic trivalent form in the presence of ferrous sulfate: $2CrO_3 + 6FeSO_4 \rightarrow$ $3Fe_2(SO_4)_3 + Cr_2(SO_4)_3 + 6H_2O$ The trivalent form is removed by the addition of lime, where it precipitates as $Cr(OH)_3$.	Reductions and oxidations are paired into *redox reactions*. Such reactions occur in the environment, leading to chemical speciation of parent compounds into more or less mobile species. For example, elemental or divalent mercury is reduced to the toxic species, mono- and dimethyl mercury, in sediment and soil low in free oxygen. The methylated metal species have greater affinity than the inorganic species for animal tissue.
Diffusion	Diffusion is the mass flux of a chemical species across a unit surface area. It is a function of the concentration gradient of the substance. A compound may move by diffusion from one compartment to another (e.g., from the water to the soil particle).	The concentration gradients within soil, underground water, and air determine to some degree the direction and rate that the contaminant will move. This is a very slow process in most environmental systems. However, in rather quiescent systems[a] ($<2.5 \times 10^{-4}$ cm s^{-1}), such as aquifers and deep sediments, the process can be very important.

Property of substance or environment	Chemical importance	Physical importance
Isomerization	A *congener* is any chemical compound that is a member of a chemical family, the members of which have different molecular weights and various substitutions (e.g., there are 75 congeners of chlorinated dibenzo-*p*-dioxins). *Isomers* are chemical species with identical molecular formulas but which differ in atomic connectivity (including bond multiplicity) or spatial arrangement. An *enantiomer* is one of a pair of molecular species that are nonsuperimposable mirror images of each other.	The fate and transport of chemicals can vary significantly depending on the isomeric form. For example, the rates of degradation of left-handed chiral compounds (mirror images) are often more rapid than for right-handed compounds (possibly because left-handed chirals are more commonly found in nature, and microbes have acclimated their metabolic processes to break them down). Isomeric forms also vary in their fate.
Biotransformation	Many of the processes discussed in this table can occur in or be catalyzed by microbes; these are "biologically mediated" processes. Reactions that may require long periods of time to occur can be sped up by biological catalysts (i.e., enzymes). Many fungi and bacteria reduce compounds to simpler species to obtain energy. Biodegradation is possible for almost any organic compound, although it is more difficult in very large molecules, insoluble species, and completely halogenated compounds.	Microbial processes will transform parent compounds into species that have their own transport properties. Under aerobic conditions, the compounds can become more water soluble and are transported more readily than their parent compounds in surface and ground water. The fungicide example given later in this chapter is an example of how biological processes change the flux from soil to the atmosphere.
Availability of free oxygen	Complete microbial processing degrades hydrocarbons by oxidation to CO_2 and H_2O when free O_2 is available (aerobic digestion). In the absence of free O_2, microbes completely degrade organic compounds to CH_4 and H_2O (anaerobic digestion).	Aerobic and anaerobic processes may need to be used in a series for persistent compounds. For example, aerobic processes can cleave the aromatic ring of PCBs that have up to three chlorines. PCBs with four or more chlorines may first need to be treated anaerobically to remove the excess chlorines before the rings can be cleaved.

Property of substance or environment	Chemical importance	Physical importance
Potential to bioaccumulate	Bioaccumulation is the process by which an organism takes up and stores chemicals from its environment through all environmental media. This includes *bioconcentration* (i.e., the direct uptake of chemicals from an environmental medium alone) and is distinguished from *biomagnification* (i.e., the increase in chemical residues in organisms that have been taken up through two or more levels of a food chain).	Numerous physical, biological, and chemical factors affect the rates of bioaccumulation needed to conduct environmental risk assessment. For chemicals to bioaccumulate, they must be sufficiently stable, conservative, and resistant to chemical degradation. Elements, especially metals, are inherently conservative, and are taken up by organisms either as ions in solution or via organometallic complexes such as chelates. Complexation of metals may facilitate bioaccumulation by taking forms of higher bioavailability, such as methylated forms. The organisms will metabolize by hydrolysis, which allows the free metal ion to bond ionically or covalently with functional groups in the cell, such as sulfhydryl, amino, purine, and other reactive groups. Organic compounds with structures that shield them from enzymatic actions or from nonenzymatic hydrolysis have a propensity to bioaccumulate. However, readily hydrolyzed and eliminated compounds are less likely to bioaccumulate (e.g., phosphate ester pesticides such as parathion and malathion). Substitution of hydrogen atoms by electron-withdrawing groups tends to stabilize organic compounds such as the polycyclic aromatic hydrocarbons (PAHs). For example, the chlorine atoms are large and highly electronegative, so chlorine substitution shields the PAH molecule against chemical attack. Highly chlorinated organic compounds, such as the PCBs, bioaccumulate to high levels since they possess properties that allow them to be taken up easily but do not allow easy metabolic breakdown and elimination.

Source: Adapted from D. A. Vallero, *Environmental Contaminants: Assessment and Control,* Elsevier Academic Press, Burlington, MA, 2004.

[a] W. Tucker and L. Nelken, Diffusion Coefficients in Air and Water, *Handbook of Chemical Property Estimation Methods,* McGraw-Hill, New York, 1982.

Appendix 2

Radiative Forcing and Global Warming

Climate change results from natural internal processes and from external forcings. Both are affected by persistent changes in the composition of the atmosphere brought about by changes in land use, release of contaminants, and other human activities. Radiative forcing (see pages 269–270) is the change in the net vertical irradiance within the atmosphere. Radiative forcing is often calculated after allowing for stratospheric temperatures to readjust to radiative equilibrium, while holding all tropospheric properties fixed at their unperturbed values. Commonly, radiative forcing is considered to be the extent to which injecting a unit of a greenhouse gas into the atmosphere changes global average temperature, but other factors can affect forcing, as shown in Figure A2.1.

The International Panel on Climate Change (IPCC) has applied a "level of scientific understanding" (LOSU) index is accorded to each forcing (see Table A2.1). This represents the Panel's subjective judgment about the reliability of the forcing estimate, involving factors such as the assumptions necessary to evaluate the forcing, the degree of knowledge of the physical/chemical mechanisms determining the forcing, and the uncertainties surrounding the quantitative estimate of the forcing. The relative contribution of the principal well-mixed greenhouse gases is shown in Figure A2.2.

For more information, see

IPCC, 1996, *Climate Change 1995: The Science of Climate Change.* Contribution of Working Group I to the Second Assessment Report of the Intergovernmental Panel on Climate Change [Houghton, J.T., L.G. Meira Filho, B.A. Callander, N. Harris, A. Kattenberg, and K. Maskell (eds.)]. Cambridge University Press, Cambridge, United Kingdom and New York, NY.

IPCC, 1996, *Climate Change 1995: Impacts, Adaptations and Mitigation of Climate Change: Scientific-Technical Analyses.* Contribution of Working Group II to the Second Assessment Report of the Intergovernmental Panel on Climate Change [Watson, R.T., M.C. Zinyowera, and R.H. Moss (eds.)]. Cambridge University Press, Cambridge, United Kingdom and New York, NY.

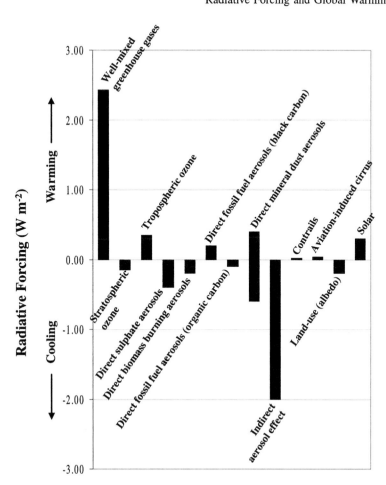

Figure A2.1 The global mean radiative forcing (watts per square meter, W m^{-2}) of the climate system for the year 2000, relative to 1750. Data from IPCC, *Climate Change 2001: The Scientific Basis,* Chapter 6—Radiative Forcing of Climate Change, 2001.

Table A2.1 Level of Scientific Understanding (LOSU) of Radiative Forcings

Forcing Phenomenon	LOSU
Well-mixed greenhouse gases	High
Stratospheric ozone	Medium
Tropospheric ozone	Medium
Direct sulphate aerosols	Low
Direct biomass burning aerosols	Very low
Direct fossil fuel aerosols (black carbon)	Very low
Direct fossil fuel aerosols (organic carbon)	Very low
Direct mineral dust aerosols	Very low
Indirect aerosol effect	Very low
Contrails	Very low
Aviation-induced cirrus	Very low
Land-use (albedo)	Very low
Solar	Very low

Source: IPCC, *Climate Change 2001: The Scientific Basis*, Chapter 6—Radiative Forcing of Climate Change, 2001.

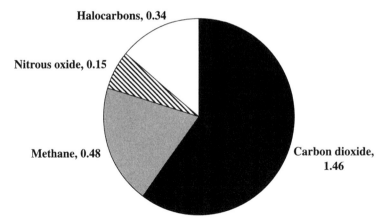

Figure A2.2 Relative contribution of well-mixed greenhouse gases to the $+2.43$ W m^{-2} radiative forcing shown in Figure A2.1. Data from IPCC, *Climate Change 2001: The Scientific Basis*, Chapter 6—Radiative Forcing of Climate Change, 2001.

Name Index

Subject Index